Across the Sahara

Klaus Braun • Jacqueline Passon
Editors

Across the Sahara

Tracks, Trade and Cross-Cultural Exchange in Libya

DAAD Deutscher Akademischer Austauschdienst
German Academic Exchange Service

Physical Geography UNI FREIBURG

المركز الليبي للمحفوظات والدراسات التاريخية

Federal Foreign Office

Springer Open

Editors
Klaus Braun
Department of Physical Geography
University of Freiburg
Freiburg, Baden-Württemberg, Germany

Jacqueline Passon
Department of Physical Geography
University of Freiburg
Freiburg, Baden-Württemberg, Germany

ISBN 978-3-030-00144-5 ISBN 978-3-030-00145-2 (eBook)
https://doi.org/10.1007/978-3-030-00145-2

Library of Congress Control Number: 2018953308

This Springer imprint is published by the registered company Springer Nature Switzerland AG
The registered company address is: Gewerbestrasse 11, 6330 Cham, Switzerland

Preface

The idea for this book goes back a long way. And, in the end, it proved to be a time-consuming and arduous affair to bring this idea to life.

It involved travelling approximately 2000 km between the Mediterranean coast and the areas south of the Sahara from the perspective of the nineteenth century, when huge caravans and numerous European travellers were on the move, transporting goods and exploring the hitherto unknown desert areas. Its geographical location alone has made Libya a hub of trade between Europe and Africa since ancient times. Right up until the twentieth century, camels formed the basis of this economic and cultural exchange, thanks to their ideal adaptation to the climate and landscape of the desert and semi-desert. Caravans comprising 1000 or more animals as well as their guides were not uncommon, travelling along established routes which offered places to rest and shelter overnight. These "ships of the desert" came from the south laden with gold, leather, ivory and ostrich feathers, while vast numbers of slaves from inner Africa were another permanent feature of the caravans over the centuries. The caravans from the north brought above all tableware, cloth, paper, spices and weapons.

Such journeys bore little resemblance to the idyllic tales of the "Arabian Nights"—traversing the Sahara was gruelling and dangerous. Travellers needed a friend and protector known as an "amidi", as well as a local guide familiar with the area and its residents. It was otherwise almost impossible to find a way through the inhospitable environment, crossing great distances devoid of water, and to withstand the attacks and diseases which were a constant threat. This was particularly true for the outsiders and "nonbelievers" from Europe who began crossing the Sahara at the end of the eighteenth century, en route to explore inner Africa for their governments. Celebrated as "martyrs of science", many of these travellers paid for their attempt to explore the world's last uncharted territories with their lives. Other "travellers" simply disappeared unremembered and unacknowledged, including the many slaves who did not survive their gruelling march through the Sahara.

What remains of the centuries of caravan trading? What has been preserved, in the landscape and in people's memories? This work sets out to answer these questions. It is based on a joint venture between scholars from Freiburg, Germany, and Tripoli, Libya, which began in 2006 and has proven to be an extremely fruitful opportunity for collaboration between everyone involved.

At the beginning of the new millennium, after a long period of separation and isolation, Libya was eager to reorient itself and open up to Europe and the Western world. Similarly, the West was interested in normalising relations with Libya. This was reflected in a political and societal shift seen, among other things, in a considerable expansion of economic ties as well as significant growth in tourism. Against this backdrop, the Libyan state and the German Academic Exchange Service (DAAD) funded a joint academic project which aimed to underscore the effects of this transition in the field of scholarship. In addition to the research questions, the project focused in particular on creating opportunities for scholars and especially students from both sides to come together and gain insights into the other culture through intercultural dialogue. The research objective of the joint project was to gain an understanding of which routes were once used by the countless people, animals and goods crossing the Sahara, in what form paths and structural remains of the historic caravan trade, such as places to rest and pass the night, oases, wells, are still recognisable in the landscape today, and to what extent the modern routes through the Sahara follow these old caravan paths.

However, this would not have been possible by merely viewing the area from above. After examining satellite images, we quickly moved on to compare our findings with other information. Historic maps and travel reports were analysed, as were Arabic sources from family archives. Most importantly, the project members searched for physical remains such as way markers and places of prayer, for the remains of places to rest and stay the night, and for wells. Interviews were held with local people whose fathers or even great-grandfathers had worked in the caravan trade. These informants were of course unable to give personal accounts of the golden age in the eighteenth and nineteenth centuries. Nonetheless, some still remembered the last caravans, as the trade only came to a complete halt around 50 years ago. They were also able to draw on the knowledge which had been passed down in their families from one generation to the next. This enabled us to connect the traces in the stony landscape with living stories.

Camels have had their day as pack animals, and goods are now transported through the desert in trucks. The modern routes follow the old trails in places, but not completely. For example, the cars and trucks used today have no need for detours to wells. While camels have lost their economic significance as robust "ships of the desert", they continue to hold a high symbolic value—they serve as bridal gifts and are used in camel-riding sports.

The project was situated at the interface between geography and history. Drawing on the different cultural approaches and points of view, and using state-of-the-art geographical methods allowed us to consider the topic from a perspective which would not have been possible without mutual exchange. We therefore also see our project as a learning process which has been productive in a variety of ways. Indeed, the spirit of intercultural dialogue which saw the project through had not just an exceedingly productive effect on our collaboration, but also a long-lasting one. On top of that, everyone involved has also gained many valuable experiences on a personal level. We have developed a deeper understanding of ways to approach and view our partners' world, which in turn has helped us to better understand our own world. We have allowed ourselves to change and to initiate creative processes.

We would therefore like to thank all of the wonderful people who have worked on this book and filled our intercultural dialogue with life. A special mention must go to Prof. Dr. Imad Al-Din Ghanem, without whom this project would not have come about, and without whose invaluable historic expertise we would have remained unaware of many things. Our thanks also go to Prof. Abdullah Zarouk, Dr. Salaheddin Sury, Dr. Said Hamad, Dr. Hassan Mughrabi, Salih al-Mahdi Khalifa, Najmiya as-Sadeq at-Tellisi, Mansour El Nayedh, Shefa El-Ghomari, Ghada Zarouk, Nureddin Al-Thani, Mohamed Al-Turki and Omar Howidi on the Libyan side, and Johannes Schlesinger, Joschua Metzger, Susanne Grimm, Aniela Jeworutzki, Sabrina Swifka, Tina Eisele, Dr. Karin Kook, Florian Lederer, Mark Grasy, Daniel Kaiser, Carina Haas, Desirée Matheis, Anna-Lena Schmidt, Anna-Maria Manz, Maria Martens, Anja Tiedemann, Sabrina Roy and Steffen Wasko on the German side. And we are grateful to our "amidi" Ali, thanks to whom we never lost our way in the desert.

Neither the project nor this publication would have been possible without the active moral and financial support of the German Academic Exchange Service (DAAD) in Bonn, Germany, the German Federal Foreign Office in Berlin, and the Centre for Libyan Archives and Historical Studies (CLAR.HS) in Tripoli. Our heartfelt thanks go to the relevant staff at these institutions. And special thanks also go to the ambassadors and staff at the German embassy in Tripoli. We have met various representatives of the Federal Republic of Germany there over the years, and every encounter has been characterised by kindness and constructive collaboration. Without this assistance, our project would not have been possible in its current form. We would therefore like to take the opportunity to expressly thank Mr Christian Much, the ambassador to Libya from 2013 to 2016, whose extraordinary commitment has made this publication a reality. In addition, we would like to thank Dr Franziska Brantner, member of the

German Bundestag (MdB), who supported us after the fall of the Gaddafi regime. We are also grateful to Isabelle Durant, Hélène Flautre and Mychelle Rieu, members of the European Parliament.

Since 2006, we Germans in particular have found our Libyan partners and contacts—who cannot all be named here—tireless in their endeavours to help. We thank them for their willingness to offer us an insight into their culture and their country. Furthermore, we have received valuable support time and time again from people in the oases who have opened their homes to us and shared their family archives with us.

A research project such as this cannot succeed without competent expert assistance. Special thanks—as mentioned previously—therefore go to Prof. Dr. Imad Al-Din Ghanem, Prof. Dr. Jörg Stadelbauer, Prof. Dr. Rüdiger Glaser, Prof. Dr. Rainer Wiegels and the long-standing expert on Libya, Hans Schlüter. The thoughtful comments and advice on our ideas, as well as the constructive criticism, were and remain especially important to us.

"Shukran" and thank you to everyone involved!

Freiburg and Tripoli
Spring 2020

Klaus Braun
Jacqueline Passon
Mohamed Alawar
Mohamed Jerary

Contents

List of Figures

List of Figure Credits

1. Across the Vast Land—Some Aspects on Libya's Geography

Fig. 1.1, 1.7–1.22, 1.25–1.27	Own photographs (= The German-Libyan Joint Venture photographs[1])
Fig. 1.2–1.6, 1.23–1.24	Maps (Design & Layout: K. Braun assisted by J. Passon and S. Grimm, A. Jeworutzki, J. Schlesinger)

2. A Journey through Time

Fig. 2.1–2.9, 2.11–12	Own photographs

3. Traders, Nomades and Slaves

Fig. 3.2, 3.11, 3.13	Historical Archives of the CLAR.HS[2]
Fig. 3.3, 3.7	Maps and charts (Design & Layout: K. Braun assisted by J. Passon)
Fig. 3.6, 3.15–3.17	Own photographs

4. Navigating and Mapping the "White Spot"

Fig. 4.2–4.4, 4.11–4.12	Own photographs
Fig. 4.5, 4.7	Maps and charts (Design & Layout: K. Braun assisted by J. Passon)

5. On Major Trans-Saharan Trails

Fig. 5.1, 5.3, 5.5, 5.7, 5.10–5.17, 5.19–5.21, 5.23, 5.26, 5.28–5.32, 5.34–5.35, 5.37–5.40, 5.42–5.43, 5.45, 5.47–5.53, 5.55–5.59, 5.61–5.69, 5.72, 5.75–5.80	Own photographs

[1]The German-Libyan Joint Venture photographs were provided by: Dr. Klaus Braun, Aniela Jeworutzky, Salih al-Mahdi Khalifa, Dr. Jacqueline Passon, Dr. Johannes Schlesinger, Sabrina Swifka, Nureddin Al-Thani, Mohamed Al-Turki

[2]Historical Archives of the CNAR.HS - Centre for Libyan Archives and Historical Studies, Tripoli, Libya

Fig. 5.2, 5.4, 5.18, 5.22, 5.27, 5.41, 5.46, 5.54, 5.60, 5.70, 5.73, 5.74, 5.81, 5.83–5.84	Maps (Design & Layout: K. Braun assisted by J. Passon and S. Grimm, A. Jeworutzki, J. Schlesinger)
Fig. 5.6	Columbia University, Department of Art History and Archaeology (https://vasari.mcah.columbia.edu/media/get/1229/r-3971830/1249/1036_ahar_libya_171_102605/)
Fig. 5.9	Historical Archives of the CLAR.HS
Fig. 5.82, 5.85	Dr. Meike Meerpohl

6. Places of Trade, Communication and Religion—Important Oases

Fig. 6.2, 6.21, 6.31, 6.57, 6.64, 6.74	Maps (Design & Layout: K. Braun assisted by J. Passon)
Fig. 6.3, 6.6–6.8, 6.12 (photo on the right side), 6.16–6.17, 6.20, 6.24, 6.27–6.30, 6.33–6.35, 6.37–6.40, 6.42–6.43, 6.45–6.46, 6.49–6.51, 6.54, 6.58, 6.60–6.63, 6.65, 6.66, 6.69–6.71, 6.73, 6.75, 6.78	Own photographs
Fig. 6.11 (photo on the right side), 6.13 (photo on top left), 6.14, 6.18–6.19, 6.47–6.48, 6.52, 6.59, 6.80, 6.81, 6.82, 6.83, 6.84	Historical Archives of the CLAR.HS
Fig. 6.41	Columbia University, Department of Art History and Archaeology (https://vasari.mcah.columbia.edu/media/get/1281/r-4789193/1286/1036_ahar_libya_60_102605/)
Fig. 6.66	LIFE Magazine
Fig. 6.67	Columbia University, Department of Art History and Archaeology (https://vasari.mcah.columbia.edu/media/get/1224/r-1941474/1180/1036_ahar_libya_167_102605/)
Fig. 6.72	photo top right: https://www.delcampe.net/en_GB/collectables/search?categories%5B%5D=996&search_mode=all&excluded_terms=&term=sirte&show_type=all&display_ongoing=ongoing&started_days=&started_hours=&ended_hours=&display_only=ongoing&min_price=&max_price=&currency=all&seller_localisation=&view=gallery&order=sale_start_datetime
Fig. 6.72	photo below: http://www.ebay.ie/itm/libya-SIRTE-SYRTE-Native-Man-Donkey-Mosque-Islam-1940s-H-Schlosser-Photo-/231742062795?hash=item35f4e740cb:g:eqUAAOSwo0JWO

1 Across the Vast Land—Some Aspects on Libya's Geography

Klaus Braun, Jacqueline Passon, and Aniela Jeworutzki

1.1 Libya at a Glance—Facts and Figures About a White Spot in Northern Africa

Klaus Braun and Jacqueline Passon

1.1.1 Topographical Outline

In a historical context, the term Libya has a really long tradition. Herodotus, for example, used it to describe an area that covers the Northern part of Africa between the Atlantic Ocean in the west and the Red Sea in the east. At that time, Libya was known to be a separate continent, the third one besides Europe and Asia.

Libya, as it is known today, is a country, which represents the central part of Northern Africa between Tunisia and Algeria in the west and Egypt in the east. In the south, Libya is bordered by Niger and Chad and in the south-east by Sudan. It covers an area of nearly 176 million km^2 and holds around 1200 km of the African Mediterranean coastline. Figure 1.2 shows Libya's main topographic features as they can be seen in the January image taken during the "Blue Marble Next Generation" (BMNG) earth observation programme in 2004 and enhanced by digital elevation data from the "Shuttle Radar Topography Mission" (SRTM) in 2000. Additional information like cities, place names and the road network are taken from other sources like the National Topographic Map of Libya or the populated places datasets from the Natural Earth archive.

With the exception of some comparatively small regions along the Mediterranean coast, Libya is mainly characterised by huge deserts and semi-deserts, which cover around 85–90% of the country. In the satellite image, those regions mostly appear in light red, brown or sandy colours, depending on whether it is a rocky, gravel or sandy desert. Belonging to the so-called Sahara which in Arabic means "the Great Desert", the most famous and well-known landscapes within this area are the tablelands of the Hamadah al Hamra, the gravel deserts of the Sarir Kalanshiyu and the Sarir Tibasti and the huge sand seas of the Idhan Awbari and the Idhan Murzuq.

Tablelands and gravel deserts form large flat plains which dominate much of Libya's topography. While the western part of the Sahara with the Hamadah al Hamra is between 300 and 600 m above sea level, the heights of the Sarir Kalanshiyu and the Sarir Tibasti in the east are mainly between 150 and 350 m. Exceptions are the depression of Sabkhat Ghuzayyil some 150 km south of Ajdabiya with the lowest point of Libya lying 47 m below sea level and the mountainous areas in the south and south-east with the highest point at the peak of Bikku Bitti in the Tibesti Mountains near the border to Chad which is 2266 m above sea level.

Comparably smaller mountain ranges like the Haruj al Aswad, the Jabal Akakus, the Masak Mastafat or the Jabal Bin Ghunaymah in the central or the south-western part of the Sahara disrupt the monotonous vastness and reach more or less a height of 1000 m above sea level. Embedded between these mountain ranges and the southern slope of the Hamadah al Hamra, the Idhan Awbari and the Idhan Murzuq form two of the huge sand seas, which are somehow a symbol of the entire Sahara, even if they do not cover more than 20%. It is not only these sand seas the Sahara is famous for, but also the significant number of oases. They indicate life in the desert. Historically, those oases are directly linked to the presence of water, either as artesian sources or as groundwater levels in reachable depths that allow for the drilling of wells. Usually, these areas can be found in deep-lying areas, in particular along wadis or in basins where fossil groundwater appears at or near the surface.

K. Braun (✉) · J. Passon · A. Jeworutzki
Department of Physical Geography, University of Freiburg, Freiburg, Baden-Württemberg, Germany
e-mail: klaus.braun@geographie.uni-freiburg.de

J. Passon
e-mail: jacqueline.passon@geographie.uni-freiburg.de

K. Braun and J. Passon (eds.), *Across the Sahara*,
https://doi.org/10.1007/978-3-030-00145-2_1

Fig. 1.1 Across the vast land: a single twisted acacia tree and certain small shrubs and herbs located within the mountains of the Jabal Bin Ghunaymah, indicating the course of a wadi system where at least some water is available at or near the surface, either as fossil groundwater or water from occasionally rainfalls captured by the river sands

With regard to the regions along the Mediterranean coastline and especially the mountainous areas of the coastal hinterland in the western and eastern parts of Libya, a quite different environment can be detected. In the satellite image, these regions appear in green and dark green colours, which indicate more fertile conditions. They belong to the mountainous area of the Jabal al Nefusah in the west and the significantly as "Green Mountains" indicated area of the Jabal al Akdar in the east. The winter rain comes mostly from west and northwest; these mountains are characterised by Mediterranean vegetation and agricultural usage. Between these productive zones, there is a stretch of around 500 km between Surt and Ajdabiya where semi-desert conditions extend northward to the Mediterranean Sea. Along this stretch, only a small zone of sparse grassland can be found which separates the Sahara from the coast. Except in the east, where the mountains of the Jabal al Akdar directly reach the Mediterranean Sea, the coastline itself is mostly characterised by an alternating system of coastal oases, sandy areas and salt flats, or Sabkhas as the salt flats are called in Arabic.

A very specific Libyan feature in this context is given by the somehow impressive but disconcerting situation where life in the big agglomerations in the north is dependent on water from the arid desert areas in the south. Due to the technical progress and the availability of sufficient financial means since 1984, a large network of pipes, the so-called Great Man-made River, has been installed to bring fossil groundwater from the Nubian Sandstone Aquifer System deep below the Sahara to the people who live along the coastline. Although this project ensures a reliable supply of water, it is proposed that the water reserves might be sufficient for not more than 50–80 years.

1.1.2 Population

According to the given conditions, the distribution and density of the population shown in Fig. 1.3 should be of no surprise. About 75% of the nearly 65 million Libyan people reside along the coastal stretch with concentrations on the urban areas of Tripoli and Benghazi and the mountainous hinterlands of these two agglomerations, the Jabal al Nefusah and the Jabal al Akdar. The other 25% are distributed amongst the different oases in the country's interior and along the main roads connecting the populated places. Although well-known oases like Ghadamis, Sokna, al Fuqaha or al Kufrah or the big number of oases along the Wadi al Adjal owe their origin to the presence of water in a way that it could readily be made available, in recent times modern technologies lead to a larger degree in independence from those natural conditions. That allowed Sabha and the region around it to be developed as the biggest and most important populated area besides the coastal stretch.

An analysis of the distribution of towns and villages that depends upon ground elevation is equally conclusive

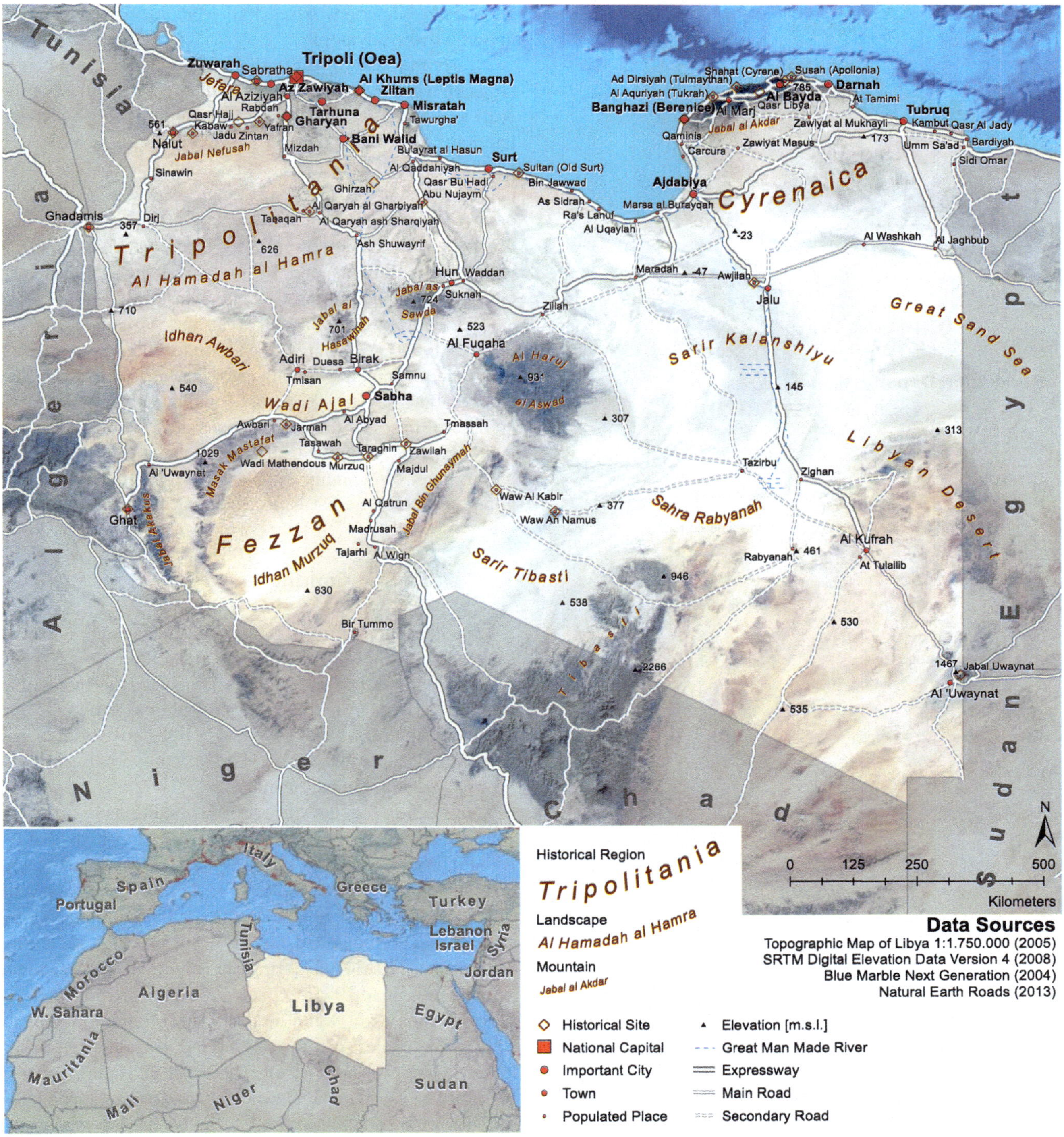

Fig. 1.2 Libya's main topographical features

(Fig. 1.4). Almost half of the population lives in settlements in the low-lying areas along the coast. These settlements are on average the biggest in the country. Most of the largest and fastest developing cities, namely Tripoli, Benghazi, Misratah and al Khums, can be found here.

The (on average and in total second largest) group of settlements are at a height between 350 and 600 m above sea level (a.s.l.). They cover, in particular, the hilly landscapes of the mountains in the hinterland of Tripoli and Benghazi. A few more settlements in this height range can be found

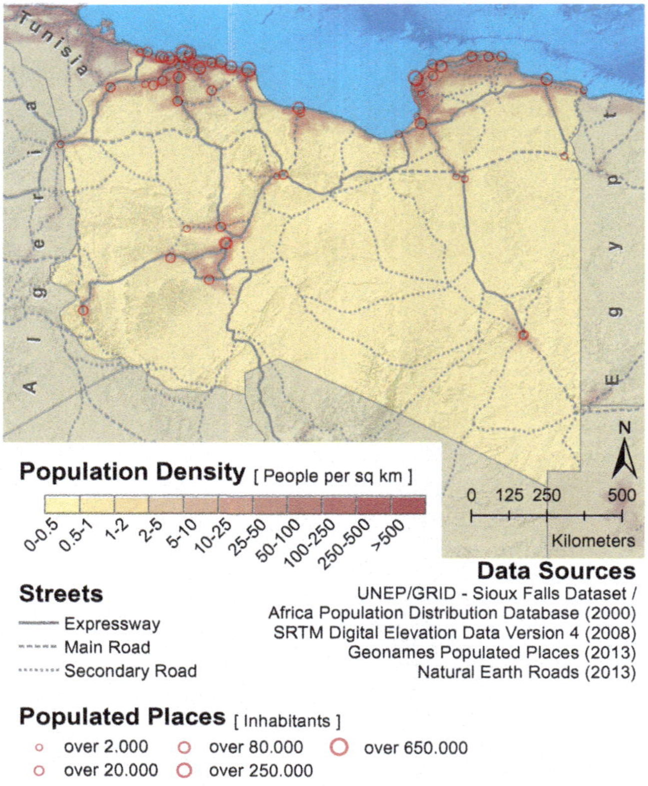

Fig. 1.3 Population density

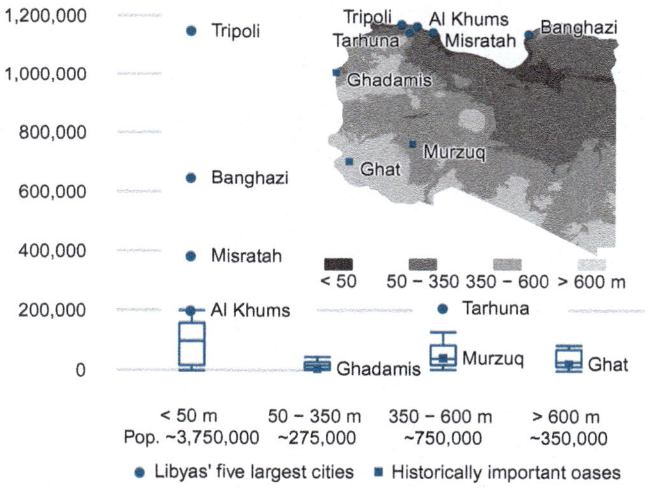

Fig. 1.4 Population analysis

along the big wadi systems in the interior. Whereas Tarhuna in the eastern part of the Jabal al Nefusah shows up as a prosperous city, there are no outliers amongst the settlements in the south. This reflects the fact that there is a considerably slower development beyond the coastal region, where even historically important oases like Murzuq do not play an important role anymore.

Nearly the same refers to the regions situated in a height range between 50 and 350 m a.s.l. that represent most of the semi-arid desert regions and some smaller depressions in the interior. Due to the natural conditions in these areas, only a small number of settlements and a limited percentage of the population can be found in this range. However, most of the oases in this height range are located in the lower parts and depressions of the desert regions. The most famous amongst them is the famous and historically important oasis of Ghadamis, situated in the Ghadamis basin.

The majority of settlements that lie in a height range of more than 600 a.s.l. can be recognised in the upper parts of the densely populated mountains of the Jabal al Nefusah and the Jabal al Akdar. They are complemented by a small number of oases in the mountainous area of the southwest. Although large in number, the settlements in this range are on average comparably small, and even the historically important oasis of Ghat is no exception to this.

1.1.3 Climatic Conditions

To understand the differences between the distinct regions of Libya such as the spots of activity along the Mediterranean coastline and the huge desert areas, one needs to look at the structures arising from the historical, political, economic and social development of Libya, which will be covered in detail in the next chapter. But there is another important aspect to this: the varying climatic conditions.

As shown in Fig. 1.5, the climatic conditions in the south are characterised by extreme temperature and high aridity. Although annual mean temperatures between 10 °C in mountainous areas and 24–26 °C in the low-lying sand seas appear to be quite tolerable, the daily variation during summer is enormous and ranges from 50 to 60 °C during daytime and 20–30 °C during the night. In winter, night temperatures up to 10–15 °C below zero are possible and not as uncommon as it would be desirable. Minimum temperatures of the coldest month with around zero degrees can be found in the tableland of the Hamadah al Hamra and the Masak Mastafat mountains, whereas maximum temperatures of the warmest month with 40 °C and more are typical for the southern part of the country and especially the sand seas of the Idhan Awbari and the Idhan Murzuq. Due to these circumstances, the temperature annual range is enormous and reaches from around 30 °C in the northern part to nearly 40 °C in the inner parts of the Sahara. Air humidity and precipitation in these desert regions cause at least as much inhospitable conditions as the ones caused by the high

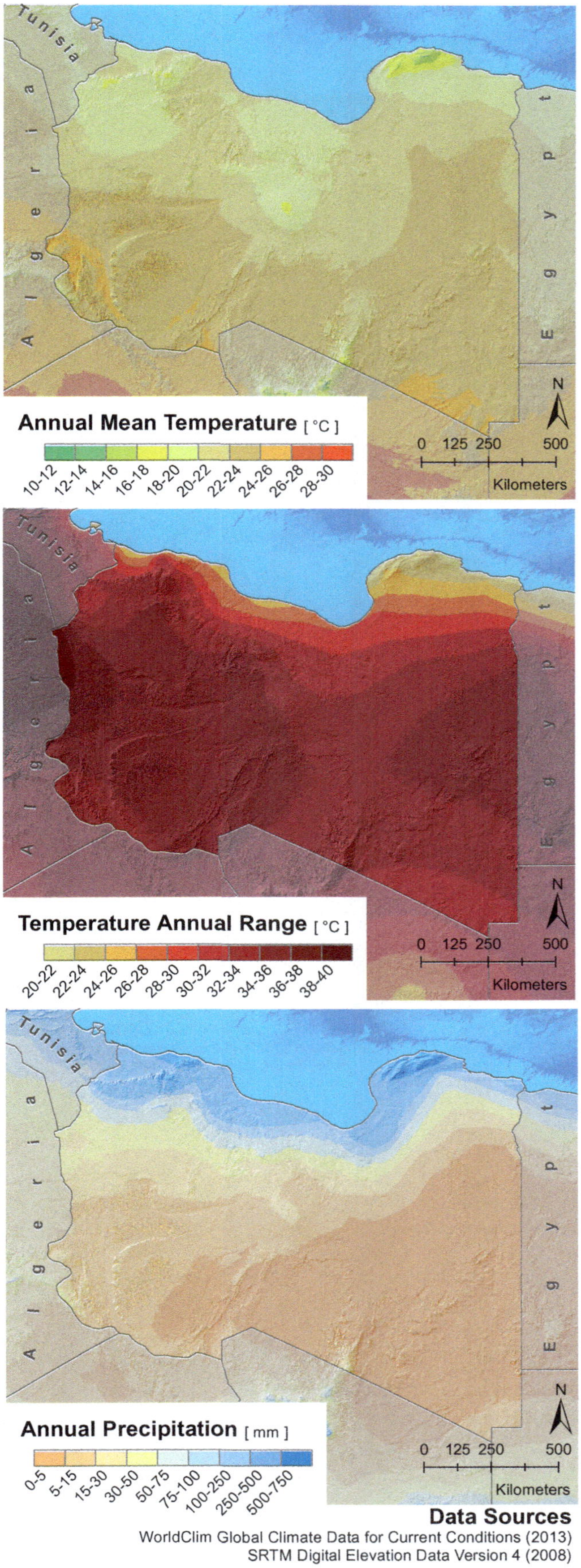

Fig. 1.5 Climatic conditions in Libya

temperature. The air humidity in the Sahara is generally very low and does not exceed 20% relative humidity with a slight decrease from north to south. Such a slight decrease can also be seen in the data for the annual precipitation. Measures indicate an amount of less than 15 mm for the most part of the Sahara. Only in the northern areas between 28 °C and 29 °C, northern latitude up to 50 mm can be expected. In some parts, no rain at all may fall over several years, but then the statistically expected amount for these years may fall in only a few hours.

The predominant wind comes from the northeast and blows nearly constantly throughout the year. Locally, hot winds often lift sand and dust particles from the desert floor, whirl them up and move them over the surface as dust devils or take them south-westward as dust storms. In situations when the wind comes from the south, air masses from the desert called Ghibli reach the coastal region and cause very hot, dry and often dusty conditions.

In contrast to these conditions, the coastal regions are influenced by a Mediterranean-type climate with hot and dry summers on the one hand and mild and wet winters on the other hand. As a result, the annual mean temperature along the coastline mostly shows around 19–20 °C with a spot of 21 °C in the area around Tripoli. Slightly lower temperatures, around 17–18 °C, can be found in the mountainous area of the Jabal al Nefusah and most parts of the Jabal al Akdar. Comparably "low" temperatures of less than 15 °C can be seen in the higher parts of the Jabal al Akdar, thus representing the coldest region of Libya overall. According to the Mediterranean-type climate, the temperature annual range is completely different from the one in the Sahara and shows values of not more than 20–22 °C in those mountains and around 24 °C along the coastline itself. Only along the stretch between Surt and Ajdabiya with its semi-desert conditions, a higher annual range of up to 28 °C can be recorded. Due to the rainy winds coming from the west and northwest, the highest precipitation can be found in the eastern parts of the Jabal al Nefusah in the region between Tripoli and al Khums and in the Jabal al Akdar. Although limited to the period between October and April, the precipitation during this time reaches up to 500–750 mm which ensure the comparably good conditions for agriculture and housing already mentioned above.

1.2 Terrestrial Ecosystems

Klaus Braun and Jacqueline Passon

In line with its topographic and climatic conditions, Libya's terrestrial ecosystems are mainly characterised by the Mediterranean Macrogroups in the north and the Saharan Macrogroups in the south (Fig. 1.6). There is no sharp boundary between these two formations, but rather a transition zone with vegetation influenced by the adjacent both northern and southern Macrogroups. Nevertheless, the boundary between the Mediterranean and the hyper-arid

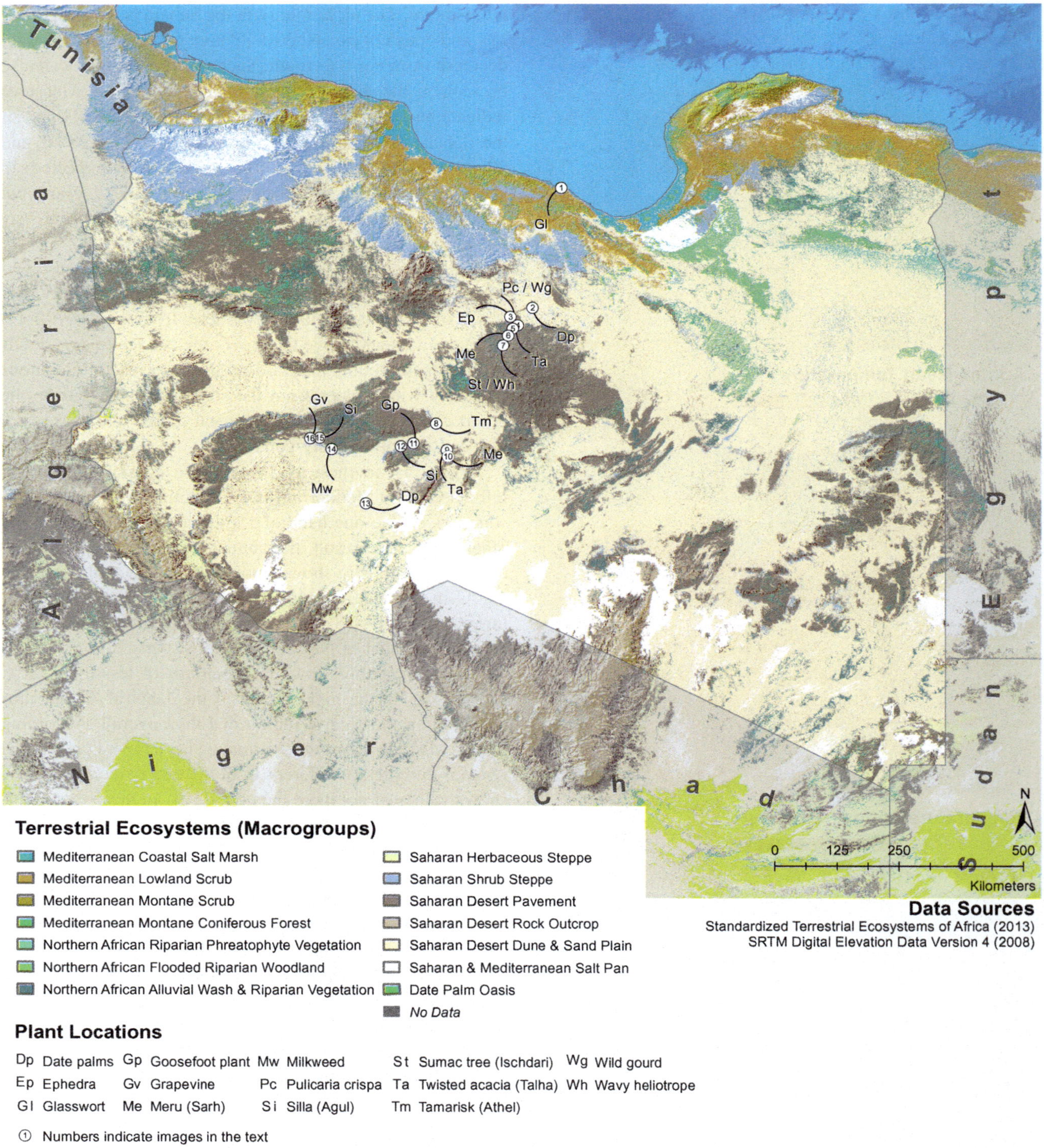

Fig. 1.6 Terrestrial Ecosystems in Libya

Sahara often follows the southern edge of the occurrence of the halfah grass.

The availability and the distribution of precipitation over the year are the determining factors for vegetation. The Mediterranean part of Libya, with its summer drought and winter rain, is marked by evergreen sclerophyllous vegetation, which can protect itself from periods of drought and heat thanks to its specific adaptation mechanisms. The plants have small, mostly firm evergreen leaves with low moisture content. Additionally, some of them have an acicular shape, are coated with a waxy layer or are partially hairy. In order to reduce water evaporation, the leaves can close their pores during periods of drought. Other plants, mostly herbs, do so by depositing essential oils. In addition, some of the plants have an expanded and deep-reaching system of roots to extract sufficient nutritional substances from the nutrient-poor soil. Known representatives of these plants are the olive and laurel trees, the cork oak and numerous herbs, such as rosemary or thyme. On the map that shows Libya's terrestrial ecosystems, these areas are classified as *Mediterranean Lowland Scrub* and *Mediterranean Montane Scrub or Mediterranean Montane Coniferous Forest*.

Low-lying areas along the coastline, where sea water occasionally reaches the surface, are exceptions to this pattern. Due to the high evaporation rates in this region, nearly all of the water evaporates, leaving flat salt pans. These areas, called *sabkhas* in Northern Africa, are characterised by particularly salt-loving varieties, including dense formations of perennial halophile grasses like alkali grasses or the halophyte grass aeluropus littoralis. Together, they form the *Mediterranean Coastal Salt Marsh*, which can mainly be found along the coast west of Tripoli, east of Misratah and in the region between Ajdabiya and Benghazi.

To the south, the *Saharan Shrub Steppe* with its shrublands and dry woodlands forms a transition zone between the Mediterranean coast in the north and the hyper-arid Sahara in the south. Wild olive and almond trees can be found here and there, but, above all, briars and undemanding grass varieties, such as halfah grass or drinn. In sediment basins without outlets, infrequent rainfalls form salt flats with no vegetation worth mentioning due to the instantaneous evaporation that leaves saliferous clay.

Despite the minimal and only sporadically occurring rainfall and a soil which, due to the slow soil-building processes, is poor in organic substances, the ecosystems of the Sahara, namely the *Saharan Desert Pavement* and the *Saharan Desert Dune and Salt Plain*, still have a flora with around 1200 different plant species. Over millions of years of natural selection processes, the plants in this region have developed specific survival capabilities. Those include extremely developed root systems, which either reach a depth of up to 35 m, or spread a closely meshed network of roots close to the surface. The root systems are large enough to compensate for the water shortage in the soil. In addition, plants have developed elaborate protection systems against evaporation, namely thick and small leaves which are covered with a waxy layer or plants which have the ability to roll up their leaves, such as grasses.

In the case of other plants, the so-called ephemerals, seeds can remain in the dry subsoil for many years until there is sufficient rainfall. They suddenly start to grow and reach the blossoming stage and seed maturation within just a few weeks. If the subsoil is sufficiently fertile, the desert can temporarily be transformed into a blooming meadow after such rainfalls. However, adapting to the climate alone still does not guarantee the survival of a plant, because the danger of being eaten is particularly great in regions with sparse vegetation. Therefore, it is an advantage to have thorns, or even to be poisonous. The colocynthis, for instance, is such a typical desert plant from the cucurbit family, which has very eye-catching, yet very bitter green-to-yellow melon-like fruits.

The most beautiful desert plants sometimes are those which tap into the root systems of other plants as parasites. They form yellow or splendidly violet inflorescences which are up to one metre tall and burst out directly from the ground. Even in the sand seas of the Sahara, vegetation might appear when rainfall exceeds a minimum of around 50 mm per year. Usually, the vegetation here consists of various grass types, which are capable of covering their water needs with their extremely long roots. Only the constantly moving, dry and unstable sand dunes, which leave plants with no support, are completely free of vegetation. Additionally, as in other regions of the Sahara, low-lying areas and basins without draining form salt flats that are nearly free of vegetation.

Besides the above-mentioned bush and shrub vegetation, some regions with extra-zonal woody vegetation can be found in the Sahara, namely where a greater amount of water is available. In the ecosystems of the *Northern African Flooded Riparian Woodland* and the *Northern African Alluvial Wash and Riparian Vegetation*, the vegetation typically does not follow the rainwater as much as the groundwater, which it reaches with its roots. Therefore, Acacia trees or tamarisks equipped with this capability do not grow just anywhere in the desert, but along the groundwater currents that often follow dry river valleys, the so-called wadis.

Extra-zonal woodland vegetation can be found in the oases, which quite often have been developed around artesian sources in the Sahara, too. These guarantee enough water supply for humans, animals and plants so that the settlements can permanently be inhabited. Despite being a cultivated plant, most probably introduced by the Arabs, the most typical plant oases are usually associated with is the date palm. An oasis without it is hard to imagine. The main reasons for this are not only the optimal growth conditions with high temperatures and a groundwater level that can be reached by the deep roots, but also the fact that the date palm is usually the main cash crop of an oasis, besides olives and other fruit trees.

Apart from the zonal succession of the terrestrial ecosystems, the mountain chains of Jabal Akakus and Masak Mastafat in the southwest, the Tibesti Mountains in the south and the Jabal Uwaynat in the south-east form the somehow isolated ecosystems of the *Saharan Desert Rock Outcrop* and the *Saharan Herbaceous Steppe*. Due to the higher elevation, these regions usually receive greater and more regular rainfalls and are characterised by cooler temperatures than the surrounding hinterlands. These conditions result in a woody and bushy vegetation of palms, acacias, myrtles, oleanders and tamarisks, as well as the occurrence of some rare endemic species.[1]

1.3 Green Spots in the Sand

Klaus Braun, Jacqueline Passon and Aniela Jeworutzki

The plants found during a journey through the Sahara demonstrate that the desert is neither lifeless, nor empty, but rather a diverse habitat. On the one hand, there are meadow plants and grasses. They are a basic resource for the extensive nomadic livestock farming and at the same time they were an important food resource for the caravans. Those plants can typically be found in the numerous oases of the Sahara such as Zillah, Tmissah or Murzuq. On the other hand, there are wild plants, typically growing on the gravel and rubble areas of the Sahara or along the wadis, where the groundwater currents are close to the surface.

There are a couple of plants throughout the Sahara which have been of great importance to the caravan trade. They supplied both humans and animals during their journey. Used as food, raw materials, tanning agents, dyes or medicine, plants have often been crucial for the success of a caravan. Others, like poisoning plants, could pose a substantial risk to the whole undertaking. Therefore, a good knowledge of plants found along the trade routes was indispensable. The following profiles of typical plants found along parts of the old caravan routes give a brief overview of the green spots in the sand. Their locations are marked in the map of Libya's terrestrial ecosystems (Fig. 1.6).

1.3.1 Glasswort ①

The glasswort is an annual succulent herb with narrow to squamous leaves and small (about 0.5 cm in diameter), white blossoms with reddish to violet colouration (Fig. 1.7). Glasswort is a halophytic plant that accumulates salts in its leaves and stems as an adaptation to its saline habitat. The young leaves of many varieties are used for food, either as vegetables or as salad. In addition, the herb is used as a diuretic and as an anti-worm medicine. In Libya, the glasswort is mainly found in the deep-lying salty areas along the Mediterranean coastline, called sabkhas.

The species once was well known due to its use in the production of potash, soda and other alkalis. However, with the use of abraum salts and the soda industry, extraction of potash fells into decline and is now economically insignificant.

Type: *Salsola divaricata*
Family: Amaranthaceae

1.3.2 Date Palms ② ⑬

With up to almost 30 m in height, roots reach down more than 20 m into the underground, and an age of up to 200 years, date palms are likely to be the best known of all

Fig. 1.7 Glasswort found along with the Mediterranean coastline. Only a number of steps away from the sea, there are ideal growth conditions within a flat salt pan or sabkha. It is of little surprise that the glasswort has settled here

plants found in Northern Africa (Fig. 1.8). Probably originating from India or the Arabian Peninsula, this cultivated plant was introduced by the Arabs.

The date palm itself is a bisexual woody plant with a frequently branchless trunk. As a cultivated plant, it requires specific attention and has to be nurtured according to guarantee good returns. It needs enough groundwater or has to be irrigated adequately. The female trees have to be pollinated to be able to produce the valuable dates. To achieve this, male trees are evenly spread over the date grove so that the male flower heads can be picked and put into the female trees by hand. This trick significantly reduces the number of male trees needed for the pollination, because one male tree can usually be used to pollinate up to 50 female trees, a procedure which sometimes is jokingly compared with a harem.

Fig. 1.8 Called the "green gold" of the desert, date palms are known to be the main cash crop of the oases and can be found everywhere where enough water is available

As the date palm not only provides dates, but also fibres, leaves and trunks used for making baskets, ropes, mats, sandals, furniture, building material and agricultural tools, it is also known as the "green gold" of the desert and has always been an important source of trade and nutrition.

Type: *Phoenix dactylifera* L.
Family: Arecaceae (palmae)

1.3.3 Ephedra ③

This light-yellow to yellowish green twig shrub can grow up to 2 m high (Fig. 1.9). The stems are rigid and hardly pliable. Sometimes, they have fine longitudinal groves similar to box stems, but without thorns or visible leaves. This form of vegetation is present in nearly the entire Sahara, as well as in dry areas of Eurasia and America.

The tea prepared from this herb is known by various names like Mormon tea, herbal dynamite, herbal XTC, ma huang or ephedra, amongst others. What makes it so attractive are the medically active alkaloids contained in this plant, such as ephedrine, norephedrine and N-methylephedrine. These are chemically related to the group of amphetamines and increase the content of the stress hormone adrenalin in the body.

The ephedrine herb was first used in 1923 to treat asthma. In Germany, ephedrine is subject to the Commodities Control Act because it can be misused to produce methylamphetamine.

Type: *Ephedra alata*
Family: Ephedraceae (ephedra plant)

1.3.4 Twisted Acacia ④ ⑨

The twisted acacia, called *talha* in Libya, is a small desert tree with an impressive umbrella crown which can reach 2–8 m in height (Figs. 1.10 and 1.11). The tree develops deep roots and therefore grows in regions where there is water at great depths, which cannot be used by other plants. In the Sahara, those regions usually can be found along dry river valleys, the so-called wadis, or lower-lying basins where the groundwater level is at around 5–10 m.

The twisted acacia forms sharp and rigid thorns, which can grow up to 6 cm in length, thus protecting its young and finely feathered leaves. Unfortunately, this does not prevent the camels from sticking their lips between the thorns and successfully biting the leaves. Mainly from October to December, and also in March and April, the tree forms light-yellow blossoms, which are arranged densely in small globules. The fruits are spiral-shaped husks.

The extremely sturdy wood of the twisted acacia is used as building material by the people in the Sahara. Other species, especially the acacia Senegal, are used to obtain gum Arabic, which is a natural gum consisting of the hardened sap of those trees.

Although the acacias are sometimes used as a symbol for Africa and are quite often described as somehow mythic trees by the early Saharan travellers, they are not only found there. Indeed, there are about 700–800 different species of acacias, with half of them growing in other regions, like Australia and the Pacific Ocean islands.

Type: *Acacia raddiana*
Family: Mimosaceae (mimosas)

Fig. 1.9 Ephedra located between the oases of Zilla and al Fuqaha. This plant can be found almost in the entire Sahara and is well known by the people who live there. It is traditionally used as medicine, or sometimes even as a drug, and plays an important role in the local lifestyle and the traditional caravan trade system. Burning of the ephedra (photograph on the right): after inhaling the smoke, mild physical and mental stimulation can be expected

Fig. 1.10 Twisted acacia found in lower-lying basins within the mountains of the Harudj al Aswad (above) and along a dry wadi within the mountains of the Jabal Bin Ghunaymah (below), where the groundwater levels can be reached by the roots of those trees

1.3.5 Wild Gourd ⑤

The wild gourd or colocynth is a desert plant native to Northern Africa (Fig. 1.12). Most probably, it was already known as a medicinal plant during the time of King Solomon. Other common names are bitter apple, bitter cucumber, desert gourd or vine of Sodom. The plant's long taproots reach deep into the ground, while its vine-like stems spread in all directions trying to find something to cling on using its auxiliary branching tendrils. The wild gourd has

Fig. 1.11 Thorns of the twisted acacia are very sharp and rigid to protect its highly demanded young leaves

Fig. 1.12 Wild gourd spread out in a small depression at the edge of the mountains of the Harudj al Aswad. Because of their bitterness, the fruits are usually not fit for human consumption, but nevertheless used for medical treatment

large yellow blossoms, and its leaves are pinnately lobed, coarse and hairy. The yellow or sometimes also green-striped fruits are usually as big as a grapefruit. As they are very bitter, they are usually not eaten, but only used as a source of food in bad years. Until today, the dried skin of the immature but full-grown fruits is used for medical treatment.

Type: *Colocynthis vulgaris*
Family: Cucurbitaceae

1.3.6 *Pulicaria crispa* ⑥

Pulicaria crispa (Figs. 1.13 and 1.14) is one of the most common plants throughout the entire Sahara. There are many locally used names, depending on the region, but not a single one which can easily be taken as a common name at all. Typically, it is used as nourishment for the camels and other domestic animals. It has an aromatic, slight taste of ginger. In the Sahara, it is used for infusions and sometimes for medical treatment.

The plant forms a thick, almost spherical bush that grows up to 60 cm in height. The upper area has many branches with lots of tiny yellow flower heads. The leaves are slim, around 10 cm long, and more or less hairy on the underside. The leaf margins are rippled.

Type: *Pulicaria crispa*
Family: Asteraceae (Compositae, Aster family)

1.3.7 Meru (Sarh) ⑤ ⑩

Meru, or Sarh, as this tree is called in Libya, is an evergreen tree with a spreading, well-branched crown, which reaches a size of up to 10 m (Fig. 1.15). Its oval and fleshy leaves grow directly from the trunk or the branches. It can be found in drier, often sandy areas all over Northern Africa.

Fig. 1.13 Overview of a small depression at the edge of the mountains of the Harudj al Aswad where mainly wild gourds, *Pulicaria crispa* and even twisted acacia can be found. One characteristic of this area is that occasional rainwater accumulates in these depressions or small basins, thus leading to comparably good conditions for plant growth. Used as pasture land, the depressions are called Grar in Libya

Fig. 1.14 *Pulicaria crispa* growing in a small depression at the edge of the mountains of the Harudj al Aswad. As it can be used as nourishment for the camels as well as for preparing aromatic infusions, traditional caravans have always been pleased to find it along their trade routes

The tree is harvested from the wild for local use as a source of food, as any type of material and as a medicinal plant. Although the wood is mostly considered as unsuitable for firewood as it produces smoke with a nauseating stench, there are tribes that live in Libya who fumigate containers with a burning stick and it produces a good smell. In terms of its medical properties, it is reported that it is used by women during their pregnancy, apparently for the induction of labour. The young branches are widely used as chew sticks, especially in Northern Africa.

In Arabic, the tree is called Meru, a name used in the eighteenth century as the source for the genus name Maerua.

Type: *Maerua crassifolia*
Family: Capparaceae (Caper plant)

1.3.8 Sumac Tree ⑦

The sumac tree, called *Ischdari* in Libya, is a shrub which is native to nearly the entire Sahara (Fig. 1.16). It can reach a height of up to 3 m, with densely ramified branches and thorny twigs at the end. Its small trunks are 3–6 cm thick, and the leaves consist of three leaf lobes.

Since the wood of this shrub is very sturdy, it is used for the production of any kind of everyday objects, from knife handles to saddle frames. Furthermore, the wood is used to produce charcoal for sometimes very cold winter nights in the desert. The fruits of the sumac tree can be eaten, and the leaves and flower buds are often chewed to banish thirst.

Even if the sumac tree could be found almost everywhere in the Sahara, there are areas in which the shrub has been overused and is nowadays quite rare.

Type: *Rhus tripartitus*
Family: Anacardiaceae (Sumacs)

1.3.9 Wavy Heliotrope ⑦

The wavy heliotrope is an herbaceous perennial plant that can grow up to 50 cm tall (Fig. 1.17). The plant is densely branched from a wooden base and presents alternate, simple

Fig. 1.15 Sarh tree in a sandy area within the mountains of the Harudj al Aswad. Especially, during dry periods the plant is an important source of nutrition. Camels are fond of the flowering shoots

and frequently bristly, hairy leaves. It is likely to be found in sandy wadis or on calcareous ridges and is sometimes abundant in poor pastures.

In some regions, the dried and powdered plant is added to water and drunk to combat fatigue; while in other regions, it is applied to treat headaches. Besides its medical use, a macerate of the plant is used as ink in the Western Sahara.

Type: *Heliotropium ramosissimum*
Family: Boraginaceae

Fig. 1.16 The sumac tree, or Ischdari, is known to provide sturdy wood which can be used for many purposes of desert life. Additionally, the green leaves are eagerly eaten by camels and other animals

Fig. 1.17 The wavy heliotrope is one of the Saharan herbaceous plants which are used for medical applications. According to the region, the local knowledge and the tradition, the use varies upon the different diseases treated by medicine based on this plant

1.3.10 Tamarisk (Athel) ⑧

The tamarisk is one of the most frequently occurring plants in the Sahara and can particularly be found in regions where the groundwater level can be reached by the deep-ranging roots (Fig. 1.18). The tamarisk, called Athel in Libya, has long, slender branches with numerous small, grey-green and scale-like leaves. Clusters of small pink flowers, hanging at the ends of branches or from the trunks, give the plants a feathery appearance.

Because of the mostly sub-optimal growing conditions in the Sahara, the tamarisk quite often develops only into

Fig. 1.18 A Tamarisk (near Tmissah) grows on top of a small elevation formed from its own falling leaves, dry and deadened branches and aeolian sand deposits

bushes or shrubs. In case of comparably good conditions, the plant can grow up to 10 m tall and forms knotty, warped trunks with strong barks.

Together with dry and deadened branches as well as aeolian sand deposits, the falling leaves form a stable, matted wickerwork which over time can reach some 3–5 m, with large clusters of tamarisks on top.

In the Sahara, the wood of the tamarisk is mainly used for carpentry or firewood, while the plant galls may be used to tan sheep or goat skins. The sugary excretions of the *Tamarix mannifera* are collected as manna and used in folk medicine.

Type: *Tamarix*
Family: Tamaricaceae

1.3.11 Goosefoot Plant ⑪

The goosefoot plant (Fig. 1.19) is a woody, stout and much-branched shrub usually 30–70 cm tall, which can be found throughout the entire Sahara, and it favours saline and sandy places (Fig. 1.18). Its leaves are sessile, small (2–4 mm in size) and rounded and often also found in a pyramidal form. They are densely grouped on the thin branches. If there is good water supply, the leaves are very fleshy. Otherwise, they can be hard and crumbly.

The juice of the leaves contains odorous substances (amines) that smell-like fish. Without this penetrating fish odour, it could easily be confused with the Mediterranean saltwort or *Salsola vermiculata*. While the latter is often used as a fodder plant, the goosefoot plant is grazed by camels, but only reluctantly by other livestock animals. According to the people living in the desert, the goosefoot plant is therefore primarily good for making fire or as a certain type of fuel.

Type: *Salsola baryosma*
Family: Chenopodiaceae

1.3.12 Silla (Agul) ⑫⑮

Silla is one of the classic thorn bushes of the Sahara (Fig. 1.20). It can mainly be found in parts of wadi beds that are covered with alluvial deposits. Known as *Agul* in Libya, this plant has light violet flowers with four identical flower petals. The leaves and the shoots have developed into sharp, hard thorns, and the fruits are chickpea-like. It reaches a height of 0.3–1 m and forms a bushy and spherical shape.

After dry years, only light to dark brown bushes remain in the landscape, often entwined by other plants.

Occasionally grazed by camels, silla is mainly used for urinary tract diseases.

Type: *Zilla spinosa*
Family: Brassicaceae

1.3.13 Milkweed ⑭

The milkweed is a 1–4-m-high flowering plant with very large round to heart-shaped leaves which can reach up to 20 cm in diameter (Fig. 1.21). Its bark is light greyish-brown, very

Fig. 1.19 The goosefoot plant can be found nearly everywhere in the Sahara and especially in those places where saline and sandy conditions exist. Grazed by camels, this plant is only reluctantly eaten by other livestock animals

Fig. 1.20 Silla, called Agul in Libya, is a typical thorn bush of the Sahara, which forms sharp and hard thorns, and is quite often entwined by other plants

thick, corky and deeply rutted. It forms clusters of violet flowers and red fruits, which can grow up to the size of mangos.

The milkweed is native to Northern Africa and can usually be found at the edge or in the middle of dry river beds. Traditionally, the milkweed was cultivated as an ornamental plant, but it has fallen out of favour because the leaves and stems contain a white, sticky and very poisonous milky juice. Already mentioned in ancient times, this plant is known by a lot of common names, for example apple of Sodom, kapok tree, rubber bush or swallowwort.

Type: *Calotropis procera*
Family: Apocynaceae

1.3.14 Grape Vine ⑯

Being known for loving warm to hot temperatures and adequate water supply, the grape vine is a common plant within the Mediterranean region. There its deep roots are usually able to get to the groundwater and help to withstand the dry summers (Fig. 1.22). The grape vine is known to be one of the oldest cultural plants of humankind. Picture representations of its cultivation are known from Egypt, dating around 3500 BC.

When Libya was an Italian colony, the grape vine was one of the most common fruit trees in the oases. In those days and until the 1950s, it was usually cultivated in mixed plantations together with olive, almond, wheat and barley. The grapes were commonly used as a fresh fruit, to produce grape juice, or dried and used as raisins for baking and cooking. While in other countries grapes have mainly been

Fig. 1.21 A milkweed tree near a house in the oasis of Murzuq. Although native to Northern Africa and traditionally used as an ornamental plant, the milkweed has fallen out of favour because of its poisonous sap and is mostly regarded as a weed

used for making wine, this has never been the case in Libya because of the prohibition of alcohol.

Using irrigation and semi-irrigation techniques, the cultivation of grape vine was not only limited to the Mediterranean region, but could also be found in oases in the south. In the 1950s, the cultivation of grape vine declined in favour of citrus plants.

Grape vine can still be found at several places in Libya today, including traditional farmland in the middle of the desert. Mostly, however, these are relicts from the past rather than representing important agricultural crops.

Fig. 1.22 Grape vine located at an old farming area near the oasis of Tasawah, west of Murzuq. Although grape vine is not common in Libya any more today, the plant reminds of former times when it was used as an important agricultural crop by Italian colonists

Type: *Vitis vinifera*
Family: Vitaceae[2]

1.4 Human Activity

Klaus Braun

A good opportunity to get an overall impression of human activity at night is provided by the use of satellite information of the earth.

Figure 1.24 shows Libya like it is seen in the "Blue Marble" images recorded by the VIIRS on the Suomi NPP satellite in 2012. This image is underlaid with the SRTM digital elevation model and supplemented by the populated places taken from the National Topographic Map of Libya and the populated places dataset from the Natural Earth archive.

As the night-time lights illustrate in this map, most of the human activity can be observed within the triangle between Zuwarah, Gharyan and Tripoli. Other spots of activity can be found in the Jabal al Nefusah and in the Jabal al Akhdar and also in the south in the region of Sabha. Most of the major oases in the desert region like Ghadamdis and Ghat in the west or al Kufrah in the east can also be detected by their visible light emission and even smaller oases like al Qatrun or Tajarhi in the south glow through the night. Even more surprising than the night-time lights that correlate to human settlements and built-up areas are the large spots of light in regions where no villages exist. This is the case in some areas along the border with Algeria in the west, in the southernmost part of the Idhan Awbari and in the almost uninhabited region of the Sarir Kalanshiyu. Like in the other areas, the night-time lights here suggest human activity. But in this case, the different "seas of light" are associated with the oil fields and the flaring of gases associated with crude oil production at those sites.

Accepting the fact that the economic development of a country is somehow reflected by changes in the emission of light, differences between appropriate satellite data taken at different times can help to detect and understand characteristics of spatial development patterns. With this in mind, composites of the NOAA DMSP-OLS Nighttime Lights Time Series from 1994 to 2009 (Fig. 1.23) have been used to analyse those changes for Northern Libya and especially along the coastal stretch where most people live. The result of this comparison is shown in Fig. 1.23 where increasing light emissions are marked by colours that turn from light to dark red. Comparably small changes in situations where high light emissions have already been recorded in the past are marked by yellow and orange colours.

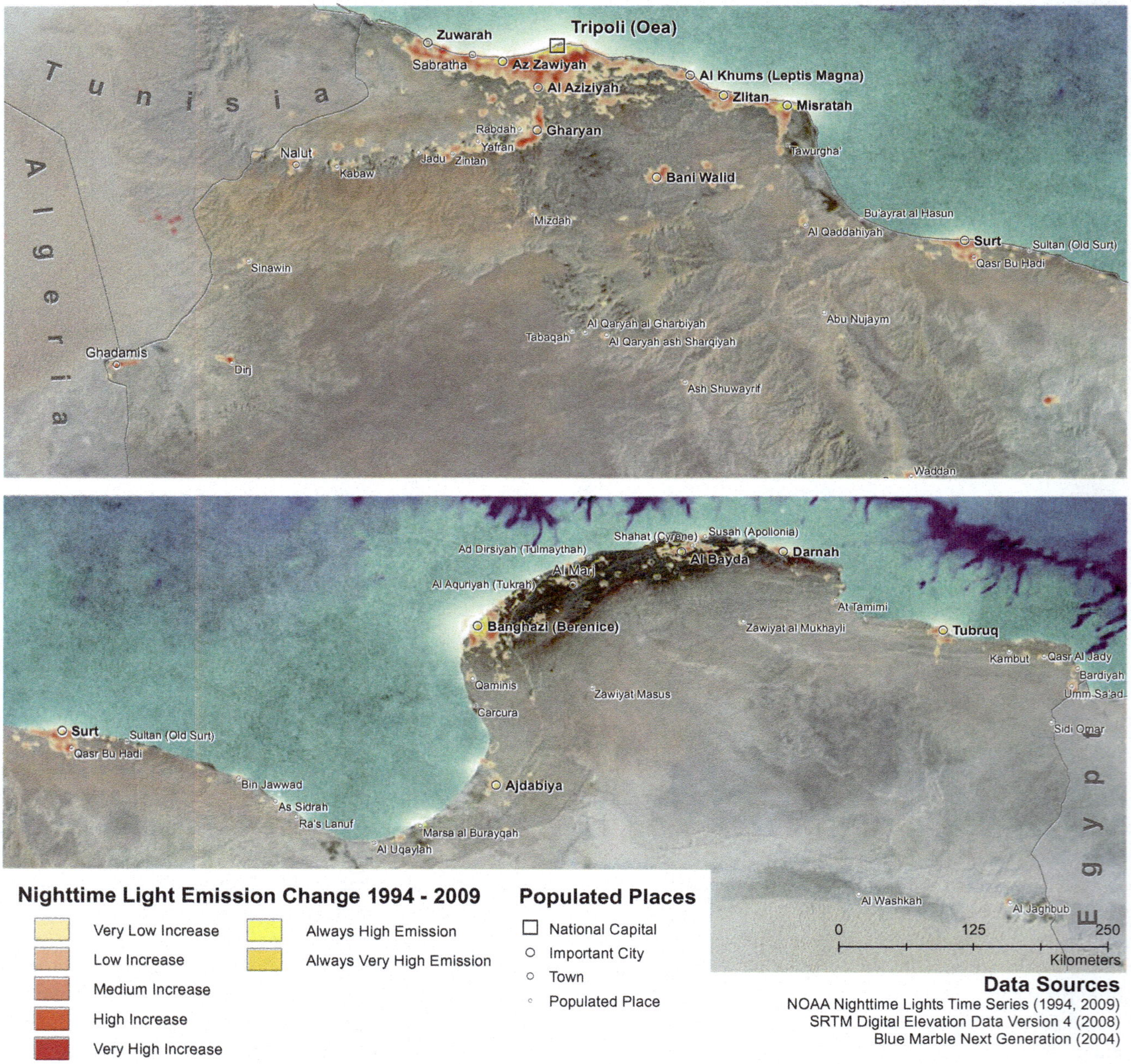

Fig. 1.23 Night-time light emission changes between 1994 and 2009 in northern Libya

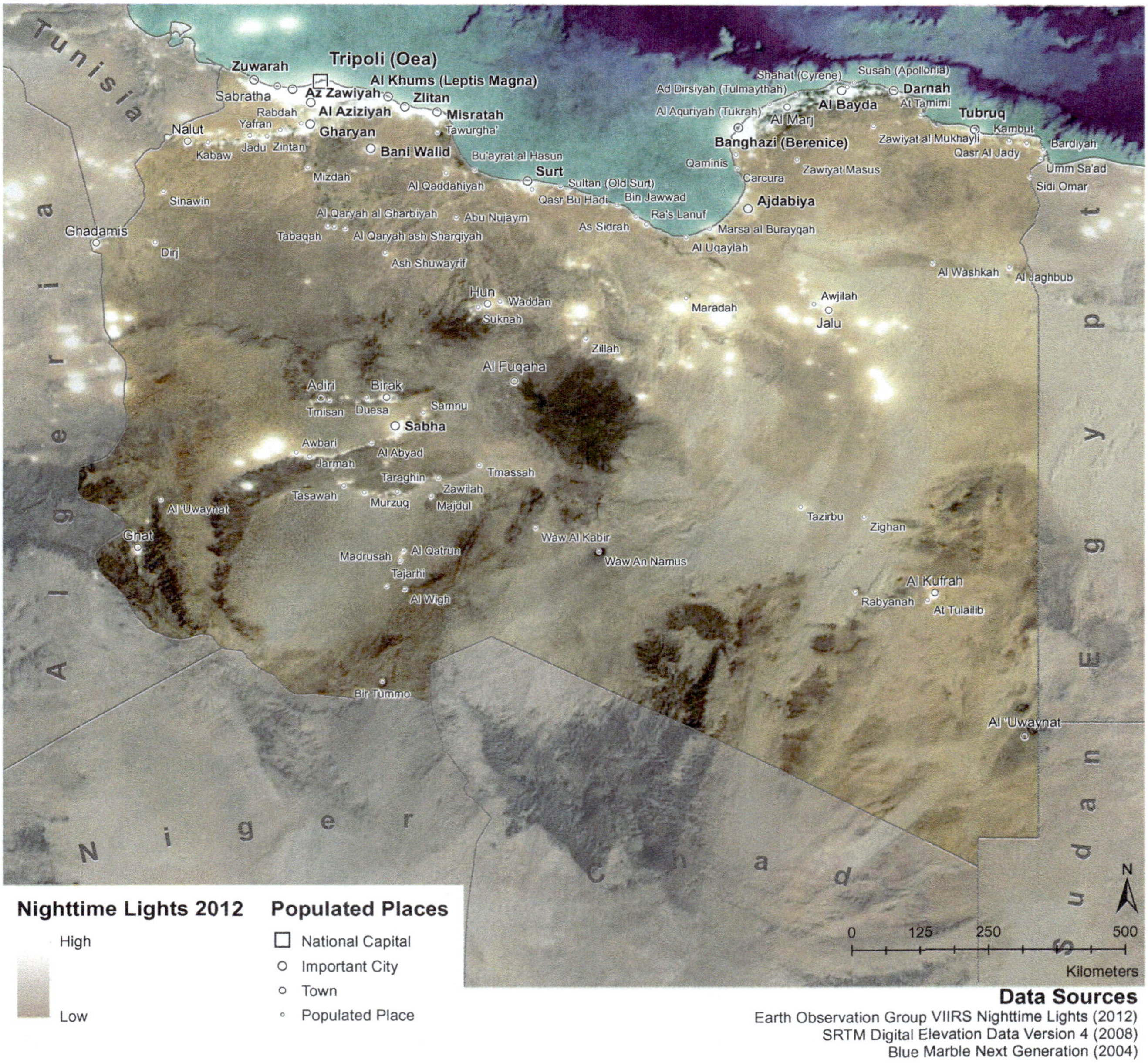

Fig. 1.24 Night-time lights in 2012

The predominant impression is that nearly everywhere where people live, light emission has increased, although the differences vary widely. Most of the increase can be seen along the stretch between Zuwarah and Tripoli, in the regions south and south-east of Tripoli and along the connection between Tripoli and Gharyan. East of Tripoli, the coastal road to Misratah appears as some kind of a secondary development axis. Beyond these axes, high increases in light emission are more or less bound to the surroundings of bigger cities along the coastline like Surt, Benghazi or al Bayda or to major oases in the interior like Bani Walid, Ghadamis or Dirj which appear as somehow isolated development spots in the otherwise almost light free environment. Minor increases in light emission can be found around nearly every populated place, equally whether it is a small village like al Qaryah ash Sharqiyah or medium-sized cities like Mizda or Ajdabiya.

Besides these indications of increasing development, the yellow spots that identify high light emissions today and in the past remain remarkable as they refer to cities and oases with a comparably long history and places of great importance. First of all, this is true for the city centres of Tripoli and Benghazi, but also for smaller centres like the ones of Zlitan, Misratah, Bani Walid, Ajdabiya, al Bayda or Tubruq. They can be detected as historical spots in an area with otherwise only few urban settlements.

1.5 Libya and Its Regions-Details About the Old Cultural Landscapes of Tripolitania, Cyrenaica and Fezzan

Jacqueline Passon and Klaus Braun

Whereas a more natural landscape-based characterisation of Libya results in a, broadly speaking, fertile and liveable north and a sparse and hostile south, the distribution of the population and the spatial differences in human activity indicate further distinctions. Those include, in particular, the ones between west and east. This leads to a three-way split of Libya into the regions of Tripolitania in the northwest, Cyrenaica in the east and Fezzan in the south, which can be understood as a result of the specific historical conditions in these regions.

With respect to designations used for these regions, it can be seen that even their names reflect somehow the historical development with its long-lasting series of discovery and colonisation. Whereas Tripolitania and Cyrenaica are denominations which are derived from Roman provinces, the name Fezzan is said to come from the Tuaregs' word Taffsania which means "the edge of the hill". Others, however, believe that the name is derived from the Roman word Phasania or Phazania, which may mean "the country of the pheasants" and had been used at the time of the Islamic conquest.

But regardless of the given designations, the historical approach of Libya as a country constituted of three main regions has been maintained to the present day and can be used to reflect in more detail differences not only in the natural landscape but also in the culture and tradition of the Libyan society.

1.5.1 Tripolitania

Starting with the north-western part of the country, the so-called Tripolitania region, the natural landscape is composed by the coastal plains of the Jefara and of the Dafnia, the mountainous escarpment of the Jabal al Nefusah and the adjacent stone desert of the Hamadah al Hamra. A steep slope towards the south leads then to the sand, gravel and scree desert of the southern region called Fezzan.

The coastal plain is a steppe. It is divided into the Jefara in the west, which stretches from Gabes in Tunisia to al Khums, and the Dafnia in the east, which comprises the narrow coastal strip between al Khums and Misratah. Although named plains, they show significant differences in height. While they are flat near the coast, the land rises gently towards the south and reaches, for example, a height of up to 380 m at the slopes of the Jabal al Nefusah. Overall, the coastal belt is defined at many places by sabkha areas in the south that extend up to 15 km inland.[3] A conspicuous morphological element is the few watercourses. In the west, there are only small wadis that end immediately after exiting from the Jabal as their water volume is too low. In the east, on the other hand, where the Jabal approaches the coast, the wadis of Raml, Mzid and Turgut reach the sea.

The Tripolitan highlands are nestled at the Jefara. The Jabal al Nefusah (Fig. 1.25), which is composed of lime, gypsum and sandstone, is a 400-km-long plateau that rises 300–400 m above the Jefara and marks heights around 600 m above sea level. Permeable and soluble rock is characteristic for such karstic areas. The water seeps, flows underground and accumulates in the mountains wherever it meets the solid, impermeable base rock and emerges then as a spring at the edge of the Jabal. These springs are centres of small settlements, where fruit trees flourish and vegetables and cereals are grown on parcels. Due to the location of the Jabal al Nefusah, changes can be determined from west to east. In the eastern part of the mountains, precipitations are higher and the ground is more fertile.

The cultivation of grain and the planting of oil, fig and almond trees have brought along with radical changes in the landscape, particularly in the eastern part of the Jabal since the Italian colonisation. In the western part, the precipitations are too low and the soil layer is too flat for appropriate use. Adjacent to the Jabal in the south, there is a roughly 100–150 km wide belt of steppes and wadis located at 600–700 m above sea level. To the east, it extends until the Greater Syrtis where it reaches up to the coast. In the south, it is joined by the Hamadah al Hamra, the red rock desert. This is a wide basin plain, which is covered with blocky, angular rubble or rock material. The region's name, red rock desert, is derived from the red soil that emerges between the rubble and rock materials. However, the Hamadah al Hamra is no wasteland. Between November and April, the area is used as a pasture for cattle.

The Greater Syrtis is the separating element between the two main settlement areas in the west and east. The "hinterland" of the Syrtis consists of vast stretches of sand with numerous sabkhas, dunes and several wide wadis. After sufficient spring rainfalls, this area is covered by a short-lived vegetation. Owing to the comparatively favourable topographic and climatic conditions, the coast of Tripolitania emerged as one of the most important regions of Libya. The capital of which, Tripoli, became the historical

Fig. 1.25 View from the Jabal al Nefusah to the Jefarah Plain

starting and end point of the trade routes to the south. It has retained the name of the late antique small province of Tripolitania.[4]

The cultural landscape has been shaped by an eventful past. Many traces of ancient settlement activities can be found. The cities of Sabratha and Leptis Magna, which have been awarded with the status of world heritage, are impressive examples. Elements of the traditional Berber cultural landscape can be found mainly in Jabal al Nefusah where the granaries of Nalut and Qasr Hajj provide impressive architectural evidence. Berber tradition can also be found in the oasis of Ghadamis. In addition to the settlement artefacts, traditional forms of agriculture such as rainfed cropping of grain, fruit-tree cultivation and the traditional oasis cultivation can still be found here.[5]

Today, Tripolitania has undergone modern modifications in many parts, especially in the coastal areas. Contributing factors to this were mainly the state colonisation by the Italians in the years between 1922 and 1940, the settlement projects during the years of the kingly rule as well as the settlement and economic policy of al-Qaddafi. First, the Italian colonists settled in the north of Tripolitania and the Cyrenaica. They established agricultural family businesses and settlements, which were adapted to southern European development standard. Apart from the large concessions with their stately homes and tenant houses, scattered settlements emerged. The Italian settlers took the entire fertile arable land into their possession and pushed the Libyan tribes into the steppe region. They were left with livestock breeding as the only form of economy. In 1959, the Libyan government founded settlement projects to resettle these nomads and landless farmers, who had once been expelled by the Italians, and to turn them into sedentary farmers. Within the wake of al-Qaddafi's decentralisation policy, these projects were expanded further. Today, the coastal strip of Tripolitania is characterised by an almost continuous urban ribbon, which reaches from Sabratha via Tripoli up to Misratah.

1.5.2 Cyrenaica

Characteristic for the second region, the so-called Cyrenaica which covers the north-eastern part of Libya, is the Jabal al Akdar, the karstic mountains called "Green Mountains" with its densely grown slopes. The entire mountainous area, which reaches a maximum elevation of 880 m, is, together with the small coastal strip, another major settlement and agricultural region of Libya (Fig. 1.26). Sufficient winter rainfalls and a relatively mild climate allow for lush vegetation and agricultural use.

The external appearance of Cyrenaica can be divided into three parts: the highlands, the southern slopes of the mountains with the western and eastern foreshore areas and the adjacent desert areas. Only a small coastal strip separates the Jabal al Akdar from the sea. East of Susah, the ancient Apollonia, this strip becomes wider and turns into gentle hills. The coastal strip narrows to an extent in some places that the mountains rise straight from the sea. The rise of the mountains takes place in two steps. The first, more northerly step, rises steeply up to 300 m above the sea and has an overall length of around 400 km. Short and steep wadis cut through the slopes covered with Mediterranean shrubs. The second, more southerly step, rises somewhat more gently. It is 300 km long and towers above the first level by 150–200 m so that it reaches a height of 500–600 m above sea level. Between the stepped slopes, a plain extends towards the sea that is characterised by depressions and sinkholes. These types of landscapes are created by water accumulation on insoluble sediments and the resulting horizontal erosion.

A very fertile zone extends over large parts of the heavily karstified plateau between the two Jabal levels. So-called poljes, large closed cavities with underground drainage, are typical for karstic areas. One must imagine those as predominantly elongated, partly valley-like winding basins with almost level ground. Poljes offer favourable conditions for agriculture. Therefore, agriculture is limited to plains and wide valleys in the Cyrenaica. In the narrow valleys of the

Fig. 1.26 The settled region of the Cyrenaica is characterised largely by the upland plateau of the Jabal al Akdar

mountains, or even on its slopes, wood is chopped or cattle are grazing. An undulating plateau that slightly slopes to the south joins the second step. South of the watershed starts the flat southern slopes of the Jabal which are covered by steppe.

In line with the common division of Libya, the Cyrenaica comprises, apart from the historic landscape in the north, also the gravel and sand deserts. They join this steppe belt in the south and consist mainly of scree and gravel plains as well as dune seas. Referred to as the Libyan Desert in the border area between Libya and Egypt, this landscape shows all characteristics of the inhospitable and almost uninhabited expanses of the Sahara desert as they are also typical for the Fezzan. Only in the remote south, the oases al Kufrah and al Uwaynat are witnesses of human existence in the otherwise still less developed area.

Particular importance is attached to the desert areas between Jalu and al Kufrah due to the fact that the region is home to a large part of the resources, which are indispensable for the Libyan economy. In addition to crude oil and natural gas, the extracted fossil groundwater, which is pumped with immense effort from the many wells with depths of more than 500 m, supplies the inhabitants of the coastal regions using a gigantic tube system.[6]

The population of the Cyrenaica mainly concentrates on the coast and in the Jabal al Akdar. The main town with the seat of the administration is the port city of Benghazi. Since the onset of the crude oil boom, the city has recorded a steep economic upswing and today is the country's second largest city. The Cyrenaica's cultural landscape has experienced modern modifications, as did Tripolitania. Throughout the area there are numerous traces of ancient settlement, traditional Berber elements, however, are rare. The creation of about 1800 agricultural family businesses in the course of the "Ente per la Colonizzazione della Libia" by the Italians greatly changed the face of the Cyrenaica between 1932 and 1940.[7] The previous landowners, nomadic and semi-nomadic tribes, who drive their herds from the steppes in the south to the pastures of the Jabal in the hot summer months, where pushed into the barren steppe areas south of the highlands. Attempts were made to revive agriculture after the independence of the country in 1951, as in Tripolitania. However, only the "Jabal al Akdar Project" implemented in the mid-1970s was able to further stimulate the agricultural development of this area.[8]

1.5.3 Fezzan

The Fezzan, as the third Libyan region, can be described as the one, which does belong neither to Tripolitania nor to the Cyrenaica. In short, it can be described as the large remaining part. As part of the Sahara, the landscape is characterised by rocky, gravel or sandy desert areas which enchant with their wide-open and constantly recurring bizarre terrains.

Amongst the most famous topographic features within this area are certainly the Ergs or Idhans, as the sand dune fields or sand seas are called in Libya. The biggest ones like the Idhan Awbari or the Idhan Murzuq in the western and south-western parts of Libya cover an area of around 50,000 and 80,000 km^2. The sand dunes themselves appear in a multitude of shapes, depending on the wind conditions and the occurrence of water and vegetation. Some of them are oriented parallel to the direction of the prevailing wind, with a height of a few metres and up to 50 km long. Others look like crescents, with a steeply sloping leeward surface, or like parabolas with convex noses trailed by elongated arms, or like a mixture of both, as they give the dunes the shape of a star. With heights between 5 and 40 m and an extent of around 5–400 m in width and 25–20 m in length, those dunes often move with the wind and tend to migrate 5–100 m a year.

Although the sand dunes are somehow an allegory of the Sahara in general and of the Fezzan, in particular, most of the surface is covered by plateaus, arid mountains, sand- and gravel-covered plains, shallow basins and large oasis depressions.

In the western part and especially along the transient area to the Tripolitanian region, impressive tablelands like the Hamadah al Hamra, which are extending of several hundreds of kilometres and partly dissected by big wadis (Fig. 1.27), shape the landscape. In the central part, huge areas are built up by gravel. In Berber language, such areas are called Serir, meaning wide flat depressions. This then describes the occurrence of those landscapes within large basins and wadis.[9]

Originally called "*Qued*" in North Africa, the wadis refer to a normally dry riverbed that contains water only during times of heavy rain. Initially modelled by large rivers in times when the Sahara region was wet and green, today those dry valleys impress by their huge extent. In cases of sudden heavy rainfall, resulting flash floods can cause dangerous situations so that the crossing of wadis always bears a certain risk. Nevertheless, wadis often tend to be associated with human activities because of the existence of sub-surface water and sporadic vegetation.

Volcanic mountains like the Jabal as Sawda, the al Haruj al Aswad and the Waw an Namus sometimes force a break in the monotony of those plains. The same holds true for the arid mountains of the Jabal Akakus and the Masak Mastafat with their bizarre arches, gorges, isolated rocks and deep ravines. Although affected by one of the harshest living conditions in the world, human activity in the Fezzan has a long history. Traditionally and even today, the basis therefore is the availability and accessibility of fossil groundwater. Especially in low-lying areas and depressions, where it can be found only some tens of metres below the surface,

Fig. 1.27 Wadi system in Fezzan

such groundwater can easily be accessed by digging wells or by using simple pumps. In a usually hostile area like the Fezzan, such places look almost like a miracle. It is of no surprise that such places like the oases of Ghat or Murzuq are associated with a specific kind of mystery and supernaturalism.

In the Fezzan region, most of the oases that depend on fossil water can be found along the three more or less parallel and west-to-east-oriented low-lying areas of the Wadi al Shati, the Wadi al Adjal and the Wadi Murzuq. Whereas the northernmost of these three areas, the Wadi al Shati along with the northern edge of the Idhan Awbari with the oases of Adiri, Tmisan and Birak is said to host nearly 75% of all wells in the Fezzan, the most impressive of all those wadis seems to be the Wadi al Adjal. This region south of the Idhan Awbari between the oases of Sabha and Awbari is characterised by a large number of little villages with green palm gardens and huge sand dunes in the north. Like a miracle in the midst of these sand dunes, a considerable number of lakes can be found, indicating a groundwater level close to the surface. It is assumed that in times of higher precipitation, a huge number of those freshwater lakes covered this area before they disappeared around 5900 (±1000) BC. Recent lakes like the picturesque ones of Gaberoun and Umm al-Maa, which means the mother of water, can be seen as relicts of gradually colder and wetter conditions. According to archaeological findings in this area, this allowed for human presence during the Pleistocene and Holocene eras.

The cultural history of the Fezzan is directly connected with the succession of humid and arid climate periods. Climatic and ecological conditions can be recorded for the early Neolithic period, around 10,000 BC, which allowed the repeated encroachment of people to the interior of the Sahara desert. The Epipalaeolithic from the northern edge of Africa is classified as hunters and gatherers just as their predecessors. The diffusion of the rock drawings suggests that these tribes with a semi-nomadic lifestyle spread from the entire

North African coast into deep in today's Sahara desert. Most of these drawings can be found in the Acacus Mountains and Tibesti Mountains in the Fezzan; however, even in Tripolitania rock, drawings were found. The development of livestock farming and later of hoe cultivation is attributed to their Neolithic descendants. Because of good storage management, the people were able to spread out in large parts of today's Sahara area. The invention of pottery and the spread of terracotta about 6–8 millennia ago marked a profound change in prehistoric times: At that time, the transition from cyclical wandering that depended upon rainfalls and food resources to sedentarism took place. The time around 5000 BC then marked again the beginning of a long-lasting drought in the Sahara desert.[10] This change goes hand in hand with the domestication of cattle, sheep and other livestock. Agriculture, however, never became a dominant form of economy. At the end of the Neolithic period, wherever there was sufficient water, a farming oasis economy was created due to the progressive desertification. Up to 3000 BC oasis, farmers had established themselves in all important parts north of the central Saharan uplands using irrigation techniques. This process led to sustainable ethnic changes and cultural differentiation. The Neolithic period marked the most dynamic phase of human development in the Sahara desert. This period is characterised by the spread of pastoral nomadism, agriculture and the emergence of urban centres. The cattle nomad culture in the Fezzan is interpreted as predecessor culture of the Garamantes.

Another living testimony to the existence of human civilisation in the Fezzan region is the oasis of Jarmah, which today acts as a central point within the Wadi al Adjal. Excavations and historical sources prove that in ancient times, an empire existed there that was able to maintain active trade relations with cultures in Central Africa and the Mediterranean and which could successfully oppose the Roman expansion around 20 BC. Towards the end of the 2nd Millennium BC, Mediterranean peoples had advanced with horses and chariots from the Libyan Mediterranean coast towards the Central Sahara and expelled the negroide population of this area. These were the people of the Garamantes, as first described by Herodotus, whose political and economic centre became the Wadi al Adjal. They are described as farmers and ranchers and worked mainly as traders. The Garamantes controlled the trade routes between the Mediterranean and Africa's sub-Saharan regions. They also built irrigation systems and cultivated date palms.

Despite the imprint left by Arab influences over hundreds of years, the Fezzan shows many traces of this previously existent golden age of the Berber. Apart from the archaeological evidence, especially the numerous *Foggara* conduit systems give evidence about the cultural achievements of the Garamantes. These "underground tunnels" are a kind of horizontal wells, which are used to tap the groundwater level in the higher mountain regions. The available drinking and service water is then guided to the lower-lying oases. This technique, dating back over 3000 years and probably originating from present-day Iran, permits the establishment of oases where the groundwater level is too low to be reached with conventional wells.[11]

In addition, there are many so-called *qsur* (desert castles, singular *qser*), castles consisting of clay bricks. The earliest examples in the Fezzan are attributed to the outgoing 1st century BC. Most seem to be assigned to the middle ages (see *On major Trans-Saharan Trails*). The traditional Arabic-Fezzan cultural landscape was strongly altered in the oases by modern Libyan cultural landscape elements, and it must, however, be added that the old settlement areas were usually not covered by superstructures. The set-up of planned villages and agricultural farmland in the course of the decentralisation forced by al-Qaddafi since the 1970s was carried out in the vicinity of the traditional settlements. However, this policy could not prevent that three-quarters of the Libyan population concentrated on the coastal strips, yet it ensured that the inland was not depopulated and not cut off completely from modern developments. An example for this is Sabha, currently the third largest city in Libya and today's centre of the Fezzan. Between 1995 and 2006, the population growth in the region around Sabha was after all almost 3% resulting in the fact that the population in this area tends to grow, although slower than on the coast.[12]

As the inhabitants of today's oases were traditionally deeply involved in trans-Saharan trade, they lived until 2011 more on direct and indirect government support. This support was provided by infrastructure projects such as the construction of schools or universities or large-scale, agricultural projects based on artificial irrigation, which should contribute to the supply of the Libyan population.

Notes

1. Houérou (2008).
2. Bayer and Ritter (2006, 32–45), Zech and Hintermaier-Erhard (2002), Strasburger (2002), Schroeder (1998), Western (1989), Schubert and Wagner (1988), Dittrich (1983), Schiffers (1971), Kassas (1967, 162–181), Suter (1967, 65–123) Interviews with elders from Zillah, al Fugah, Murzuq, al Qatrun and Tajarhi.
3. Hecht, Fürst Klitzsch (1964, 413–470).
4. Baume and Sager (1992, 208–213), Schiffers (1975, 87–121), Ahmad (1969, 24–51), Hecht et al. (1964, 413–470).
5. Weis (1966, 17–24).

6. Braun and Passon (2009, 56–66), Schiffers (1975, 26–133), Ahmad (1969, 24–27), Suter (1967, 65–123), Hecht et al. (1964, 413–470).
7. Ahmad (1969, 24–27).
8. Schiffers (1975, 126–133).
9. Brentjes (1982, 14), Ahmad (1969, 24–27).
10. Brooks et al. (2004, 35).
11. Klenkler (2005, 110–127), Brooks et al. (2004, 35–39), Brooks et al. (2003, 73), Mattingly et al. (2003, 337), Brentjes (1982, 16–34), Ziegert (1969, 49–58), Ergenzinger (1969, 59–81).
12. Braun and Passon (2009, 56–66).

2 A Journey Through Time

Jacqueline Passon and Said Hamid

2.1 Caravan Trade in Ancient Times

Jacqueline Passon

2.1.1 Ancient Geography

According to Herodotus, the description *Libye/a* was understood in ancient times to mean all of North Africa from Egypt to the "Pillars of Heracles" (Herodotus, Book IV), the straits between Spain and Morocco, which are now called the "Straits of Gibraltar".[1]

The coastal landscape of present-day Libya is divided into three regions: Tripolitania in the west and Cyrenaica in the east, between which the approximately 700-km-wide strips of the Syrtis desert stretch out. In ancient times, these geographical conditions had already led to different developments: Tripolitania was more oriented to the west, and Cyrenaica more to the east. In the flat coastal plains of Tripolitania, the ruins of Sabratha and Leptis Magna constitute particularly impressive evidence of the Roman culture that was tied in with the traditions of Punic *emporia* (commercial settlements).

In contrast, the hilly coastal area of Cyrenaica has been populated by Aegean Greeks since the seventh century BC. Although this region was integrated into the Roman Empire, the connections of Cyrenaica to the Aegean region were still stronger than those to Tripolitania.[2]

J. Passon (✉) · S. Hamid
Department of Physical Geography, University of Freiburg, Freiburg, Baden-Württemberg, Germany
e-mail: jacqueline.passon@geographie.uni-freiburg.de

K. Braun and J. Passon (eds.), *Across the Sahara*,
https://doi.org/10.1007/978-3-030-00145-2_2

2.1.2 Phoenician Trading Posts in Present Day Western Libya

At the beginning of the first millennium BC, trade in the Mediterranean was in the sole hands of the Phoenicians. The endeavours of the Phoenician merchants led to the establishment of trading posts along the Mediterranean coasts in western Maghreb, on the Iberian as well as on the Italic peninsula, on Sicily and Sardinia. Over the course of the colonisation wave of the first millennium BC, the Phoenicians established trading posts in eighth century BC on the northern coast of Africa, which were initially only used seasonally.

The African coast between the Gulf of Gabès and the Gulf of Sidra is almost all flat, sandy and rather unwelcoming. The Phoenicians established trading posts, where their ships could anchor safely in the lee of a cape, where the trading routes from the African interior ended and where there was fertile land. Three of these outposts, which later carried the Roman names Leptis Magna, Sabratha and Oea, were developed into bustling urban centres. It was presumably the actual foundations of Carthage that dominated this coastal area for centuries. During this time, the Phoenicians came across nomadic inhabitants, who lived in tribes and belonged to one of the Hami Haitic ethnic groups.[3]

2.1.3 Greek Colonisation in Present Day Eastern Libya

The Greeks settled in eastern Libya and invaded the interior countryside in search of farmland. In 631 BC, the new settlers, who came from the island of Thera, won a larger settlement area around Cyrene. In this territory, which was later called Cyrenaica, they established several flourishing colonial cities, the capital of which was Cyrene. In response to this, the Phoenicians of North Africa gathered around Carthage and prevented further Greek settlement attempts. In the middle of the fourth century BC, the borders to each of the spheres of influence were contractually determined. Hence, the Gulf of Sidra became the dividing line. As a result of the training of the successor states of Alexander's Empire, Cyrenaica was affiliated with Egypt and later with an outlying possession of the royal house of Ptolemies, the successor of Alexander in Egypt (323–30 BC). As the sole ruler of this region, Ptolemies Apion eventually bequeathed Cyrenaica to Rome in 96 BC.[4]

2.1.4 Roman Display of Power in North Africa

The rise of Rome to a Mediterranean power triggered the so-called Punic Wars, which ended with the fall of Carthage in 146 BC. The Carthage region was a Roman province (*Africa*) right up to Fossa Regia. It marked approximately the border between the province of Africa and Numidia. The *emporia*, including Leptis Magna, Sabratha and Oea, ceded the Numidian allies to the Romans, who were content with just collecting taxes, so that the *emporia* actually enjoyed a factual independence. Within the escalating conflict between Rome and Numidia, three Tripolitanian cities formed an alliance with Rome.

During the Civil War between Caesar and Pompeius, the *emporia* made a pact with the Republican side. After the massacre of Thapsus (46 BC) and the associated defeat of the Pompeians, Numidia became the province *Africa nova* and former Africa became *Africa vetus*. As part of *Africa nova*, the *emporia* presumably fell under direct Roman rule at this time. Over the course of the Augustan provincial reform (starting in 27 BC), *Africa vetus* and *Africa nova* united to become the province *Africa proconsularis,* and the senate was presumed to be under the leadership of a proconsul.[5]

2.1.5 Pax Augusta and Economic Prosperity

During first century AD, the ancient cities and settlements (*civitates*) began to achieve privileged Roman legal status (*municipia, coloniae*). Leptis Magna and her two sister cities, Sabratha and Oea, were subjected to a massive urban transformation in early imperial times, which strongly connected their vision of cities to a Roman city model. By building temples, public squares and houses, the inhabitants established organised municipalities, according to this Roman model. The changes brought about by this process affected not only the political organisation, but also the economic system, social structure and spiritual life. For rich and ambitious members of the upper class, it opened up the possibility of embarking on a political career in their hometowns. Through positions as the decurionate, aedileship, chief magistrate and provincial flaminate, one could attain even more influence, power, prestige and wealth.[6]

During the first three centuries, the emperor's rule (*Pax Augusta*) in North Africa was hardly challenged. It was only the region of Cyrenaica that suffered greatly from 114 to 118 AD under the attacks of an uprising by the Egyptian Jews.

2.1.6 Caravan Trade as a Source of Prosperity

It was within this context that Oea, Leptis Magna and Sabratha developed into prosperous cities. Their locations at the various trading routes between the Mediterranean coast and Africa's interior made them important settlements for trade and communication. The trans-Saharan trade became a significant source of wealth for the inhabitants. The cities functioned as stockyards and loading terminals. It was especially the trading of slaves, ivory, gemstones and wild animals, intended for games in the amphitheatre that played a central role.

The mediators of the trans-Saharan trade between Libya's Mediterranean coast and the African interior were an ancient people from the Fezzan region, the Garamantes. They inhabited the present-day Fezzan region since the fifth century BC, around the capitals of Zinchecra and Garama in the Wadi al Ajal (Wadi al Hayat). Horse breeding and the use of chariots enabled them to defeat the surrounding tribes. After initial military conflicts with the Garamantes, the Romans had given preference to diplomatic actions since the middle of the first century BC and facilitated the expansion of north–south trading. For their services, the Garamantes received Mediterranean luxury goods, *amphorae* filled with wine, oil or garum (fermented fish sauce) and glassware, as proven by the archaeological finds discovered in their settlements (Fig. 2.1).[7]

The caravans that crossed the Fezzan region brought the goods from Central Africa to the Mediterranean. There, ships took over the transport of the valuable products to the most important centres in the Mediterranean region. Intensive trading was practised with cities such as Rome. Many African cities had a trading station in Ostia, Rome's port.[8] This is verified by pictorial representations and inscriptions on numerous mosaics at the Forum of Corporations in Ostia, which are partially still in good condition. The traders of Sabratha maintained a trading office in Ostia, as shown in the inscription *STAT (io) SABRATENSIVM* (Fig. 2.2). The picture on the mosaic of an elephant suggests that elephants or wild animals were traded too.

The Garamantes controlled the caravan routes from Africa's interior all the way to the Tripolitanian limes. Crossing the extremely arid landscapes was dependent on the few watering holes. The direct and shortest connection lead from Oea or Leptis Magna via Gharian to Mizdah and

Fig. 2.1 Traces of the early settlements of Greeks and Phoenicians in Libya: The temple of Zeus in Cyrene is one of the best-preserved examples of ancient Greek art of architecture. In contrast, Punic traces like the grave in Sabratha are rare. Due to a directive issued by the Italian colonial authority, only remains of the Roman epoch could be excavated and restored

Fig. 2.2 Mosaics in front of the Sabrathian trader's office at the "Forum of Corporations" in Ostia. Ostia was the port of Rome, where many African cities maintained a trading station

to the military camp of al Qaryah al Gharbiya, and from there southwards through the Fezzan region towards Lake Chad and further into Africa's interior. Monitoring was carried out by the individual military outposts. There was also a western caravan route that ran from Sabratha via Ghadamis, or ancient Cydamus, through the Hoggar Mountains to Niger. An eastern road began at Leptis Magna and headed southwards past the fort Abu Nujaym. The paths between these fortresses could be relatively easily blocked (Figs. 2.3, 2.4, 2.5, 2.6 and 2.7).[9]

2.1.7 Developments in the Late Antiquity

Under the rule of the emperor Diocletian (284–305), an imperial administration was reorganised through the reclassification of the Roman territory into smaller provinces. The ancient *emporia* region in western Libya formed the independent province of Tripolitania. Cyrenaica was released by Crete and divided into two provinces: the western *Libya Superior*, which included all of the Pentapolis, and the eastern *Libya Inferior*. In the fourth century, the

Fig. 2.3 Arch of Septimius Severus in Leptis Magna: Symbol of the prosperity of the ancient town during the second century

Fig. 2.4 Roman theatre in Leptis Magna: Entertainment nearby the sea

Fig. 2.5 Roman theatre in Sabratha

Fig. 2.6 Market place in Leptis Magna

Fig. 2.7 Detailed view on Cyrene

Tripolitanians, like Cyrenaica, were subjected to increased pressure by the Libyan tribes. The effects of the mass migration also reached North Africa by the fifth century. Within a few years, the Vandals of the west had seized the North African coastal region. However, they did not cross the natural barrier of the Gulf of Sidra. They held their ground in Tripolitania for about one hundred years. Repeated attempts to win back the East Roman central power, Tripolitania and Cyrenaica, were finally successful in 534 AD under the leadership of Justinian. As with the Vandals, the Byzantine ruler only came into power along the coast of Libya. Byzantine fortresses in the cities elucidate the reduced size of the ancient centres. One Byzantine dux resided in each of Leptis Magna and Apollonia. The control of the country's interior was assumed solely by the strengthened Libyan tribes, who also increased their pressure on the coastal region.[10]

2.2 The Light of Islam Penetrated the North of Africa

Said Hamid

Islam appeared in Mecca as a result of the mission of the Arab Prophet Muhammad ibn Abd Allah, who preached monotheism and condemned the worship of statues. The Islamic religion spread across the Arabian Peninsula, and the Prophet was the nucleus upon which the Arab Islamic state was established. After the death of the Prophet, in the year 632 AD, he was succeeded by Abu Bakr as-Siddiq, the caliph who strengthened Islam in the Arabian Peninsula. He sent armies to take over Syria and Iraq; however, this was not achieved until the caliphate of Umar ibn al-Khattab. Control over Palestine and Egypt was taken later in the year 640 AD.

2.2.1 The Arabs Take Over Libya

When the Arab leader Amr ibn al-As took control of the city of Alexandria, it was incumbent upon the Muslim forces to control the Cyrenaica region which, historically and administratively, depended upon Egypt. This became imperative in order to secure Egypt's western borders from any possible Byzantine attacks and also to spread Islam. The Muslim forces were encouraged in their endeavours by the quick and stunning victories which they achieved against the Byzantines. Moreover, they were motivated by the situation of the people of Egypt, who did not consider the Muslims as invaders, but rather liberators who had put an end to Byzantine oppression. Supplies were guaranteed because of the Muslims' control over Egypt and the absence of any form of Byzantine resistance. Most Arabic sources mention different dates for the Arab victory in Cyrenaica. Despite Alexandria surrendered to the Muslims in the year 642 AD, Amr ibn al-As was able in the same year to triumph in the Cyrenaican cities of Tobruk, Derna, Susa and Tolmeita without any resistance. After the fall of the city of Marj, Tukra, Benghazi, Ajdabiya and Sirt shared the same fate, Amr ordered a campaign, under the leadership of Uqba ibn Nafi, which headed towards the interior regions. This campaign reached as far as the city of Zawilah. Another campaign, this time under the leadership of Busr ibn Abi Artat, was able to conquer the Waddan and other neighbouring areas, as al Hakam, a ninth-century historian, reports. It should be noted that there is doubt on the historicity, in particular, to al Hakam's account of the expedition to Fezzan.

Amr ibn al-As headed for Tripoli, and on his way, he took control of Leptis Magna. He then besieged Tripoli and was able to take it over in 643 AD. In Tripoli, Amr established his mosque, which is Africa's second mosque. He then sent his army, under the leadership of Abd Allah ibn az-Zubayr, to take control of Sabratha. Consequently, Amr led his army to the city of Sharus, one of the capitals of the Jabal al Nefusah region and brought it under his control. Before the overtaking of the remaining North African cities, Amr considered it prudent to consult the Caliph, Umar ibn al-Khattab. He wrote him a letter, saying: "*God allowed us to be victorious in Tripoli. There are not more than nine days between this city and Africa. If the Commander of the Believers thinks Africa should be won, then this will be done*". However, the Caliph did not agree to Amr's proposal. The latter was anxious about a possible return of the Byzantines to Alexandria and, therefore, hastened his return to Egypt. He left behind Busr ibn Abi Artat at the head of a garrison in the Sirt area and Uqba ibn Nafi heading a garrison in the Cyrenaica region. As far as the other regions are concerned, Amr kept no garrisons behind because he considered his conquests of such regions as an initial step towards a permanent Arab Islamic conquest to follow. It should be stated that no reference was made to any aggressive action on the part of the Libyan tribes against the Arab Islamic conquest. The Libyan tribes might have considered the Muslim Arabs as liberators from Byzantine hegemony which had long usurped their resources.

During the ensuing periods, there was a succession of Arab Islamic campaigns which sought to strengthen the conquest. The result of these campaigns was the conquest of Ghadamis in 662 AD. Moreover, Abd Allah ibn Sad ibn Abi Sarh dispatched campaigns against Ifriqya (Africa). According to the accounts of the inhabitants of the city of Awjila, he was the one who established a mosque in that city. When he died, he was buried in Awjila, where his tomb still can be found today. Ruwayfa ibn Thabit al-Ansari also raided Africa. He was the Amir (Prince) of the region when he died in Cyrenaica in the year 674 AD, and his tomb is still to be found in the city of al Bayda. By the year 674 AD, Libya was subject to the Arab Islamic state. It became a Muslim military centre from which Arab armies advanced towards the battles further inside Africa. It was towards this military centre that the Muslim armies retreated to reorganise themselves and, later, to resume their attacks.

After the martyrdom of the leader Uqba ibn Nafi in Tehuda in the year 683 AD, that same year Zuhayr ibn Qays al-Balwi, who fought within Uqba's army, retreated with the troops to Cyrenaica and was appointed governor of inner Africa. Upon receiving the necessary supplies, he raided inner Africa and made resounding victories. He then moved eastwards and, upon reaching Cyrenaica, found out that the Byzantines had recaptured it. He proceeded with 70 of his troops towards the coastal regions and engaged the Romans in an unequal battle. About the year 689 AD, he and all of his companions fought till martyrdom, and their graves, known as "the graves of the Companions", are still to be found in Derna.

Arab administration in Libya did not establish itself firmly before the time of Hassan ibn an-Numan, who had stayed for 5 years in the region of Sirt, waiting for supplies. There he built palaces, known as Hassan's palaces, remains of which can still be seen up to this day. During the 3 years of his stay in inner Africa (701–704 AD), Hassan dedicated his time to spreading Islam and the Arabic language and the establishment of Arab Islamic administration.

For eight centuries, Libya was subject to numerous Islamic petty states, such as those of the Fatimids and the Zirids. It was also part of the Almohad state and, subsequently, of the Hafsid state, before succumbing to the Spaniards in the beginning of the sixteenth century.

2.2.2 Period of the Spanish Occupation (1510–1530)

The end of the fifteenth century coincided with the rise of a Christian power, namely Spain. It managed to oppose the Arab Muslims and to bring about the fall of the last Muslim outpost in al-Andalus, Grenada, in 1492. Spain headed the fanatical Christian Crusades against Islam, and the Muslims, who had previously reached the zenith of their political expansion, were considerably weakened.

Soon thereafter, Spain's attention shifted southwards towards Morocco, and many cities, such as al-Marsa al-Kabir, Oran and Bejaya, succumbed to it. Subsequently, Spain set its eyes upon the conquest of Tripoli. With this aim in mind, the king of Spain prepared a military campaign entrusted to Pedro de Navarra who managed to invade Tripoli on the 25 July 1510, following a strong resistance from the city's local inhabitants. When describing the battle in his report, the Spanish leader wrote the following: "*There was no room to set foot, if not on corpses*". Some sources state that the number of the Tripolitanian deaths reached five thousand, and another six thousand were enslaved.

The inhabitants of Tripoli soon organised themselves and adopted the nearby oasis of Tajoura as the centre for popular resistance. Since then, this city acquired a special place in the history of Tripoli. As a result of the local people's abandonment of Tripoli, as well as their strong resistance and lack of co-operation with the occupying authorities, the Spaniards tried to encourage the Christians, especially the Sicilians, to settle in Tripoli. To achieve this aim, the Spaniards offered the settlers many incentives, such as the provision of proper housing, the allocation of agricultural lands, exemption from all taxes for a period of 10 years, as well as civil and criminal impunity. Nevertheless, the call for settlement fells on deaf ears.

The Spaniards sought to fortify the fortress and the city of Tripoli, yet this task proved to be very difficult due to the resistance put up by the local inhabitants. In 1530, they preferred to cede Tripoli to the Order of the Knights of Saint John. In this manner, the Spaniards relieved themselves of a liability which was not to be overlooked and which had burdened the Spanish budget very much. Moreover, this move allowed the Spanish court to concentrate on its European problems.

The Spanish occupation was very damaging to the people of Tripoli. Many of them were killed, and huge numbers were enslaved and sold in the slave markets of Sicily. In addition, the commercial situation deteriorated very quickly, due to the lack of ships which sailed to the city's harbour. The popular resistance movement and the city's isolation from the coastal centres and from the interior resulted in a deterioration of the caravan trade. In addition, Tripoli lost many of its architectural landmarks, among them the Mustansiriya school and its grand mosque, along with many of its houses.

2.2.3 Tripoli Under the Knights of Malta (1530–1551)

The Order of Saint John hesitated to accept the Spanish offer of assuming the defence of Tripoli and its fortress. That was because, at that time, the order's weakness prevented it from assuming the responsibility of defending Tripoli and its fortress, especially since they were located amidst territories fuming with anti-Christian resistance. It was difficult for help to reach the order in this isolated area, especially since Ottoman naval units had started navigating in that region. The report prepared by the order's delegation, which visited Tripoli to scout the city's condition, was not encouraging. However, the order had no choice. The Spanish offer to cede Tripoli to the order was bound to the order's acquisition of the two Islands of Malta and Gozo, both of which could not be renounced. Thus, Tripoli was subjected to the Order of the Knights of Malta on the 24 March 1530. When they took over the city, they found themselves in a serious dilemma and in the midst of many dangers. For on the one hand, they faced popular resistance which was organised by the local inhabitants in Tajoura, and, on the other hand, they were exposed to the threat from the Ottoman fleet in the Mediterranean Sea. The means to defend Tripoli and its fortress were very limited, and during their occupation, the Order of the Knights of Malta did not leave any clear marks in Tripoli. This might be due to the fact that only fortified some sections of the fortress and of the city's bastions. In addition, the best halls of the saraya al-hamra (red castle), that is the main fortress in Tripoli, were turned into a church. In 1551, the Ottomans prepared a fleet which, with the help of the local inhabitants, managed to take over the city of Tripoli, and, consequently, Libya came under Ottoman rule.[11]

2.3 Libya in Modern History

Said Hamid

2.3.1 The First Ottoman Period (1551–1711)

After extending their imperial influence across Europe, the Ottomans diverted their attention towards the North African territories, aiming to occupy them and add them to their empire. Libya was well within this sphere of interest, and, as a result, the Ottoman sultan dispatched a fleet, led by Sinan Pasha and Dragut Pasha. These sought to take advantage of the inhabitants' revolt, which was based in Tajoura, against the Knights of Saint John. Ultimately, the knights were expelled, and Tripoli was liberated.

Since 1551, Libya became part of the Ottoman Empire, and it was ruled by a succession of governors. Among the most prominent, was the first governor Murat Agha, who sought to fortify the city and its fortress, and who turned the church inside the fort into a mosque. He also built a mosque in Tajoura which still carries his name till this very day. Governor Dragut Pasha (1553–1565), in turn, also fortified Tripoli, and a mosque (Fig. 2.9) built a fortress which bears his name. He also established the gunpowder depot. The Governor Othman Saqizli (1649–1672) was responsible for the building of the markets, the public baths and a religious school. He also strengthened the naval fleet and entered into trade agreements with some European monarchies.

The last years of the first Ottoman period (1551–1711) were characterised by the control of the Janissaries over the ruling circles in the country. Their influence increased to the extent that they appointed and deposed governors according to their whims and interests. The country became corrupted, causing the local people, who were exhausted by heavy taxes to revolt against governors. This was a time characterised by great political instability and the struggle for power.

Fig. 2.8 Map of the fortified old town of Tripoli, the harbour and the red castle. The wrecks of the frigate USS Philadelphia and USS Intrepid, that went up in flames, can also be seen

2.3.2 The Karamanli Period (1711–1835)

The eighteenth century witnessed the beginning of the reign of Ahmad al-Karamanli, who had managed to realise his dream and become the ruler of Libya that he achieved by taking advantage of the weakness of the preceding rulers. He took control of the affairs of the country and established a hereditary type rule which lasted from 1711 till 1835 and is referred to as the Karamanli period. This historical phase was characterised by its independence from the Ottoman state and the establishment of diplomatic and commercial relations with some European countries, engaging in direct interaction with them without resorting to the Ottoman authorities. During this period, fortresses were built to oppose the periodical European attacks from the sea, and shipyards were established to build a Libyan fleet. Within a short period of time, this fleet became quite important in the Mediterranean. Moreover, Arabic became the language of the state's institutions. The Yusif al-Karamanli period (1795–1832) witnessed the Libyan-American conflict in the Mediterranean Sea, during which the Libyan people resisted the American attacks and attempts aimed at securing the passage of American commercial ships. History has recorded the story of the capture of the American vessel Philadelphia by the Libyan navy in 1830 (Fig. 2.8). This was the second largest American naval unit and carried 307 officers and troops. America took advantage of the presence of Ahmad al-Karamanli, the brother of Yusif Pasha, in Egypt and encouraged him to return to Libyan rule, supplying him with weapons and facilitating his attack against Tripoli, starting from Derna. Events followed in quick succession until a treaty was signed between Libya and America. On the basis of this treaty, Libya handed over the American captives and, in return, received compensation. Furthermore, America turned its back on its agent Ahmad al-Karamanli.

Towards the end of the Karamanli rule, Libya faced an economic crisis which came as a result of its weak financial resources, particularly after the decline in maritime activity

Fig. 2.9 Mosque of Dragut Pasha, who was a governor of Tripoli during the first Ottoman period between 1556 and 1563. This mosque is one of the largest mosques in the medina of Tripoli

and the abolition of slave trade. In an effort to curtail the budgetary deficit, Yusif Pasha borrowed money from foreign traders, burdened the population with taxes and changed the currency eleven times. On their part, the foreign consuls intervened in matters concerning the rule of the country and, as expected, a revolt broke out against Yusif Pasha in al-Manshia, calling for his deposition and for the appointment of his grandson Muhammad as ruler of the country. The Ottoman state sought to calm the situation in the governorate, whereas the French and English consuls seized the opportunity to interfere in the struggle for power. Conscious of his inability to handle the events, Yusif Pasha had no alternative but to renounce power in favour of his son, Ali Pasha (1832–1835) who managed to secure the approval of the Ottoman sultan for his appointment as governor. Nevertheless, the revolt of the inhabitants of al-Manshia continued unabated.

2.3.3 The Second Ottoman Period (1835–1911)

The Ottoman state decided to settle the Libyan question and restore its sovereignty over the governorate of Tripoli. It was anxious about the European ambitions concerning the occupation of Tripoli, particularly after the French occupation of Algeria in 1830. For this reason, the Ottoman sultan sent a fleet to secure Tripoli, which became, once again, an Ottoman governorate.

At this time, the country witnessed a number of revolts against Ottoman rule. Foremost among the Libyan resistance leaders were Ghuma al-Mahmudi and Abd al-Jalil Sayf an-Nasr, but these revolts were suppressed. Some governors, in an attempt to improve the conditions of the governorate, were able to introduce a number of reforms. For example, civil and military schools, hospitals and post and telegraph offices were built, wells were dug, and a printing press was imported for the publication of newspapers and magazines. Furthermore, it was a time in which Libya witnessed a considerable architectural boom (Fig. 2.11).

For its part, Italy's ambitions to occupy Libya were evident. In order to pave the way for such an endeavour, Italy took a number of economic and political measures. Notwithstanding the attempts of some governors to resist these colonialist ambitions, Italy exploited the weakness and problems facing the Ottoman state and managed to occupy Libya on 7 October 1911.

2.3.4 The Italian Occupation

By the end of the nineteenth century, Italy's occupation ambitions became clear. The governorate of Tripoli was the last governorate under Ottoman control. The Ottoman state had become unable to protect the integrity of its territories. At the same time, Italy prepared for the occupation of Tripoli. At exactly 2.30 in the afternoon of 28 September 1911, the Italian minister of foreign affairs, di San Giuliano, submitted a memorandum to the Turkish grand vizier which stated: "*The Italian government ... has decided to occupy Tripoli and Cyrenaica militarily*". Following this communication, Italy entered into a state of war with Turkey and took the necessary military steps to execute the occupation. On 3 October 1911, the Italian navy started to bombard the Libyan coastal cities, and, on 5 October, Italian forces landed on Libyan soil. The Libyans called for *jihad*, a war of resistance, against the invaders. Epical battles were fought in

Fig. 2.10 Italians and their idea of a new urban design for Tripoli: At the seaside, they built the Lungomare Conte Volpi, a promenade lying towards the sea with the Miramare theatre in the foreground. On the right side: for the old Sidi Hamuda mosque, situated at the corso Vittorio Emanuele III, the Italians designed a Moorish facade (Laronde 1997)

Jiliana, al-Hani, Shari ash-Shatt, al-Murgub and elsewhere, in an attempt to defend religion and the Libyan soil. Faced with such fierce opposition, Italy tried to evacuate the land of its inhabitants. To achieve this aim, it started to carry out the most brutal deportation in history. Thousands of Libyans were deported to remote Italian islands. The first of such deportation campaigns was put into effect on 26 October 1911. In the face of the huge Libyan resistance, and having suffered a number of humiliating defeats, the Italian forces were restricted to an area which did not exceed the shooting range of its fleet. In the meantime, on 18 October 1911 in Lausanne, Italy signed a treaty with the Ottomans by which the Ottomans renounced Libya. The Libyans had, by now, lost all hope of Ottoman support and had to rely on their own resources in their fight against the Italians.

The Italian forces occupied several Libyan cities, such as Gharian, Sirt, Zawia and Ujailat, and managed to reach as far as Marj, in the east. The Libyan freedom fighters engaged the Italians in many intense battles, such as the battles of Jdabia, Sidi Krem al-Girba, in the Green Mountain. The Libyan fighters also resisted the Italian forces in the Fezzan. The battles of ash-Shabb, Ishkda and Mahruga were heated battles in which the Italian troops were forced to retreat, particularly after the battle of al-Gara, in Sebha. Italy tried to re-occupy the Fezzan, the region which controlled trade movement and the most important caravan roads linking the Libyan ports with the African interior. However, Italy suffered a great defeat in the Gordabiya battle on 28 April 1915.

In 1922, the Fascist party came to power in Italy. This party adopted a policy of extreme brutality towards the Libyans and resorted to executions without trials. It sets up concentration camps in Libya, most renowned of which were those of al-Agila, al-Magrun and Slug, in which thousands of Libyan prisoners were crammed, in an attempt to isolate the population from the freedom fighters. The most prominent of these fighters was Sheikh Omar al-Mukhtar, who was arrested and later executed on 16 September 1931.

Italy confiscated Libyan land and distributed it among the Italian colonists (Fig. 2.10). The Libyans were recruited to take part in Italy's colonialist wars in Ethiopia. They also formed part of the Italian army during the Second World War. Libyan lands were also converted into minefields which took their toll on man and greatly hampered the area's development. At the end of the Second World War, Italy was forced to withdraw from Libya. Italy left a destroyed and devastated Libya behind, which at the time became one of the poorest countries in the world.

2.3.5 After the Second World War

After the Second World War, Tripoli and Cyrenaica came under the British mandate, whereas the Fezzan came under French mandate. In the meantime, the Italian community retained its control over the country's economic resources. In the wake of the declaration of independence on the 24 of December 1951, Libya entered a new phase in its history as the United Kingdom of Libya. However, the country lacked state economic recourses and expertise. Poverty and illiteracy among the people were widespread. In fact, the Libyan state had to rely on foreign aid. The country witnessed a number of crises which left their mark on the country.

Fig. 2.11 Clock tower, a landmark of the medina, is the only surviving tower from the Ottoman period and was built during the second half of the nineteenth century

Among them, was the assassination of the minister for Palace affairs, Ibrahim al-Shalhi, by a member of the royal family in 1954. This event had considerable negative implications on the state. Another issue facing the Libyan state problem derived from the lack of a direct heir.

The economic situation in Libya improved considerably with the discovery and exportation of petroleum in 1961. This development greatly influenced various sectors of life in Libya. Soon, a five-year plan was put in place. The oil revenues enabled Libya to support the Palestinian cause, on the one hand, and the Arab frontline countries after the June 1967 war, on the other.

A number of factors led to the fall of the Libyan monarchy. In the year 1964, popular protests were organised against the foreign bases on Libyan soil. Other demonstrations took to the streets, as a result of the conflict between Israel and Egypt in 1967. The internal situation within the Libyan regime, characterised by various crises and the proliferation of political blocs, resulted in the alternation of eleven governments in eighteen years. A number of military officers, belonging to the lower ranks, carried out a military coup on 1 September 1969. They soon set up the Revolutionary Command Council to administer the country, and in their first declaration, they stated that the official name of Libya would be the Libyan Arab Republic. They specified their guiding principles, namely freedom, socialism and unity. They also stated that they would be eradicating corruption, bribery and favouritism.

Libya was greatly influenced by the internal and external policies adopted by Muammar al Gaddafi as President of the Revolutionary Command Council. One could consider his speech in the city of Zuara in 1973 as a decisive juncture, which expressed the domination over power. In that speech, he declared the suspension of all Libyan laws, as well as the so-called Cultural Revolution. Subsequently, on 2 March 1977, al Gaddafi declared the ushering of the People's Power, and the establishment of the Socialist Peoples Libyan Arab Jamahiriya (Fig. 2.12). Libya suffered great political and economic problems, among them, the suffocation of liberties, the control over commerce, the involvement in external wars, such as the war in Chad, confrontation with Egypt and the interference in internal affairs of other countries. A series of executions were announced via the mass media in the holy month of Ramadan, in addition to the infamous Abu Salim massacre.

On 17 February 2011, a number of factors led to the Libyan people rising up in peaceful demonstrations which were brutally confronted by the regime. This led to the people taking up arms and defending themselves. Faced with the people's will, the regime had no other possibility but to give in. Tripoli, the capital city, was liberated on 20 August 2011. Subsequently, on 23 October 2011, the National Transitional Council, from the city of Benghazi, declared the total liberation of Libya.[12]

Fig. 2.12 Al Gaddafi left various imprints all over Libya, many of which were eradicated

Notes

1. Herodot (2004).
2. Schmidt-Colinet (1979), Kirsten (1961).
3. Di Vita et al. (1999).
4. Di Vita et al. (1999), Kirsten (1961).
5. Di Vita et al. (1999), Mattingly (1995); Reynolds and Ward Perkins (1952).
6. Alföldy (1987).
7. Di Vita et al. (1999), Mattingly (1995).
8. Squarciapino (1966).
9. Weis (1996).
10. Di Vita et al. (1999), Mattingly (1995), Teutsch (1962), Charles-Picard (1962).
11. Al Hakam, Futuh misr wa-l-magrib, Al-Tijani (1958), Ghalbun (1960); al Ansari (1967) al manhal al adb fi tarihTarabuls al-Garb, Rossi (1974), Abbas (1970), Thiry (2004), at-Tillisi (1985), Hamid (1978), Vireau (1973), Az-Zawi (1963), al-Barghuthi (1973), Bernia (1969).
12. Al Ansari (1967) al manhal al adb fi tarih Tarabuls al-Garb, Rossi (1974), Abbas (1970), Thiry (2004), Tillisi (1985), Hamid (1978), Vireau (1973), Az Zawi (1963), al Barghuthi (1973), Bernia (1969).

3 Traders, Nomades and Slaves

Trans-Saharan Trade

Jacqueline Passon, Said Hamid, Hassan Mughrabi, and Sabrina Swifka

3.1 Cross-Cultural Network

Jacqueline Passon

Since ancient times, goods have been transported from Africa's interior to the Mediterranean ports. For centuries, the economic life of the Mediterranean coast, from the Sahara to Sudan, was determined by a network of trade relations, which consisted of traditional trans-Saharan caravan trading. The flow of traffic crossed through all of former Libya. The traders had their collection points and loading terminals in the interior and on the coast; in this regard, the caravan cities of Ghadamis and Murzuq played a central role for a long time. Besides trade in goods and merchandise, thoughts and ideas were also exchanged within a cross-cultural network. Furthermore, merchants were the first to bring the light of Islam to inner Africa. Others have followed them such as scholars, diplomats, men of religion or plain travellers. Indeed, Islam was a creed that developed in an urban, mercantile setting. The diffusion of Islam throughout the population took many centuries.[1]

Of the three large trans-Saharan routes, which rank among the oldest trunk roads in the world, two crossed present-day Libyan territory: originating from Tripoli, one eastern route ran via Murzuq and continued on the one hand into the Sokoto Caliphate and the Bornu Empire, both of which lie in present-day Northern Nigeria, and on the other hand to Kanem and to the Sultanate of Wadai in present-day Chad. This route marked the shortest path to Africa's interior, which has been used continuously since ancient times. It would become the main slave-trading route of the Sahara. The western route ran from Tripoli via Ghadamis to Ghat and from there on into Niger and to Mali. These routes are assumed to have been used regularly until the beginning of the twentieth century, even when political events, in Europe as well as in North Africa, significantly influenced caravan trading and subsequently also the intensity of the usage of the routes (Fig. 3.1).

J. Passon (✉) · S. Hamid · H. Mughrabi · S. Swifka
Department of Physical Geography, University of Freiburg, Freiburg, Baden-Württemberg, Germany
e-mail: jacqueline.passon@geographie.uni-freiburg.de

K. Braun and J. Passon (eds.), *Across the Sahara*,
https://doi.org/10.1007/978-3-030-00145-2_3

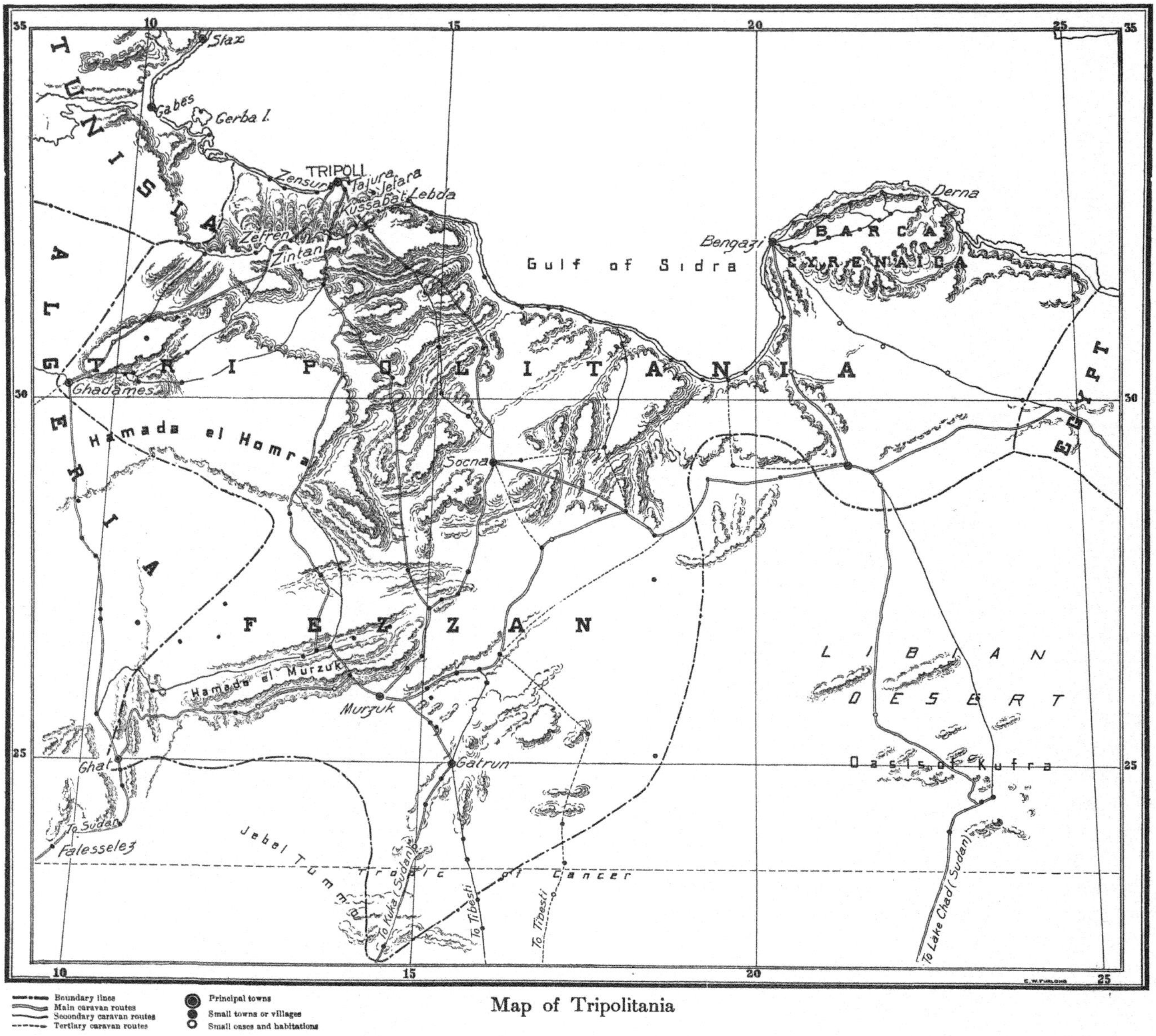

Fig. 3.1 This 1909 map by the American traveller Charles Furlong shows trans-Saharan trading routes. Features identified on the map include dunes, dry valleys as well as Arab caravan routes, and towns and oases. Hundreds of researchers have attempted to explore the landscapes as well as the different cultures that are found within this part of the world. Many of them lost their lives (Furlong 1909, 174)

3.2 A Terra Incognita for Europe

Jacqueline Passon

Up to the end of the eighteenth century, Africa's interior was still considered a *terra incognita* for Europe. The descriptions of ancient historians and geographers as well as reports from Arabian travellers lay the groundwork for that which was known in Europe about this region. The motivation of the systematic collection of this region was primarily trade policy interests, next to the scientific exploration by the Europeans over the course of their travels. With the independence of North America and the increasing industrialisation in the second half of the nineteenth century, Europe's interest in Africa grew as did its resources. The African back country was largely unoccupied up until 1880. The continent was the primary objective of the new imperialistic expansion of European powers. In particular, explorers from Europe embarked on expeditions over the course of the nineteenth century. These led them into the interior of the continent. Many explorers travelled on behalf of so-called African societies. The *African Association* in London, which was absorbed in 1831 by the *Royal Geographical Society*, is worth mentioning as a driving force in this process. Apart from a geographical thirst for knowledge, there were primarily political and commercial interests among the members.[2] Sir John Sinclair, member of the committee of the *Association,* said about the legendary city of Timbuktu: "*... gold is there so plentiful as to adorn even the slaves ... If we could get our manufactures into that country we should soon have gold enough*".[3] So next to travellers being dispatched, the *African Association* also tried to win over the British government in order to facilitate the development of the continent. As a result, European travellers consciously or subconsciously became the forerunners of the colonial annexation.

The era of the European scientific exploration was initiated by the German Friedrich Hornemann, who travelled on behalf of the British *African Association*. He succeeded in laying down the first paths according to an itinerary through the deserts. The starting point of his journey was Cairo. In order not to stick out or come under the suspicion of being a spy, he took on the identity of a Mameluke. He then established contact with caravan traders and pilgrims so that he could join a caravan on 5 September 1798 that used the north–south route, a significant pilgrimage route between Cairo and Timbuktu. The route he took ran from Cairo via the famous oases of Siwa and Audjila. He then crossed the basalt volcanic area of Harudj al Aswad via Tmissah into the trading metropolis of Murzuq. From there, he travelled southwards and reached Bornu and Nupe, where he presumably died. The findings of his journey from Cairo to Murzuq are recorded in a diary, in which Hornemann described the geographical location of the Fazzan as well as the environmental conditions. In his diary, he also included a historical overview of the political situation of the Fazzan at the end of the eighteenth century and the beginning of the nineteenth century as well as a description of the population in this area.[4]

Despite the tragic end of the first travellers of the *African Association*—next to Hornemann, the Englishman John Ledyard and Major Houghton, and the Scotsman Mungo Park also paid with their lives—the British government decided to make a new attempt and sent the Britons George Francis Lyon and Joseph Ritchie to Tripoli in 1818. They had the task of making a new attempt to penetrate Africa's interior. In the process, both men chose the classic caravan route from Tripolitania into Sudan and first travelled from Tripoli via Suknah to Murzuq, where they met on 4 May 1819 and where Ritchie died a few months later. Lyon made a journey through the oases in Fezzan and arrived at Tajahri via Zawilah and al Qatrun (see *From Murzuq to al Qatrun*). Lyon's recordings about Murzuq rank among the most detailed ones of the early nineteenth century. His journey provided new insights into the cartography of North Africa.[5]

With the detailed recordings from Hornemann and Lyon, which not only described travel routes, Murzuq and Fezzan, but also recorded all information about Sudan that was available at the time, the British travellers Walter Oudney, Hugh Clapperton and Dixon Denham embarked on their journey. The travellers also chose Tripoli as a starting point and set out to Murzuq in February 1822. Clapperton and Oudney made an attempt to reach the oasis of Ghat, which had not yet been accessed by any Europeans.

In winter 1822, all three explorers set out on the difficult march southwards with a stately caravan of over 100 men. They reached the Tummo springs via the oases of al Qatrun, Tajarhi and the Musciuru wells. They made their way via Bilma and Agadem to Lake Chad, upon the banks of which Oudney died in June 1824. Denham travelled on to the Sultanate Bagirmi alone. After the death of his companion Oudney, Clapperton moved on westwards to the Sokoto Caliphate. He returned to England in 1825 with Denham. Thus, Sudan had been reached through crossing the desert, and Lake Chad had been discovered.[6] The secret of Timbuktu, however, where still no European had set foot, continued. The zeitgeist of romanticism offered a fertile ground for the writings of Ibn Battuta or Leo Africanus. In Europeans' minds an idealised vision of a rich and mysterious city, that could easily have emerged from the tales of the "Arabian Nights", was born. That is why the British government sent out a subsequent explorer, Alexander Gordon Laing. As with his predecessors, the young man also chose to begin his journey through the Sahara at Tripoli. In August 1826, he entered Timbuktu, where he was murdered near the city, two months later.

Timbuktu was undoubtedly a focal point of the caravan traffic and a centre of Arab culture. However, the existence of magnificent palaces with gold-covered roofs belonged to the realm of myth. It is easy to imagine that there was a great disappointment when Réné Caillié, the first European to return alive from Timbuktu in 1828, reported of a rather decrepit town of mud brick, which was located in a sandy desert. The French explorer was accused by the critics of having never even been in Timbuktu. Finally, all doubts were swept away only decades later when Heinrich Barth, a German historian, visited the town. He finally was able to verify the account that Caillié had sent back. This anecdote shows the high expectations on the part of national governments, and incidentally, it illustrates how much the explorers were tied to colonial ideas.[7]

Due to its topographical location, Tripoli seemed to be the best entry point to the Sahara. An important milestone for the explorers on the way to Africa's interior was the oasis of Murzuq, the capital of Fezzan. However, often months went by before a good opportunity arose to continue the journey. Many explorers utilised this involuntary waiting period for an in-depth exploration of the landscape. In 1850, one of the most famous research groups was stuck in Murzuq too. James Richardson, who had already travelled in 1845 from Ghadamis to Ghat and had brought valuable news regarding trans-Saharan trading, was sent out again by the British government to explore the caravan routes of the Sahara. At the same time, people in London hoped that Richardson may be able to acquire potential trade associations for Great Britain. On this expedition, Richardson was accompanied by the German explorers Heinrich Barth and Adolf Overweg. The first milestone of the journey was Murzuq. But the travellers used a different route than the one that Richardson had used four years earlier on his way back to Tripoli. They decided against taking the easier way via Suknah and chose to cross the Hamadah al Hamra, a barren stone desert instead. They were the first Europeans to travel this route, which was shorter but much more difficult. After a very difficult one-week march via Mizdah and al Qaryah al-Gharbiya (see *From Tripoli to Mizdah*), the travellers reached Murzuq, where they immediately began preparations for their journey to Ghat and into the still unknown Aïr massif, which was approximately one thousand kilometres south of Ghat as the crow flies. The negotiations with the Tuareg for a safe escort on their journey progressed slowly. It was not until four months after their departure from Tripoli that the explorers reached Ghat and continued their journey into the Aïr massif. But only Barth returned home; both of his colleagues died on the way.[8] At first, Barth's fate in Europe was unknown. For this reason, the German astronomer Eduard Vogel was commissioned with the clarification of the matter. Via Murzuq, al Gatrun, Bir Tummo and Bilma, he reached Lake Chad at the beginning of 1854. This was his third time crossing the central Sahara. On his journey, the young scholar performed numerous astronomical localisations, which were of great benefit for cartography. He also made meteorological, magnetic and geological observations. Finally, at the end of 1854, he met up with Heinrich Barth. Then they parted their ways forever. Vogel was murdered in Wadai in 1856.

Heinrich Barth returned to Germany, where he published a comprehensive work consisting of 3500 pages. He had gathered such an enormous amount of information about North Africa that his sponsor and promoter Alexander von Humboldt stated that Barth opened up a new continent to the European scientific community. His meticulous descriptions of the journey from Tripoli via Murzuq and Ghat into the Aïr Mountains were pioneering in the field of research of African cultures, and he was the first European to characterise them, extensively and mostly without prejudice. He delivered a comprehensive history of the region, detailed descriptions of the social structure of the Tuareg and completed his descriptions with sketches and regional maps. In his last few years of life, the German acted as the advisor for the young generation of explorers such as Gerhard Rohlfs or Henri Duveyrier, who were to continue his initiated scientific exploitation of the black continent.[9]

The Frenchman Henri Duveyrier built on Barth's research of the northern Tuareg and collected a great amount of ethnographical and historical data. Since 1859, Duveyrier had undertaken many trips to Algeria, Tunis and the Libyan Fezzan. He succeeded in making friends with the leaders of the Tuareg in the Tassili n'Ajjer (southern Libya), namely with their political leader Ikhenukhen and the religious leader Sidi Uthman (see *From Ghadamis to Inazar and Adiri*). Together with them, he moved from Ghadamis to Ghat, reached Murzuq in 1861 and then travelled in small segments to Tripoli. In doing so, he followed a route that no explorer had yet travelled before, but which also proved to be a long and difficult one.[10]

Barth and Duveyrier's journeys received a huge response in Europe. The Fezzan, the great axis of Tripoli to Chad, and the legendary Timbuktu oasis were now known and described. Between 1864 and 1880, new researchers emerged, especially including the Germans Moritz von Beurmann, Gerhard Rohlfs and Gustav Nachtigal. In 1862, Moritz von Beurmann travelled from Benghazi to Murzuq via Audjila and Zillah, where he was the first European to set foot (see *From Zillah to Murzuq*). In Murzuq, endless negotiations hindered the continuation of his journey. He used the waiting period to design a map of this caravan melting pot and became the first white man to enter the mysterious oasis of Wau al-Kebir. A year later, he was murdered in Wadai.[11]

Gerhard Rohlfs has the merit of having walked the greatest width of the desert from the Atlantic Ocean to Egypt and the greatest length from Tripoli to Lake Chad. His travels in the region of present-day Libya led him from Tripoli to Murzuq via Ghadamis in 1865. He used his five-month-long stay to thoroughly study the political and economic situation of the Fezzan region. Upon continuing his journey, he reached Sokoto via Kuka, the capital of the Bornu Empire. Finally, in May 1867, he reached Lagos on the Atlantic coast. In 1869, he began his journey from Cyrenaika and travelled through the Libyan Desert to Egypt. The Libyan Desert, which makes up the eastern region of the Sahara between Fezzan and the Nile Valley, belongs to the hyper-arid regions of the earth. There are no nomads here. Human life can only be found in the few oases. In the heart of this desert, there is a group of oases, the most significant of which is al Kufrah. For a long time, the unfavourable climatic conditions constituted an insurmountable obstacle for European researchers. The caravans that travelled through this region had to go without water for up to ten days on some routes. In 1878, Rohlfs embarked on another expedition on behalf of the *African Association*. This time he succeeded in actually becoming the first European to reach al Kufrah. With that, another white spot on the map was filled.[12]

Fig. 3.2 Famous Saharan explorers: Dr. Heinrich Barth, James Richardson, Dr. Wolf Overweg, Henry Duveyrier, Dr. Gustav Nachtigal, Alexandrine Tinne and Gerhard Rohlfs (from top to bottom)

Another important person in researching the Sahara is Gustav Nachtigal, who travelled through the Sahara between 1869 and 1874. He had set himself the goal of reaching Wadai. First, his travels led him from Tripoli to Murzuq. There, he, like many of his researcher colleagues, used the waiting period and took a trip into the mountainous country of Tibesti, which had never before been travelled by a European. He gathered countless amounts of data, described the city of Murzuq, the geographical nature of the Fezzan region, its history and demographic structure, climate and diseases, etc.[13] On his journey, Nachtigal continuously ran into an extraordinary woman. Alexandrine Tinné, a rich Dutch woman, who had made travelling her life's fulfilment and who was dreaming of exploring the sources of the River Nile. Her first expedition (1862–1864) started in Cairo and ended tragically and without the discovery of any sources. On that expedition, her mother died as did her aunt, one scientist and two servants. At the beginning of 1869, she arranged a new expedition in order to become the first female European to cross the Sahara. Together with two Dutch seamen, a few released slaves, Algerian women and a young German, Adolf Krause, she left Tripoli with the goal of reaching Lake Chad. Next, Tinné and her travelling companions reached Murzuq, where she waited for a caravan with which she could travel through the Sahara. Krause left the group and returned to Tripoli, which was a stroke of luck for him. As a result, the young German was able to create a detailed historical evaluation of the Fezzan region. He also described the city of Ghat as well as its inhabitants and history. The meticulous descriptions of the Tuareg tribes and the drawings, illustrating the trade situation in the nineteenth century, are attributed to him. Tinné's second expedition was not very lucky either. On her journey towards Ghat, she took the same route that Heinrich Barth had taken before her. But only a few days after departure, Tinné was murdered (see *from Murzuq to Ghat*) (Fig. 3.2).[14]

Towards the end of the nineteenth century, the time of great discoveries by individual pioneers was over. The Sahara entered a political phase. The era of mavericks, who usually had a scientific goal, was now replaced by "state expeditions", which were systematically organised and equipped with a substantial amount of resources.

3.3 From North to South and from East to West

Jacqueline Passon

By analysing maps and descriptions of Arab as well as of European historians, geographers and travellers of the Middle Ages and the eighteenth and nineteenth centuries, two routes would be ranked among the most important connections through the desert, both of which started in Tripoli: the western route led from Tripoli via Ghadamis, Ghat, Agadez and Zinder to Kano, a province of the Sokoto Caliphate (northern Nigeria). The other more eastern route ran from Tripoli directly southwards via Murzuq and Kuwar into the Bornu Empire and Kanem (Lake Chad). Numerous back roads and connecting paths run through present-day Libya.[15]

The "Bornu route", as the great trans-Saharan caravan route between Murzuq and the Lake Chad region was called in the nineteenth century, played an extraordinarily significant role in the transport network. In the middle of the second century AD, Ptolemies had already provided the first reference to this route in his report about an army of the king of the Garamantes in the sub-Saharan country of Agisymba. According to the records of the ancient chroniclers, between 83 and 92 AD, the king attempted to bring the black African inhabitants of the country under his rule again, who he regarded as his subjects. Punic slave hunters from Carthage had probably set up one or more bases in the Chad Lake region long before this, from which Agisymba eventually arose. Due to its relatively short and comparably easy access, the Bornu route developed into the most durable and most actively used trade route between the Mediterranean and sub-Saharan Africa.

According to current research activities, it can be assumed that trading had already developed between the Mediterranean and central Africa in ancient times.[16] From the early Middle Ages, a series of powerful empires arose in the sub-Saharan area, which is called Sudan. Trade brought Sudanic rulers the means of consolidating and extending their powers. Existing trade relations had been intensified. A consequence of this development was a dense network of connections between the Mediterranean world and Sudan by the end of the eleventh century. Traffic along the different routes continued then over the centuries. It always depended on the security of the Fezzan and the central desert. Until the end of the medieval Islamic period, trade was under the control of Kanem and then switched to Bornu, of which Kanem became a province.[17] After the fourteenth and fifteenth centuries, the rising power of Bornu needed to ensure the security and continuity of the trans-Saharan trade on which the prosperity of the state and the prestige of its rulers depended. Intensive contacts with Tripoli were established.[18] During the nineteenth century, when wars raged among Bornu and Wadai, many merchants were forced to shift their operations to other routes across the Sahara. Consequently, part of the traffic along the Bornu Road was diverted into a new route which ran between Wadai, al Kufrah and Benghazi. Trade flowed also along the route between Kano.

The European explorers went directly on the trade or pilgrim routes. They did not explore any new routes, but rather stuck to the already established routes, primarily used by traders and caravans but also used by pilgrims and herdsmen. This was mostly due to the fact that the travellers were dependent on guides. The routes understandably followed the topographically most appropriate conditions, led from well to well and connected the most important oases (see *Navigating the white spot*).

3.4 Trade Between Different Worlds

Jacqueline Passon

3.4.1 Ibadi Trade

The transport of goods across the Sahara has been organised thoroughly. Between the ports of the Mediterranean Sea, the oases of Fezzan, so-called inland ports, and the political and economic centres of Sudan an intense network of relations had emerged. According to the writings of early Arab geographers, trans-Saharan trade in the Middle Ages was operated by trading enterprises of Ibadi Berbers from Fezzan. Their main entrepôt was situated at the oasis of Zawilah. Its favourable situation on the main road northwards to Tripolitania, which here crossed the important pilgrim route from Cairo to Timbuktu, made the market site develop soon. It became the centre of trade activities from the eighth until the twelfth centuries. The Ibadi Berbers of Fezzan created a trading network that was firmly linked with Ibadi communities of the Jabal al Nafusah as well as with centres of Ibadi communities in central Sudan.[19] Al Yaqubi provides in his *Book of the Countries (Kitab al Buldan)*, which he completed in 889–890, the earliest insight into the political history of Sudan. For Zawilah, he reveals that various people from Khurasan, Basra and Kufa were living there.[20] The book conveys the image of a vibrant oasis gathering people from different parts of the Islamic world. Those people might be Ibadi traders from the east, who sought refuge among the Ibadi communities of the Maghrib.[21] This Ibadi fusion has in turn provided the basis for a successful long-distance trade network and thus for the rapid economic expansion. Ibadi commercial expertise was combined with the indigenous tribesmen's knowledge. This resulted in a monopoly of the eastern and central Saharan routes, which obviously crossed Ibadi towns. This ensured the economic underpinnings of the Ibadi community.[22] Apart from the Berbers, the Jewish community were relevant players in trans-Saharan trading. From a widely extended network of far-flung trading posts such as Tripoli, Misratah, Sirte, Ajdabiya, Darna, Tolmeta, Ghadamis and the Jabal al Nafusah, they coordinated their commercial activities.[23] As members of a trans-territorial diasporic society, Jews played an important role, both as traders and intercultural mediators, since antiquity.

The thirteenth century proved to be a time of enormous political and economic change. With Kanem's expansion into Fezzan, the Africans took control over the central

Saharan caravan routes. This domination gave control of the north–south trade and the east–west trade between Egypt and the west Sudanic empires. In the fifteenth century, Bornu, situated west of Lake Chad, became a leading power. To ensure security and continuity of the trans-Saharan trade, the Bornu kings maintained cordial relations with Tripoli, which controlled the northern trade routes.[24]

As Wright points out, the Sudanic empires relied on the Saharan nomads for the organisation, guidance and protection of the trading caravans. The economies of the nomads, the oasis dwellers and settled communities north and south of the desert are largely complementary. This is because each group was able to supply necessities the others lacked. In the central Sahara, the nomadic Tubu families, who have long maintained complex trading relationships, provided the essential contact between these different markets and demands. In the northern Sahara, trade activities were controlled and protected in particular by the Tuareg. By the Arab Middle Ages, patterns of international trade had evolved across the central Sahara. It was based on nomadic knowledge, practices and contacts extending from Sudan to Fezzan and to the Mediterranean coast.[25]

Looking back in history, it can be said that since the times of antiquity, a dense network of trade and pilgrim routes had formed in the central Sahara, on which trading stations were strung together one after the other. Trans-Saharan trade was a guarantee for the well-being of these places. If trade disappeared, these commercial centres would suffer a decline, as was the fate of the important centres of Garama (Jarmah) and Zawilah. Goods from central Africa were accumulated in Garama (Jarmah), located in the Wadi al Ajal and were then brought to the ports in the north, where first the Phoenicians, and later the Romans, carried on trading the goods. The Arabic invasion in the seventh century probably spelled the end for Garama (Jarmah). At this time, another location, Zawilah, rose to become one of the most important inland ports of the central Sahara. Zawilah advanced to become a multicultural centre of the medieval Fezzan. The merchants of Zawilah, Ibadite Berbers (Ibadis) from the Nafusah Mountains and from Khurasan, Basra or al Kufah, had a monopoly over trade in the central Sahara between the eighth and thirteenth centuries. Due to close contacts—and family ties as well—with the Jabal al Nafusah, Ghadamis and the Levant, as well as the cooperation with the Jewish merchants who, on their part, had widely ramified contacts in the entire Mediterranean area, it became possible to control the import–export business.

This changed after the kingdom of Kanem put Fezzan under its control and made Taraghin its capital. Zawilah gradually lost its position and finally sank into insignificance in the sixteenth century with the rise of the Awlad Muhammad dynasty. Murzuq now grew into a trading hub in Fezzan. Into the early twentieth century, the inland ports of Ghadamis, Murzuq and Ghat remained the distribution centres. Goods from Africa's interior were stockpiled here and later transported to the ports of the north, above all to Tripoli, where they were traded further in the Ottoman Empire or in Europe. In doing so, the traders acted as mediators between different worlds.[26]

The basic conditions throughout the Sahara were characterised by a lack of political stability and regional security. For ages, the Sahara has been portrayed as a more or less unstable area where economic activity was based on trans-Saharan or regional caravans. These caravans have been described as an ephemeral entity, formed on an ad hoc basis, by a group of travellers, merchants and others in response to seasonal trading opportunities.[27] It is often thought that little, if any, technological change has occurred in the organisation of camel caravans across the centuries.

What are the transitional questions that arise in this context? Hence, it is interesting to ask who were the traders and how could long-distance trade be conducted successfully when political stability and regional security could not be guaranteed?

3.5 Business Structures of the Caravan Trade

Jacqueline Passon

3.5.1 The Ghadamis "Trade Diaspora"

Since the Middle Ages, the Ghadamis traders had established themselves not only in Timbuktu, but also in other strongholds along the important caravan route to Kano, with branches and secondary residences and had created a so-called trade diaspora. This can be defined as dispersed communities of traders living as minorities in a foreign culture. Their purpose was to facilitate cross-cultural trade.[28] Duveyrier noted that Ghadamis firms "*have branches at Kano and Katsina in the Sudan, in Timbuktu on the Niger, in Ghat and in Sallah in the central Sahara, and in Tripoli and Tunis on the Mediterranean shore*".[29] Al Hachaichi, an Arab traveller, who had visited Ghadamis in the 1890s, adds to his descriptions a detailed list with residences of the Ghadamis traders,[30] who lived and worked in a "trade diaspora" outside of Ghadamis. The advantage of a "trade diaspora", which was also open to other North African traders, consisted of the creation of a common trading culture in areas where the traders were strangers. This made access to new source and target markets considerably easier.[31]

To finance desert trade, the Ghadamis traders, who, according to Barth, had invested their entire capital in trading companies,[32] cooperated with merchants from Tripoli. The wholesale merchants based in Tripoli, which were represented mainly by Jews of Italian origin and had been established for centuries in the Medina of Tripoli, managed the goods of the large trading firms from Europe.[33] According to the British consul Fraser, Christian and Jewish agents in Tripoli imported trade goods and sold them on to "*the Moorish merchants of Tripoli who transported them In Land*".[34] The wholesalers from Tripoli as well as their colleagues from Ghadamis as middlemen closed the gaps between two different worlds. The merchants from Tripoli advanced the goods they received from European companies to the Ghadamis traders on a credit basis. They, in turn, had to repay their debts with products from Sudan. Until the end of the nineteenth century, the merchants from Tripoli and Ghadamis controlled a considerable part of the import–export business between western Sudan and the North African coast. In doing so, they had established a system similar to that of a bank, which permitted financial transactions between the coast and the African interior on the basis of checks and letters of credit.[35] Major Alexander Laing reported: "*They calculate with profound nicety the expense of carriage to distant countries, duties or customs risks, trouble and the percentage that their goods will bear, and even do business by Bills or written agreements or promises*".[36]

When one specifically looks at the route between Tripoli and Kano, on which the traders from Ghadamis controlled the trade, the settlements of Ghadamis, Ghat and Agadez served as so-called inland ports. At the seasonal markets that took place there, goods were either sold, exchanged, or the freight was unpacked and repacked and transported further to the south or to the north, respectively, by newly formed caravans. The task of the Ghadamis traders consisted first in arranging the transportation of the goods from Europe to Ghat.[37] Richardson reports in 1845 that a large caravan departed in November from Ghadamis in order to trade goods from Kano. For this purpose, the Ghadamis traders went to Ghat, where they met the caravans travelling from Kano to the north, which primarily transported slaves, senna leaves and ivory. At the market in Ghat, there was always a colourful hustle and bustle.[38] Besides the caravans from Kano, there were also other caravans arriving in Ghat from all the different cardinal points, as Richardson wrote in the winter of 1845: "*Caravans from Soudan, including all the large cities, but especially from Kanou, from Bornou, from the Tibboo country, from Touat, from Fezzan, from Souf, from Ghadames, and from Tripoli, Tunis, and the North coast, visited the Ghat Souk of this winter. The number of merchants, traders, and camel-drivers was about 500, the slaves imported from Soudan to Bornou about 1000, and the camels employed in the caravans about 1050. Provision caravans from Fezzan also were constantly coming to Ghat during the Souk. The main commerce of these caravans consisted of the staple exports, of slaves, elephants' teeth, and senna, the united value of which, at the market this year, was estimated at about 60,000 lb, which value would be doubled, on arriving at the European markets*".[39] Ghat was, compared to the other centres in the Sahara, a relatively small settlement. It had a population of a few thousands. However, during the market times, the population of the oasis increased several times over. The market flourished in the nineteenth century because Ghat was an ideal trans-shipment point. This can be seen not only in relation to the good connections, which were maintained with the trade source markets and the Tuaregs, but also due to the fact that the oasis, until its occupation by the Turks in 1875, was not subservient to any foreign power and as a result thereof not subject to taxes.

In Ghat, there were two forms of a market. The seasonal markets, which usually took place twice a year, namely in the winter and in the summer, were held in front of the gates to the oasis and lasted for three months each. The inside of the oasis did not have sufficient space for the traders and their goods, and this is why the seasonal markets were held in front of the eastern door of the oasis. Ghat was one of those inland ports where the caravans exchanged their goods, stocked up their supplies and took a rest for a while before continuing on. As a rule, the caravans were divided here. Part of the caravans continued further to the north towards Tripoli or Murzuq, and the other part went back to the respective areas of origin after the goods had been exchanged with the numerous traders from the north. The Ghadamis traders now operated from here in two ways: they either transferred the goods on a credit basis to employed persons or agents, so that they could transport them farther to the south, or they exchanged or sold the goods in order to return to Tripoli and settle their debts there. The procedure depended on the liquidity of the traders. Since the travel to and from Ghat to the source markets in Sudan and back to Tripoli could last from one-and-a-half to three years, the Ghadamis traders themselves had to possess sufficient financial resources if they wanted to send the goods to Sudan.[40]

It is clear from the descriptions as well as from interviews with elders that the market in Ghat became a place where intensive processes of negotiation took place. It was necessary to negotiate the formation of new caravans and consider the conditions for the onward journey. The oasis was a meeting place for the tribes that had settled in the area. The Tuareg rented or sold the camels they did not need and offered camel herders and guides. The price for these services, as well as the amount of the "protective duty" that had to be paid, was negotiated in Ghat. An important part of the success of the traders from Ghadamis is based on the close cooperation with the Tuareg, in particular with the Adjer Tuaregs or the Kel Ewey.[41] Geographical proximity, and the acting within eyesight, led to the formation of common business practices and trust. For the negotiations, this meant that they could take place more smoothly and also more quickly. Over time, it became evident that the Adjer Tuareg were responsible primarily for the route segments between Ghadamis, Ghat and Iferwan in the northern Aïr Mountains. Contracts with Kel Ewey were then negotiated in Iferwan. Based on them, they ultimately accompanied the caravans to Kano or to other towns of the Hausa countries—predominantly Hausa-speaking communities were scattered throughout West Africa. Overall, the Tuaregs mainly worked in the services sector of the caravan trade. However, there are also indications that some of them had been engaged as traders and in the micro-financing business of the trade, especially between Zinder and Ghat. They received credits for these ventures on the market in Ghat.

The middlemen in Tripoli tended to be a thorn in the side of some European wholesalers. With the expansion of trade since the 1870s, it appeared that there were attempts on the part of some European companies to bypass the wholesalers from Tripoli and to establish their own branches there. Some companies established themselves directly in Tripoli in order to be able to control the import–export business. Generally, it should be emphasised that nothing had changed in the trading network, which had been established for centuries, and had spread from Europe through the stations of Tripoli, Ghadamis, Ghat and the target and source areas in western Sudan (Bilad as-Sudan).[42]

3.5.2 Murzuq—Diversified Market in "Blad Elhemah", Country of Fever

Murzuq functioned both as an entrepôt and also as the terminus of the trans-Saharan trade. However, unlike Ghadamis, there was no layer of traders domiciled here who called Murzuq home. Traders from different ethnicities came together here, in the oasis, in order to trade, but not to invest their money: in the eighteenth and nineteenth centuries, there were Arab merchants from Tripoli, Suknah, Jalu, Aguila and Egypt, traders from Bornu and the Hausa countries, as well as Europeans.[43] The traders from Suknah represented the largest group. They managed the trade traffic between Tripoli and Murzuq. Just like the other traders, they had established secondary residences here, but their capital was elsewhere.[44] It is not surprising when, in the 1850s, Heinrich Barth reported that Murzuq was "*rather the thoroughfare than the seat of a considerable commerce, the whole annual value of imports and exports amounting, in a round sum, to 100,000 Spanish dollars; and the place, therefore, is usually in great want of money, the foreign merchants, when they have sold their merchandise carrying away its price in specie, the Mejabera to Jalu, the Tebu to Bilma and Bornu, the people of Tawat (Tuat) to their respective homes*". He then added, "*[...] few of the principal merchants of the place are natives*".[45]

Between 1550 and 1812, Fezzan was under the supreme authority of the Awlad Muhammad dynasty, which had established its capital in Murzuq. From the sixteenth century on, Murzuq served as a hub for the trans-Saharan trade on the Bornu route between Lake Chad and Tripoli. According to the description of Ibn Ghalbun, a local historian, as well as the German traveller Friedrich Hornemann, a large part of the economic foundation of the Fezzanian state was to collect taxes from the caravans which crossed Fezzan. As compensation, it offered a secure market and accommodation for the night. In addition, the traders could get new camels or hire a new guide in Murzuq. The tribe of Awlad Sulayman made one more attempt to participate in this successful domination between 1830 and 1842, but it was only of short duration. After the chief of the clan was murdered in 1842 by the Ottomans, this domination over Fezzan ended and the Turkish Pasha in Tripoli took control over Fezzan.[46]

The caravan season ran in the same way as in Ghat or in the other oases: every winter, one large caravan from Bornu reached the seasonal market of Murzuq. The arrival of the winter caravans from the south coincided with the pilgrim caravans, who had come via the Tuat oasis from western Maghreb, and marked the beginning of the winter market which, as a rule, took place between October and February. During this time, large caravans from Tripoli, Benghazi, Cairo, Ghadamis and Sudan arrived. The smaller entrepreneurs travelled in the desert for the entire year. As has already been indicated, the north–south traffic between Tripoli and Murzuq was primarily in the hands of the traders from Suknah, while the connection to Bornu was controlled mainly by the Tibus from Bilma. The Tibus themselves were also active as traders, and they financed their own ventures between Zinder and Bilma or between Zinder and Fezzan.[47]

Hornemann, who reached Murzuq in November 1798, reports on the diversity of the trade. Besides slaves, there was a wide palette of goods being traded in Murzuq: "*From Bornou copper is imported in great quantity. Cairo send silks, melays (striped blue and white calicoes), woollen cloths, glass, imitations of coral, beads of bracelets, and likewise and assortment of East India goods. The merchants of Bengasi, who, usually join the caravan from Cairo at Augila, import tobacco manufactured for chewing, or snuff, and sundry wares fabricated in Turkey. The caravan from Tripoly, chiefly deals in paper, false corals, fire-arms, sabres, knives, and the clothes called abbes, and in red worsted caps. Those trading from Ghadames, bring nearly the same articles. The smaller caravans of Tuaricks and Arabs, import butter, oil, fat, and corn; and those coming from the more southern districts, bring senna, ostrich feathers, and camels for the slaughter-house*".[48]

Hornemann also writes about the sum a caravan has to pay in Murzuq, whereby the amount of the sum to be paid could vary: "*The sultan derives further income from duties on foreign trade, paid by the several caravans. That from Cairo pays from six to eight dollars for each camel load. The caravans from Bornou and Soudan pay two matkals, for each slave on sale*".[49] However, it was not only the traders who had to make payments; the rulers in Murzuq were also obliged, on their part, to pay a tribute to the Ottomans and later to the Karamanlis in Tripoli. Depending on the stability of the rulers in Tripoli, this task was performed once per year by a senior official from Tripoli who arrived personally in Murzuq in November. He collected the sum either in monetary form (Hornemann mentions 6000 dollars, which was later reduced to 4000 dollars) or the corresponding value in the form of gold, senna or slaves.[50]

During the nineteenth century, the security situation on the Bornu route worsened considerably. In addition, after the official abolishment of the slave trade, it became increasingly more difficult to ship slaves at the port of Tripoli. As a result, the trade moved farther to the east on the so-called Benghazi–Wadai route. At the port of Benghazi, the shipping of slaves could continue to take place in a relatively undisturbed manner.

3.5.3 The Benghazi–Wadai Trade

The Benghazi–Wadai route (see *From Benghazi to al Kufra* via *Jalo*) was initially controlled by two tribes: the Majabra from the Jalu oasis and the Zuwaya from the al Kufrah oasis. They worked as guides or as porters and rented their camels to the traders.[51] Due to the nomads' close connections with the Sanussi brotherhood, which had established itself with the opening of its first monastery (*zawiya*) in al Bayda, situated in the eastern part of today's Libya, in 1843, this order was soon able to play a major role in trade, promoting also sedentary behaviour, and education. In 1856, the brotherhood built the new centre of the order in the remote oasis of al-Jaghbub, about 500 km south-east of al Bayda. The town was fortified, and an Islamic university and a new mosque were built. In the following years, a state was established by the order which, at the apogee of its might, dominated not only the Libyan Desert, but also the entire eastern Sahara. Control over the important caravan route from Benghazi via al Kufrah to Wadai, which was the only important trans-Saharan route that was completely outside the field of vision of the European powers, was of great importance. Hence, this route was used primarily for bringing slaves to the Mediterranean Sea, mainly in return for older European firearms, which were traded in the South. Until the end of the nineteenth century, the order was also represented in Fezzan, Tripolitania (Fig. 3.3) and Egypt, as well as in the western region of Sudan and even in India.[52] The approach was always the same: *zawaya* were created along the trade routes and at important crossroads, in particular along the Benghazi–Wadai route. Traders found not only safe accommodation in the monasteries in the desert, but it was also a place where news was exchanged, since the monasteries were connected to each other by a postal system. In addition, the brotherhood took care of the maintenance of the wells along the trade routes, which made travel through the desert much easier.[53] However, the Sanussi brotherhood was also able to build up its discipleship in another way. In a region where there was no strong central authority, and the right of the stronger reigned, the brotherhood tried to mediate disputes and maintain calm. Arbitration courts were established for this purpose. In addition, they offered education and religious instruction to the local population in their monasteries. In some monasteries, they added arbitration courts, in which law scholars prepared guidelines for the creation and the successful maintenance of business relationships, rules for granting credit or general commercial contracts. For these reasons alone, it was advisable for the traders to associate themselves with the monastery; it was the only way they could use the full scope of the services, and it made them exempt from paying customs dues. In return, they gave tea, sugar or other goods to the order. The integration of the different groups in the brotherhood fundamentally simplified trade.[54] The outlook on the world shared by the members led to a kind of cooperative identity which enhanced solidarity and on whose basis possible problems could be solved more easily. As a result, the spreading of the common doctrine became an essential guarantee for the safety on the route between Benghazi and Wadai. In this way, the Sanussiya created trust between the traders, the agents and the other groups participating in the trade and established a new culture of social interaction.

With the opening of a *zawiya* in Ghadamis, the brotherhood tried to operate on the Tripoli–Kano route, which they achieved successfully. However, the Sanussiya could not play the dominant role here as they had in the East.

3.5.4 Summary

The forming of a caravan was an expensive venture which, was not financed by European creditors alone. The most influential agents and traders included, besides the wholesale merchants from Tripoli in the form of Jewish family companies, primarily the trader dynasties from Ghadamis or

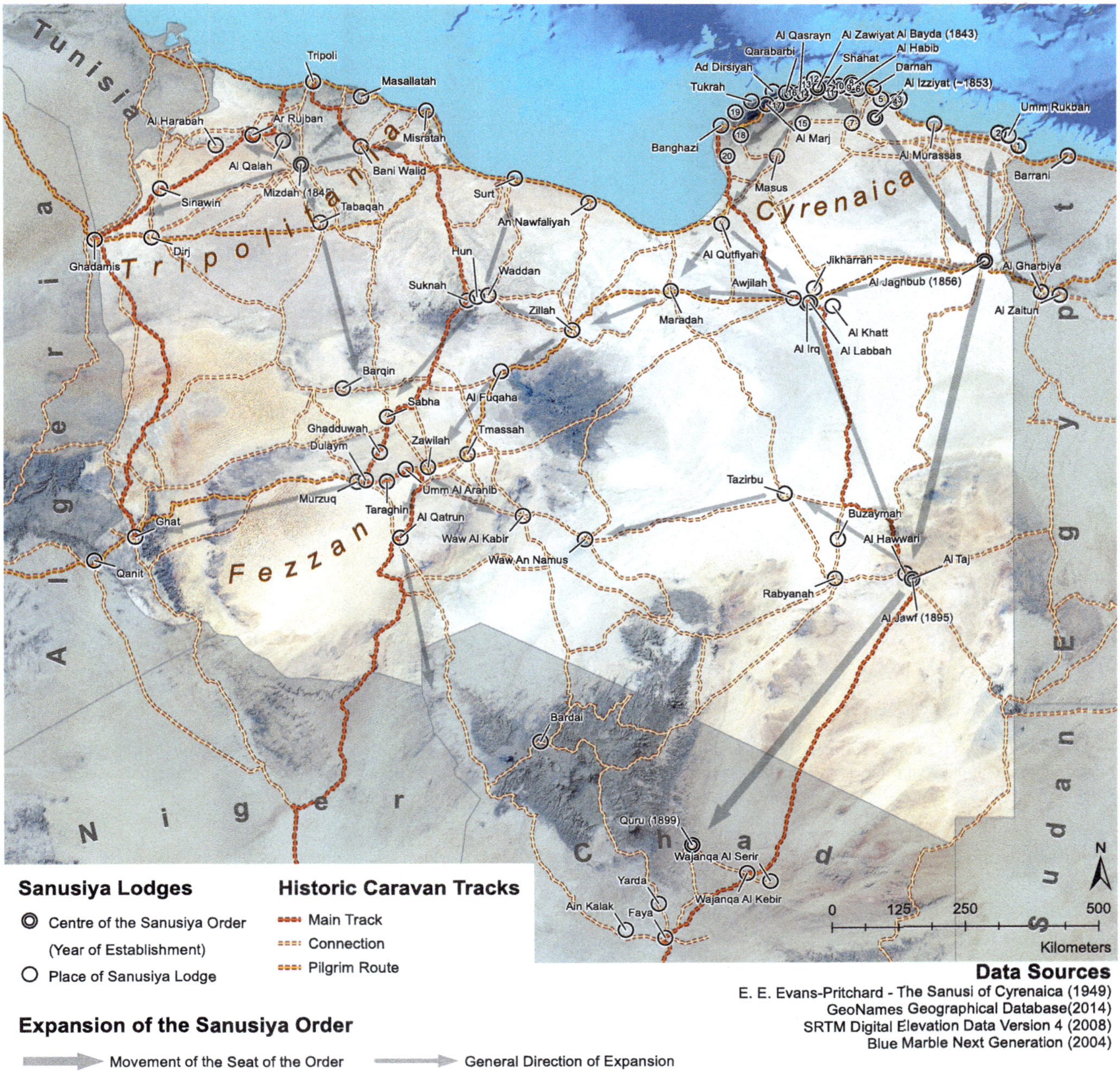

Additional Place Names of Sanusiya Lodges (Cyrenaica)

1 Al Jirfan	5 Martubah	9 Bisharah	13 Al Haniyah	17 Al Qusur
2 Zanzur	6 Mara	10 At Tart	14 Al Urqub	18 Umm Ash Shakhnab
3 Umm Hufayn	7 Al Mukhayli	11 Al Faidiyah	15 An Niyan	19 Daryanah
4 Umm Ar Rizam	8 Zaigh	12 Al Hamamah	16 Sirat Mirad Masud	20 At Tilimun

Fig. 3.3 Map of the distribution of Sanussiya lodges throughout the Libyan territory in the nineteenth century

Suknah. While these groups of traders were responsible for the transport inside the country, the Jewish entrepreneurs organised the import and export of the goods. In addition, they took care of the preparation of the Sudanese products for export—ostrich feathers, for example, had to be processed extensively. For this purpose, these traders received special protection from the European consulates and enjoyed a special legal position on the part of the Ottoman Empire. Both the Jewish merchants and the traders from Ghadamis and Suknah acted with far-reaching

Fig. 3.4 Sanussiya lodge in al Kufrah. The illustration is taken from Rohlfs book „Reise von Tripolis nach der Oase Kufra“ from 1881. Control over the important caravan route from Benghazi via al Kufrah to Wadai, which was the only important trans-Saharan route that was completely outside the field of vision of the European powers, was of great importance

relationships in the commercial network and used their close relationships in central Sudan, in the Levant or to Europe, which they had formed from time to time since the sixteenth century. So, for example, some Jewish families maintained family connections in Great Britain, which considerably facilitated the processing of transactions with the local suppliers and merchants there. The strong capital-related connections with the influential trading enterprises in Europe and in the Ottoman Empire allowed the traders to financially cover risky ventures. In the final analysis, the European and Ottoman trade enterprises, with their global networks, ensured that the local exchange of goods was connected with the world trade (Fig. 3.4).[55]

Along with commercial companies in Europe, Ottoman investors also participated in the caravan trade. In addition, sources contain many documents showing the involvement of Turkish officials, particularly in the slave trade. The posts in Tripoli were considered unattractive, and so the officials tried to make as much money as possible during their term of service. In this respect, the British Consul Crowe reported in 1847 to London that a shipment with 60 slaves on board was on its way from Tripoli to Constantinople. Half of the slaves, according to the consul, would have belonged to the director of customs, Mehment Sherif said. “*The official position by this person afforded him great facilities for carrying on the detestable traffic with peculiar advantage and profit. There can be little hope of the cessation of the trade while those most deeply engaged in it are appointed to important stations*”, relayed Crowe.[56] In addition, he reports that, in June 1848, two ships belonging to the governor of Tripoli were on their way to Smyrna and Constantinople with 288 slaves onboard.[57] Until the early nineteenth century, most goods, including slaves, were transported over the Mediterranean Sea with European ships, in particular French ships. The situation in the Mediterranean Sea had changed since the 1830s because the Ottomans now had a merchant fleet with the respective ships. Ship owners from Tripoli, Benghazi, Misuratha and Derna were active in the shipping business from this time on and chiefly brought slaves to the markets in the Aegean Sea and the eastern part of the Mediterranean Sea.[58]

3.6 Negotiation and Trust

Jacqueline Passon

Ibn Khaldun, who was the first to systematically analyse the functioning of an economy in his book *Muqaddimah*, stated with regard to the long-distance trade that "*(...), it is more advantageous and more profitable for the merchant's enterprise, and a better guarantee (that he will be able to take advantage of) market fluctuations, if he brings goods from a country that is far away and where there is danger on the road. (...) Therefore, the merchants who dare to enter the Sudan country are the most prosperous and wealthy of all people. The distance and the difficulty of the road they travel are great. They have to cross a difficult desert which is made (almost) inaccessible by fear (of danger) and beset by (the danger of) thirst. Water is found there only in a few well known spots to which caravan guides lead the way. The distance of this road is braved only by a very few people. Therefore, the goods of the Sudan country are found only in small quantities among us, and they are particularly expensive. The same applies to our goods among them*".[59] To operate successfully in this business the Arabian scholar provides advice: "*In the attempt to earn the increase (of capital) that constitutes profit, it is unavoidable that one's capital gets into the hands of traders, in the process of buying and selling and waiting for payment. Now, honest traders are few. It is unavoidable that there should be cheating, tampering with the merchandise which may ruin it, and delay in payment which may ruin the profit, since (such delay) while it lasts prevents any activity that could bring profit. There will also be non-acknowledgement or denial of obligations, which may prove destructive of one's capital unless (the obligation) has been stated in writing and properly witnessed. The judiciary is of little use in this connection, since the law requires clear evidence*".[60]

Khaldun advised commercial entrepreneurs to document their economic activities in writing already in the fourteenth century. That was the theory, but what did that mean in practice?

The preparation for a caravan season required careful planning. Before the start of the enterprise, the organisers of caravans first had to think about what kind of goods they would need to transport to which target markets. Then, they had to decide upon which routes the goods would be transported. Initially, this required building networks with trading partners. Afterwards, the different parties had to agree among themselves on how many participants and camels the caravan should have and on the financial means they would have to contribute. Ghislaine Lydon, who has made important contributions to our understanding of the caravan trade in Western Sahara, has pointed out that in the nineteenth century, the most widely spread form of cooperation in long-distance trade was the contractually sealed partnership (commenda type). The following aspects were regulated:

- The specific capabilities, as well as the work performance, of one of the contracting parties were settled with the financial resources of the other party.
- The locations and the objects of the trade were defined.
- The shares, into which the profits and the losses were to be distributed, were defined.
- Precise liability provisions, as well as protection measures, were included for the partners, which had been indebted during the cooperation.

In addition, there were contractually regulated forms of cooperation.[61] Further planning required thinking about who was to guide the caravan. After an agreement had been reached, the caravan guide (*saykh*) was now tasked with hiring local and loyal companions (*khabir, taksif*) who knew the places well. He also had to make contacts with the various groups, through whose area the caravan would pass. Travelling into Africa's interior consisted, as has already been mentioned, of stops at different stations, at which the passage always had to be negotiated anew. Not only did one have to rely again and again on new rulers, tribal chiefs or companions at these locations, but the composition of the agents, traders, camel owners and helpers could also change. Although the size and the composition of the caravan could change during a journey, the preparation and execution of the trip were always subject to careful planning.[62] In this way, it also becomes apparent that a commodity, which has not yet been considered in the previous deliberations, namely mutual trust, always had to be negotiated anew, both outside and inside the caravan. "*Like long-distance trade throughout the early modern world, organizing camel caravans involved resources and above all trust in people*".[63] Trust is a phenomenon which appears in insecure situations or in the case of a risky outcome: those who are sure about something do not need to trust. However, trust is also seen more as belief or hope; it always needs a foundation, the so-called *basis of trust*. To what sort of trust could the travellers in the Sahara refer to?

The *basis of trust* was, on the one hand, guaranteed on the basis of kinship relationships, and on the other hand, it was created on the basis of common belief, as well as by contracts. Ghislaine Lydon illustrated this for the trade in Western Sahara: "*The access to literacy and faith-based institutions provided support, laws, and incentives that structured the organisation of early modern trade. The practice of Islam structured both the organisation of long-distance caravan trade and the operation of trade networks.*

Type of Documents	Family Archives in the Western Sahara	Family Archives in Libya
Accounting		
Accounting Books and Ledgers	X	X
Contracts		
Agency Contracts	X	
Commenda-type Contracts and other Partnership Agreements	X	
Commission Contracts	X	
Debt and Equity Contracts	X	X
Forward Purchase Contracts	X	
Labour Contracts	X	
Leasing Contracts	X	
Marriage and Petition Contracts		X
Real Estate Buying and Selling Contracts		X
Storage Contracts	X	
Correspondence		
Business Letters	X	X
Private Letters		X
Shopping Lists	X	X
Financial Instruments		
Bills of Exchange, Money Orders, Dept-Swapping, Traveler's Checks	X	
Waybills		
List of Goods to Establish Ownership of Dispatched Parcels and Camel Loads	X	
Other Documents		
Tax Farming		X
Folk Medicine Prescriptions		X
Family Genealogy		X
Codicology Tented Quran with Remarks		X

Fig. 3.5 With regard to their quantity, the sources accessed in the al Qatrun region cannot be compared with the sources accessed by Lydon in Morocco

Muslim merchants and traders used their Arabic literacy and access to writing paper to draw contractual agreements and dispatch commercial correspondence, while depending on their mutual trust in God. At the same time, they relied on an Islamic institutional framework defined by local scholars versed in legal doctrine and local customs".[64]

3.7 Al Qatrun—Trading Post on the Bornu Road

Jacqueline Passon, Said Hamid and Hassan Mughrabi

For parts of the central Sahara, in the area of today's Libya, interviews were conducted and family archives in the al Qatrun area were processed and examined. These contained handwritten Arabic sources from the nineteenth and early twentieth centuries (Fig. 3.5) and provided an insight into daily life, as well as into the business relationships of the population in this area during this period of time.

The oasis of al Qatrun appears as a remote green spot surrounded by desert. It occupies a depression in the south-western part of Libya. The area along the so-called al Qatrun depression is very sparsely populated within an area of about 60 km. The surrounding area of al Qatrun encompasses the small villages Tajarhi, al Bakhi and Madrusah. As in the nineteenth century, the inhabitants are still mainly of Tubu tribes. The other inhabitants descend from the mixture of Garamantes and the Arabic tribes which once settled in the area.

Several artesian wells provide the only source of water for drinking and irrigation. Like in other oases, farming had been essential for the residents. The date palm is the main cash crop of the oasis, hence dates acted as the primary source of food supply. The palm tree provides fibres, leaves, trunks, agricultural tools, furniture, baskets, mats, sandals and numerous other items. From its wood and leaves, the oasis dwellers prepared most of their household items. It had been a culture based on and supported by the date palm. Some vegetables were cultivated for local consumption, as well as grasses for fodder for the livestock. Chickens and livestock, especially goats, sheep and camels, were raised. In addition, the residents from al Qatrun were engaged in trade as the oasis was a meeting point on a major caravan route that crossed the central Sahara; the north–south one was called Bornu road that linked the Mediterranean shores and the heart of Africa via Zawilah and later Murzuq. Several other paths ran, for example, between al Qatrun and Ghat.[65] In this area, the population concentrated around the few available water sources, connected by trade routes. A number of unsurveyed sites punctuated the portion of the Bornu road between Tajahri, respectively, al Qatrun and Murzuq as well. These desert castles appear to have been outposts guarding the route, providing water and taking tolls. Others may have been also trading centres or still await an interpretation (see *from Murzuq to al Qatrun*).

Data on social life is very limited. For that reason, the following excerpts, derived from family archives in the al Qatrun area, offer a vibrant insight into the daily life of traders and the population in this area. They help us to make a next step towards understanding life and trade in a Fezzan oasis. They also illustrate the use of written documents such as business letters, contracts or accounting books which allowed for a certain commercial transparency.

3.7.1 Chronological Dating and Type of Sources

Chronological dating has determined that the majority of the texts originate from the second half of the nineteenth century. The other documents date back to the first half of the twentieth century. The sources show that written documents, undoubtedly, formed part of life in this region of the world. They were used to govern daily business as well as private life.

The range of content featured in the texts includes numerous legal texts, which, predominantly, outline sales contracts or provide information on private inheritance arrangements. One text describes the correct distribution of inheritance, similarly to a religious manuscript which also exists for this purpose. Furthermore, excerpts from an accounting book exist that recorded, for example, commercial transactions and financial services that were provided in the form of loans. Several letters related to trade and a document written by someone presumed to be a tax farmer offer insights into the world of trade too. Another more extensive series of documents comprise various religious manuscripts which were circulated in the area studied. Other texts refer to endowments (*waqf*) and reveal personal correspondence between people from the area studied. Private letters were exchanged, discussing information on different people and networks, relaying messages of gratitude for gifts received or contacting relatives to request the sending of a certain commodity from the north. Finally, some of the texts are associated with traditional medicine (see below).[66]

3.7.2 Written Correspondence

Letter writing was an opportunity to conduct business transactions and the most efficient way to coordinate information flows, as Lydon has already pointed out.[67] Like in the Western Sahara, there was a need to communicate business across distance. And it was the only way to maintain business relations in international commerce in the nineteenth century. In business communication, private letters also formed an element in the organisation of long-distance trade by helping to understand current events and developments and to identify reliable business partners. It is not surprising that letters were filled with a variety of gossip. For example, information about agent misbehaviour was conveyed to family members. This kind of information helped to update the reputation and reliability of distant counterparts. As Trivellato pointed out, letters were not only valuable for circulating commercial and financial information, but they were also a crucial instrument in maintaining informal cross-cultural networks.[68] To coordinate their activities, traders from the al Qatrun area engaged in letter writing with family members and network partners, requesting and supplying information about business partners, market trends and prices or current developments on routes.[69] Since the Sanussiya monasteries were connected to each other by a postal system, they provided merchants with a postal service.

3.7.3 Legal and Commercial Documents

As mentioned above, a selection of documents shed light on the type of legal transactions conducted in the area studied. To this effect, a series of sales contracts exist, which certify the sale of land and/or date palms. The documents generally record the location of the object of purchase, details of the seller and buyer, the price of the transaction and also all witnesses present.

Example text: Sale of a group of date palms

Like all the letters, it begins with warm salutations and religious invocations before proceeding to business:

Grace and thanks to Allah and peace upon our prophet Mohammed, his family and his followers. With Allahs help and power al-Haj named son al-Murabit al-Haj Ibrahim al-Gatrouni bought from named son of Abd al-Qadir Colcot a group of trees in the area of Aiyat al-Ajaais originally bought from Ghazala daughter of al Tuarggi and the daughter of her sister Mabarkah. These groups of trees are situated to the east and north of sheik Muhamed Ben Omar Wishka's group of palms and also to the east of Ajaaiz al-Brawnah which is in the Maghmag area. This group of palm trees is sold with all the produce and its surroundings. The sale is final and immediate which is worth 150 Turkish Girsh.[70] *The whole amount was paid and received complying with the Islamic Sharia. The witness certified that the sale took place last year. The witness of this sale knows both parties who are both mentally fit. They have both given him the permission to witness this sale.*

Issued on the 11 Shawwal (=tenth month of the lunar Islamic calendar) in 1298 Hijri (=1881)
Ibrahim Ali son of al Haj Ibrahim al Jamai the servant of the Sannussya[71]

With regard to the settled areas of Tripolitania and the urban areas of Cyrenaica,[72] it seems that private ownership and registration in landholding were introduced in the al Qatrun area during the second half of the nineteenth century too. It appears that tribal lands were divided and ownership was assigned to individuals. Several sources also indicate that residents from al Qatrun or smaller villages nearby owned trees or animals, whose care they contracted out to other residents or shepherds.

The Quran is considered to be the most important base of coexistence in the Islamic sphere. Sura 2, verse 282 (*surat l-baqarah*) instructs believers how to write down their contracts. It also emphasises that all transactions require testamentary evidence.[73] It is important to note that the before-mentioned contract only was witnessed by one person and that happens to be a member of the Sanussi monastery in al Qatrun. Other documents frequently refer to members of the Sanussiya too. This suggests that towards the end of the nineteenth century, the order had also become highly influential in this area of the Sahara.

The sources further indicate that parts of the privately owned land had been converted by the end of the nineteenth century into *waqf* (plural *awqaf*). Some of the palm groves in the outskirts of al Qatrun were *awqaf*, dedicated to funding mosques.

3.7.4 Excerpts from an Accounting Book

During the nineteenth century, it appears that more people used paper to record their business transactions. The documents at hand (Fig. 3.6) are excerpts from an accounting book belonging to two people, presumed to have been traders working in various markets (*aswaq*, sg. *suq*) around al Qatrun. The documents list an assortment of traded goods as well as the prices paid for them, and transactions exacted in lending businesses. Money lending in the form of credit was a common practice at the market, and this remains the case today. The *suq* was and still is the centre of commerce and selling, as well as a hub for financial activities and money lending. It spread out along the city streets, typically in a linear pattern, and the location of the different markets was determined by the type of goods sold in them. The same commodity or craft was grouped together, and the most profitable and prestigious crafts or traders were situated close to the political and religious hub.

The documents record the name of the person and the amount lent, together with a note that the sum comes in the form of a letter of credit. The documents are each marked with a stamp which lends them validity and legal force. It is not impossible that the excerpts refer to a document which has been looked at by a tax farmer.[74]

Tax farming, known as "*malikane*", was introduced in the Ottoman Empire in the seventeenth century. The right to collect a certain tax was sold at auction, where the winning bid was given directly to the fiscal authorities. Of course, the tax farmers, the *mültezims*, tried to squeeze considerably more tax out of the area allocated to them than they had paid for it in the auction.[75] This form of taxation was also practised in the area studied, together with a second form of taxation, where 10% went directly from the people to the central authorities. This was possible in monetary form as well as in natural produce (though money was preferred). One document in the collection of sources which was signed and stamped by a man named Mamour Shakir, likely to have been a tax farmer, becomes relevant in this context. The document contains the names of people who own date palms. There are also entries regarding the number of date palms belonging to any individual person.[76]

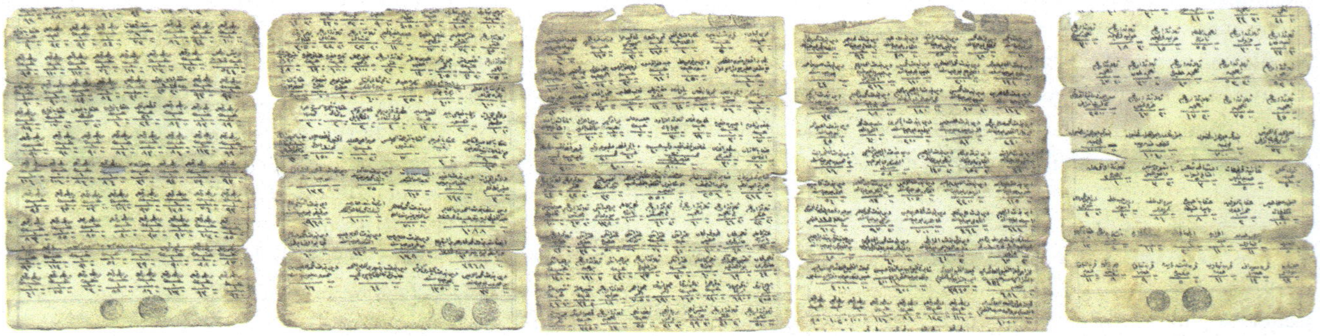

Fig. 3.6 The documents seem to be excerpts from an accounting book belonging to two people, presumed to have been traders working in various markets around al Qatrun. The documents list an assortment of traded goods as well as the prices paid for them, and transactions exacted in lending businesses

The excerpts from the accounting book list the goods and lending businesses in what is probably the order of sales transacted or money lent. The price demanded or the credit sum lent is specified in the row beneath the transaction. The first document conserved contains a list of leather skins produced in varying sizes which were used, for example, for manufacturing shoes. The second document features a diverse range of products including camels, sheep or goats which were sold and also two books or nails. Alongside this, records of money lent to people who are listed by name are kept: Mohammed Tahir bin al-Haj, for example, took out a loan of 700 Girsh. There is also a note indicating that an original document was issued for this transaction. The third document also lists disparate products in entries featuring copper products, items of clothing, leather skins and also a book. In addition, there are also entries for animals (camels, goats, sheep) which were sold either as a whole or in pieces. The fourth document predominantly details lending businesses but also includes earnings made from the date harvest. A wide variety of products are listed again in the final document, including items of clothing, combs, weapons and animals which have been sold.[77]

3.7.5 Excerpts from a Manuscript on Traditional Medicine

As they do today, illnesses, injuries or psychological ailments formed part of everyday life of people in Europe and Africa and elsewhere in the nineteenth century or even earlier. They constituted a real threat and people were eager to improve their physical and psychological welfare, so they sought ways to rid themselves of suffering. Islam, just like Christianity, did not provide people with adequate resources for this. The majority of the population did not have recourse to "scholarly medicine", so traditional medicine developed in this region as well. This is confirmed by manuscripts discussing traditional medicine which were circulated in the area studied and in other areas. They describe curative medical knowledge passed down from generation to generation.

The texts indicate that popular beliefs and superstitions, as well as the belief in the effects of magic, were widespread in the population. The handbook of superstitions defines magic as the art of exercising power over nature and people with the aid of incantations, sorcery and rites. According to Richard Kieckhefer, the defining difference between religion and magic is that in the case of the religion, people supplicate God or the gods, while magic coerces spiritual beings or forces. Belief in magic is also at least in the early modern period a point of intersection where the common people's culture collides with the scholarly world. Magic was therefore a part of people's reality, and they were not indisposed to trust in healers and magic in the case of illness. Magic is a major aspect of certain fields of traditional medicine which probably have their origins in pre-Islamic traditions and customs. Magical fields of medicine would have included natural magic or herbal healing methods.[78]

The text at hand, written in the form of a handbook, is part of a more comprehensive manuscript. The curative methods described were used in a religious Islamic context. It describes various rituals performed by healers to drive the genies out of the bodies of the sick and strengthen them against new attacks. This manuscript includes a detailed description of a ritual to be employed by women who suffer from headaches or heart problems: Firstly, *surahs* from the Quran were written on a piece of paper, and the names of genies were then added to them. The paper was then folded and burned together with essences and inhaled. The protecting or burning healing power was then breathed in together with the smoke of the burning words.[79]

People learned in the Quran, teachers of the Quran or Islamic religious leaders who were knowledgeable in the holy writings, were able to become religious healers. They were mostly active within an Islamic community and, additionally, worked as non-medical healers. There were, however, healers who were not taught in the Quran and had not received any particular training and mainly engaged in practices that combined magic and religion. The boundaries between healers using magic and people with acquired knowledge of the holy texts were fluid.

It was accepted that the healers had armies of genies on their side which would take action against the "evil spirits". Genies are a central part of Islamic beliefs; however, Islam rejects belief in magic and sorcery in accordance with the Quran. Magic and theurgy are considered to be bringers of bad omens and are strongly connected to anxieties. Psychiatric illnesses, in particular, are often attributed to possession by an evil genie or the influence of the "evil eye". People are also discouraged from interfering with such things so as to avoid nurturing the superstition. However, the traditional Islamic beliefs have held a special place for the existence of genies. Lucky charms are also rejected, as the supernatural power attributed to them is held to function in an automatic manner. According to Islam, God is omnipotent and the single creator of all things.[80]

3.7.6 Summary

The picture that slowly emerges from these sources as well as from interviews highlights the importance of literacy among the people living in the al Qatrun area—not only in terms of trade but also life in general. With respect to the functioning of the trade, it underlines the thesis provided by Lydon that both the regional and the transnational caravan trade were based on organised networks. It appears that the organisational framework of the trade was based on oral, as well as written, agreements for processing the trade transactions. The trading business was based on Islamic law. It has to be assumed that, as a result, in particular, of the spread of the Sanussi monasteries, which functioned both as centres of trade, as well as education, the ability to read and write in the desert (Fezzan) increased. The Sanussi order spread through Mizdah, where a monastery was founded in 1845, to Tripolitania, and it spread through the monasteries in the backcountry of the Great Sirte to Fezzan.[81] Through the monasteries, access to education and other trade-relevant knowledge was facilitated. By means of Islamic education, the trans-Saharan traders could acquire key competencies, which they used in the processing of their business transactions.

The oasis of al Qatrun cannot be regarded in isolation from the other villages along the al Qatrun depression as well as the desert castles. From a strategic point of view, the installation of a chain of fortified settlements along a major caravan route of the central Sahara appears to represent a large-scale effort to control trade and travels in this area. In this respect, al Qatrun retains its role as a part of the Bornu road. It is far from being an isolated spot on the map.

To sum up the above, it can be said that the success or the failure of a given trade activity in the central Sahara depended upon various aspects:

Negotiation processes: Whether a given venture ran successfully depended on the practice and dexterity with which the traders and other staff in the organisation of a given caravan could facilitate communication and negotiation processes among different groups and stakeholders.
Faith/trust: The Islamic religious practice had a conjoining and trust-forming effect. Faith provided a structure to the trade and had a stabilising effect on the functionality of the trade networks. The Sanussi brotherhood unified many tribal, regional and ethnic loyalties. This religious ideology integrated these ethnic groups under the banner of trade and Islam.[82]
Paper industry: For the success of the trade, the possession and the exchange of written documents were decisive. As Lydon shows, trans-Saharan trade could only be successfully organised in a paper economy favouring reliable information flows, contractual accountability and proper accounting methods.[83]
Dealing with dangers *en route*: Conflicts and turmoil between different ethnic groups, tribes or clans of the desert could easily break out at any time. Consequently, the organisers of a given caravan needed the ability to respond to unforeseen crises by either negotiating with those groups or by identifying alternative trails to move on. Furthermore, the merchants, who fitted out a caravan, had to be capable to compensate for the potential losses.

3.8 "Moving Cities"

Jacqueline Passon

3.8.1 Build-up and Social Structure of the Caravan

Both, large trans-Saharan caravans and smaller interregional caravans crossed the Sahara. The trans-Saharan caravans were not only large economic enterprises, but they were also travelling social communities. In their social composition, they were a collection of different cultures and identities. In them, hundreds of people from different regions, from the coast and from all parts of the Sahara, including sub-Saharan regions, could meet with each other. They spent weeks and months together, sometimes under extreme conditions. While doing so, they had to cooperate in various matters related to everyday life, to share their knowledge and to negotiate their positions. However, they also had to deal with dangerous situations in unity in order to be able to defend themselves against the outside world.

Within the caravan, a strict hierarchical order reigned. The composition of the caravan depended upon its size. No matter, whether it was large or small, the caravan always needed a leader who made the decisions. That is why the caravan was led by a so-called *saykh* who was supported by his assistants. The *saykh*, who possessed great authority, also had, in addition to the practical guiding of the caravan, the task of making the decisions on any legal issues. For that, he had disciplinary powers. Guiding the caravan also included the *saykh* determining the route. Furthermore, he established the first contacts with the tribes when he planned to pass through their areas. In doing so, the *saykh* had to act as a skilled negotiator in order to be able to negotiate the security issues of the crossing as well as the road tolls. For the protection of the caravan, he could recruit an armed escort. In this respect, the *saykh* also needed to understand human nature. On the one hand, the armed escort served to protect the caravan participants and had to ensure that the valuable goods were not looted. On the other hand, the escort was employed to discipline the "employees" in case they deserted or became mutinous. In fact, there were often disputes in the caravans when the porters wanted better pay, better supplies or reduction of their loads.

In order to satisfy these diverse requirements, the *saykh* had to have profound geographic knowledge. In addition, a certain knowledge about the political and cultural structures of the societies in the desert, as well as in the sub-Saharan areas, was necessary. As a rule, the *saykh* was advised by a *khabir* and/or a *taksif*. They were hired for certain sections of the route which they knew particularly well. Along these sections, they were the navigation experts (see *Navigating the Desert*).[84]

A trans-Saharan caravan could be viewed as a "moving city", in which people from different social fabrics and cultures coexisted. Along with the personnel for the navigation, it also included, above all, porters, cooks and kitchen personnel. In addition, there was also a muezzin or an imam and, for the entertainment of the travellers, there were musicians or prostitutes. The porters, which could be hired through agencies in the more important caravan towns, travelled back and forth, with only seasonal interruptions. These were mostly young men originating from inland or from the coast; they could be free people, but also slaves. The slaves were partially rented by the owners and had relative freedom, or they acted by themselves and gave part of their salary to their owner. Women always travelled with the caravans too. Besides the prostitutes, they could be relatives, wives, slaves or concubines of the caravan participants. In any case, they were also a supporting workforce. They helped in carrying the loads whereby they transported the personal belongings of the porters or the military personnel and served the food during the breaks.[85]

A camel caravan could comprise on average around a thousand camels, although the historian Ibn Khaldun mentions caravans of up to 12,000 loaded camels travelling between Egypt and Sudan. He writes in his universal history *Book of Lessons, Record of Beginnings and Events in the history of the Arabs and Berbers and their Powerful Contemporaries (Kitab al Ibar)*, which was completed during the last decade of the fourteenth century that an informant had told him that: *This year there passed through our city on the way to Mali a caravan of merchants from the east containing 12,000 camels. Another [informant] has told me that this is a yearly event*".[86]

The Arabian geographer al Idrisi gives an insight into how a caravan worked. The trans-Saharan caravans normally crossed the desert in autumn and winter. Usually, the camels were loaded at sunrise, and the caravan started to move along the route. Normally, they walked until the heat became unbearable. Then the camels were unloaded and their feet were tied together, so they could not run away. At the same time, the tents were set up and everybody tried to find protection from the sun. Early in the evening, the caravan started again and travelled until night came. On average, the caravan travelled nine hours per day. Sometimes, however, the caravan had to travel for 12 hours per day; this happened when a long section of the route between two wells had to be covered.[87]

3.8.2 Goods Traded Across the Sahara

The goods traded across the Sahara had to be of an intrinsic value justifying the costs of the desert crossing. Sudan's main exports by value over the centuries, gold and slaves, met these conditions (Fig. 3.10).[88] Next to the trading of gold and slaves, leather, ivory and ostrich feathers from inner Africa were also traded; they ranked among the modest according to their market value. At the same time, Europe supplied mainly industrial products such as arms, glassware, perfume, tea, spices, paper and later textiles (Fig. 3.7).[89]

The trans-Saharan trade in slaves and ivory dates from Roman times, while the trade in ostrich feathers began in the twelfth century. Due to the lack of adequate data trade cannot be quantified for that period. Based both on consular reports from Tripoli and the Arabic and European explorers' observations on the trade, it is possible to generate statistical information for the nineteenth century. The data on ivory, ostrich feathers and tanned skins are relatively complete for the last decades of the nineteenth century.

In the 1860s, commodities other than slaves began to become statistically important. As Lovejoy pointed out, gradually at first and then dramatically three of these other goods ivory, ostrich feathers and tanned skins became very important. A comparison of the estimates for the slave trade with those for ivory, ostrich feathers and skins shows that in the last quarter of the nineteenth century, ostrich feathers, ivory and tanned skins did surpass slaves in value (Fig. 3.8). The trade of slaves appears to have dropped off dramatically in the 1870s (Figs. 3.8 and 3.9). At that time, British-Ottoman agreements to fight the slave trade were better coordinated, so that the trade had fallen off considerably. Less can be said about the value of many other commodities, particularly African textiles, leather goods or kola nuts which were distributed in the northern Sahara and along the North African coast. These imports are assumed to have been relatively constant.[90]

If one considers the flow of goods that were transported on the Bornu route, for example, it is apparent that in contrast to gold trading, the West African slaves were the most significant "commodity" on this route. Next to them, the consular reports from Tripoli also registered small amounts of gold, senna leaves, animal products and woven fabrics. From the north, mainly horses, materials, cowry shells and weapons were imported.[91]

The Kanem Empire and later, the successing Bornu Empire had the ruling power over this important trade connection in the Middle Ages as well as later on. In succession of the occupation of the Central Saharan oases in the north by the Ottomans in 1574, Bornu lost its outpost in the Fezzan region. At the time of the travellers of the nineteenth century, Bornu had degenerated to a regional power and trade on the Bornu route was less common than in earlier centuries.

The desert traffic is closely tied to the camel (Fig. 3.13). Much has been written about the introduction of that beast of burden into North Africa. Most scholars assume that the camel was introduced into Tripolitania by the Romans probably around the first century AD. But there seems to be

Fig. 3.7 Map of the caravan trade features and their spatial distribution

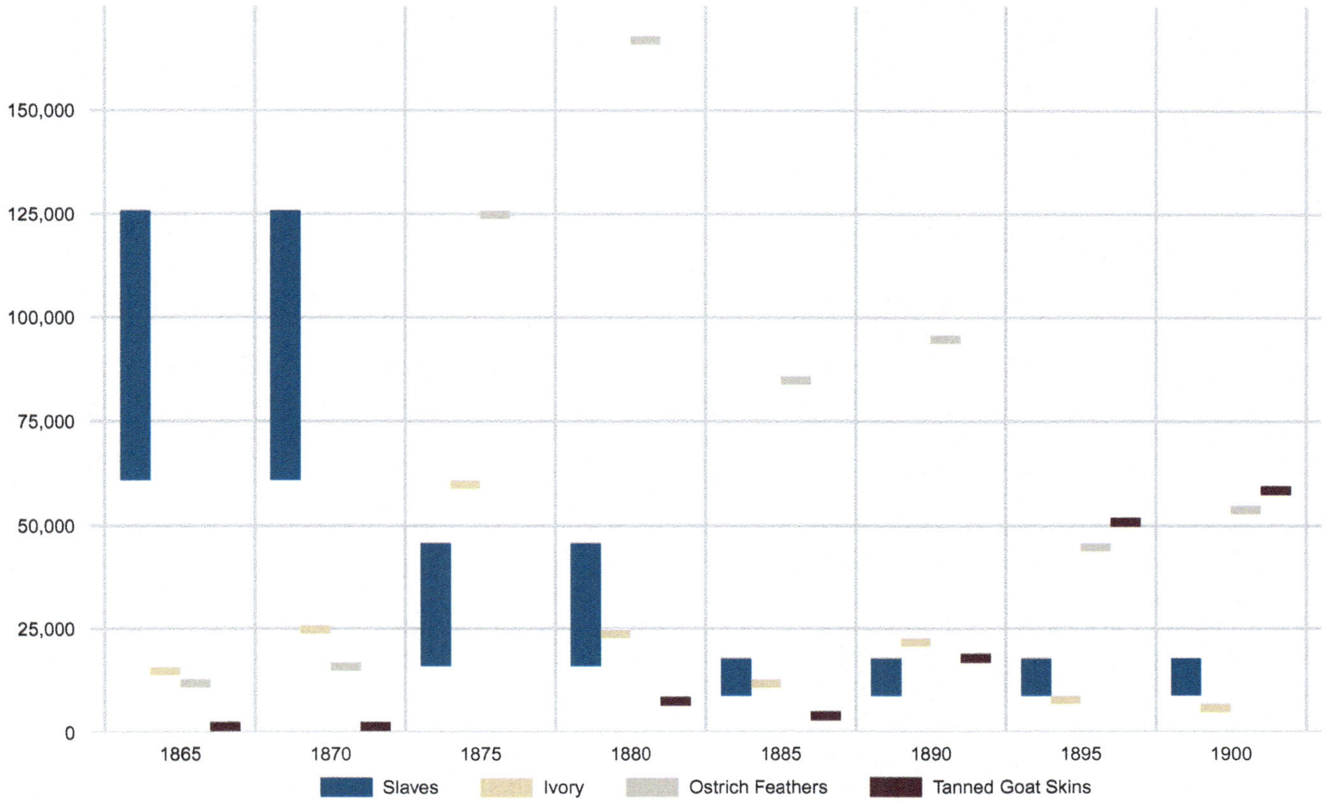

Fig. 3.8 Sudan exports in transit through Tripoli between 1862 and 1900 in £ (see Lovejoy 1984)

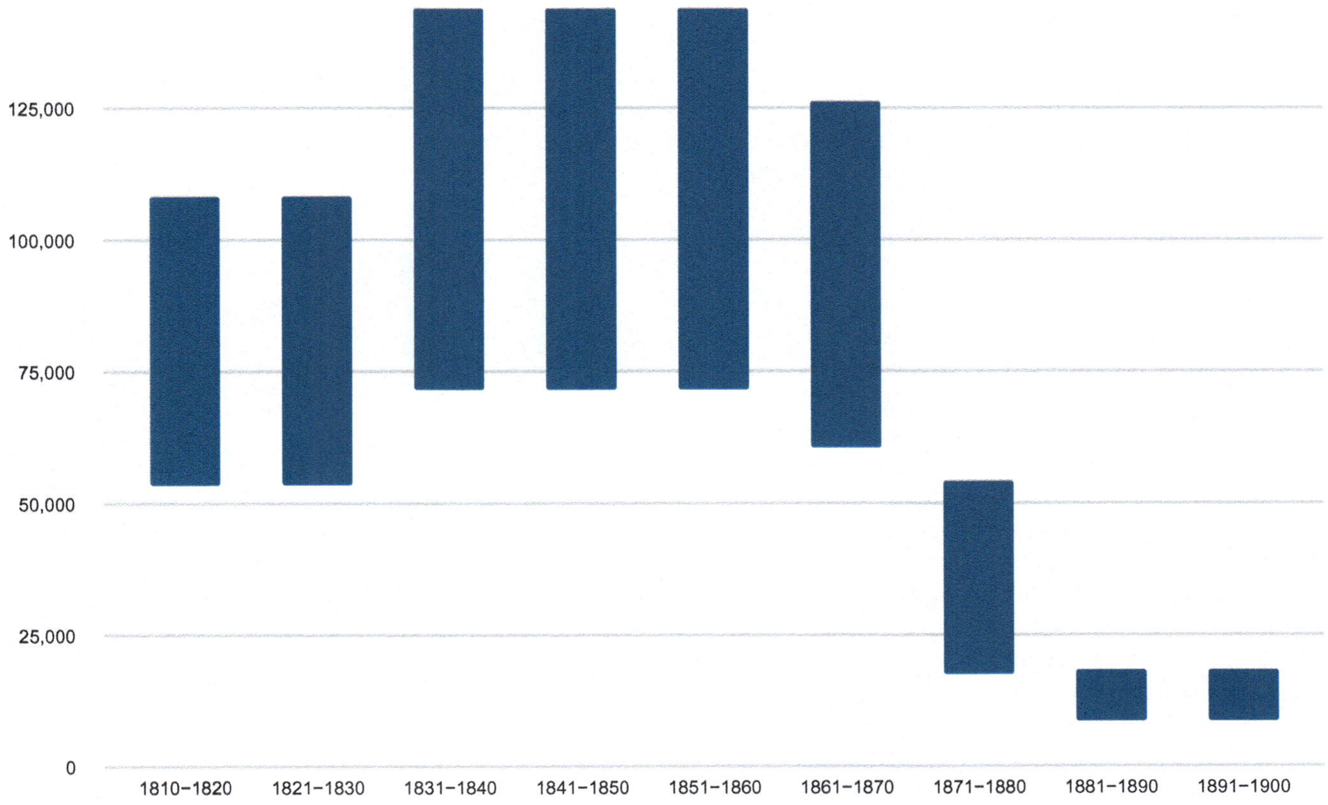

Fig. 3.9 Trans-Saharan trade in slaves (estimated annual value by decade in £; see Lovejoy 1984)

	650–800	800–1400	1400–1500	1500–1600	1600–1700	1700–1800	1800–1900	Total
Atlantic			81,000	328,000	1,348,000	6,090,000	3,466,000	11,313,000
East Africa		1,400,000	100,000	100,000	100,000	400,000	442,000	2,542,000
Red Sea		600,000	100,000	100,000	100,000	200,000	450,000	1,550,000
Sahara	150,000	3,690,000	430,000	550,000	700,000	700,000	1,200,000	7,420,000

Fig. 3.10 Estimated slave exports from Africa between 650 and 1900 (see Lovejoy 2000; see also Austen 1988, 1992)

some evidence to suggest that this assumption is just in agreement with the common opinion that trans-Saharan trade, even if already existing in the first millennium BC, became important only in Islamic times. New historical and archaeological data allow to assume that the trans-Saharan trade could have been a relatively substantial and organised affair as early as around 500 BC. Trade was in the hands of Libyan (mainly Garamantian) chiefdoms clustered in the Fezzan area. It seems unbelievable that the distance from the upper Nile valley to Niger could have been traversed without the use of camels. Archeozoological evidence is needed.

It can be taken for granted that other itineraries than the Bornu route were in existence. During the sixth century BC, the coastal *emporia* in Tripolitania were built and simultaneously the Greek colonies in Cyrenaica underwent a notable growth. The coastal area offered substantial markets and trading posts not only for the Garamantian. Caravan roads, presumably, were leading to the coast of Tripolitania.[92]

3.9 Lost Tracks in the Sand: Slave Trade Across the Sahara

Jacqueline Passon

The very earliest external slave trade was the trans-Saharan slave trade. Although there had long been some trading with the Roman Empire, the transportation of large numbers of slaves did not become viable until the Muslims arrived in North Africa in the seventh century.

During the Middle Ages and the early modern period, Africa has been a politically fragmented continent which was marked by recurrent instability and tensions. The most obvious feature was the lack of political or military leaders to consolidate large areas into centralised states, apart from Bornu or Songhay. Systems of servitude and slavery were common in many parts of the African continent. It was already fundamental to the political, social and economic pattern of African societies in the northern Savanna. Consequently, enslavement was an organised activity, sanctioned by law and custom.[93] Following the ancient tradition, slaves were in demand throughout pagan and Islamic Africa as well as the wider Islamic world.[94] The major external markets for slaves were North Africa and the Middle East. For the North African merchants, it was therefore not difficult to get males and females in response to the strong market demand. Throughout the Islamic conquest of North Africa, slaves were prisoners captured in the holy wars. After the early caliphates were established, slaves then came from the frontier areas. This practice was mainly based on the justification of enslavement and on the basis of religion; hence, a non-Muslim could be enslaved legally. Due to the given social context, a Muslim could not be enslaved. However, this commandment was often violated in practice. Furthermore, slavery was perceived as a means of converting non-Muslims. Although conversion did not automatically lead to emancipation, it was normally some guarantee for better treatment. The German traveller Nachtigal could bring himself to remark that Islam generally brings with it a mild administration of the institution of slavery. Although this perception is insufficient, one aspect of the Islamic tradition was that emancipation of slaves was clearly defined (Fig. 3.11).[95]

Within the Islamic cultural sphere, slaves did not come necessarily from Inner Africa. Although a great proportion of the slave population were black Africans, they also came from Europe and southern Russia.[96] Unlike in the Americas, in the Arab world enslaved people were mainly deployed as army recruits since the end of the ninth century. So-called Mamluks were slave soldiers who converted to Islam and served the Muslim caliphs and the Egyptian Ayyubid sultans during the Middle Ages. In the markets of the central Islamic

Fig. 3.11 Slave coffle in Sudan

lands, slaves were sold as eunuchs for harems and government service, and as concubines and domestics. Despite the lack of comprehensive sources and the literature, it is also to assume that slaves were used as agricultural labour.

Berbers, who in the days of the Arab conquest had been a licit booty, too, eventually became the masters of trade during the Islamic Middle Ages. Towards the middle of the eighth century, they were gaining control of the black slave trade along the eastern and central Saharan routes. According to the writings of early Arab historians and geographers, it is common to assume that the Ibadi merchants made the slave trade a predominantly Ibadi monopoly from the mid-eighth century onwards. Al Yaqubi and al Bakri described the slave trade as an important source of revenue for the Ibadi economic system. The entrepôt of this economy lays in the heart of Fezzan, in Zawilah. The marked place soon developed into a bustling economic centre and was well connected with Ibadi branches in the Jabal al Nefusah in the north and Kawar in the south. From al Yaqubi's writings, it becomes apparent that slaves were captured from people of Sudan, who lived close to Zawilah, as well as bought from the kings of Sudan.[97] By the tenth century, most of the black slaves sold in Muslim countries were imported via Zawilah. The town was particularly noted for the re-export of eunuchs taken mainly from the area west of Lake Chad (see *Ibadi Berbers of Zawila masters of trad*e in: *From Zillah to Murzuq*). From Zawilah, the slaves went north to the ports of the Mediterranean Sea like Tripoli, Benghazi or Cairo.[98]

All the important towns in the Islamic countries had their own slave markets, where the majority of slaves were sold. The specially trained ones were purchased at private houses like the famous Berber female singers, *qayna*, who entertained the caliphs. While Berber women were also highly regarded for housework, sexual relations and childbearing, black women had a reputation for being obedient, robust and excellent wet nurses. Male slaves worked domestically as cooks, storekeepers, porters, boatmen or keepers of private libraries.[99] Overall, there was a strong demand for women and children. Indeed, they were wanted in greater numbers than men. This certainly has to do with the fact that women and children were more likely incorporated into Muslim society. Throughout the Islamic cultural sphere, slavery was not a self-perpetuating institution. Those born into slavery formed a relatively small proportion of the slave population. Most children of slaves were assimilated into Muslim society, only to be replaced by new imports. The institution of slavery was not called into question. As already noted, this was a common practice and belonged to the political, social and economic fabric of African as well as Muslim societies.[100]

Different patterns of slave supply can be observed. Slavery was not always a result of violence, including wars or kidnapping. Although a large proportion of people were captured and enslaved during armed conflicts or large-scale slave raidings, sale by relatives or superiors, sale to repay debts and punishment for legal offenses were reasons for losing one's freedom. Estimates of the number of slaves taken across the Sahara vary greatly, particularly for the period prior to the nineteenth century. During the Middle Ages, it is assumed that between 3000 and 8000 slaves were taken through the Sahara every year by six main trade routes. According to Wright, exports perhaps reached a peak annual average of 8700 in the tenth and eleventh centuries falling thereafter to between 4000 and 5500 a year down to 1600.[101]

In view of profound power shifts during the fifteenth and sixteenth centuries, trans-Saharan trade was subjected to a structural and spatial transformation. The European seafaring powers landed on West African soil to make the treasures of the continent their own. As a result, the nature of trade was changing. Trade became a global phenomenon: beyond the great seafaring nations of Europe, Spain, Portugal, England, France and Holland, the whole European continent was, through a far-reaching network of commercial and financial links, integrated into the triangular trade between Europe, Africa and America. At the same time, the Ottomans gained control over the Mediterranean coast of North Africa. With regard to politics, there were also changes in sub-Saharan Africa where Bornu became a leading power.[102] Together, these developments firmly grounded the eastward shift of trans-Saharan trade. The trade of Western Sahara was directed towards the west coast. This contributed to the decline of the west Saharan routes and to the rising prosperity of the central and eastern roads. The slave trade continued to provide the main bond of the different powers of Tripoli with the Bornu Empire and also with the Hausa states, the main centres of slave collection and distribution on the Niger. Most slaves primarily followed the Bornu–Murzuq–Tripoli road (so-called Bornu Road), with the rest coming from Niger via Ghat and Ghadamis. On both routes, the traffic varied considerably from year to year. In the early 1700s, slaves were imported into Tripoli at an average rate of 500–600 per annum, increasing to 2000 by the 1750s and falling to an average of around 1500 a year by the end of the century.[103] In the first half of the nineteenth century, the trade volumes of the Bornu–Tripoli route consisted of roughly 5000–8000 slaves per year who passed through the Fezzan region. After the wars within the Sokoto Caliphate (1800–1830), the revolutions in Tripoli (1830–1842) and the increasing caravan robberies after 1830 between Bornu and Murzuq, trade fell to a third of the volume around 1869.[104] As the robberies surmounted, many traders resorted to using other routes. Similarly, trade on the Kano–Ghat–Ghadamis route (and further on to Tripoli or Tunis) increased considerably at this time to 2500 slaves per year (around 1850).[105]

Despite the manifold dangers that lurked over the slave caravans, the traders expected their "victims" to bear the extremely difficult journey through the Saharan region. The result of the week-long marches was a high death rate among the slaves, who had to suffer throughout the journey from heat, hunger, thirst and stress. On the 2700-km-long stretches between Lake Chad (Bornu) and Tripoli, a 50% loss of slaves was expected. On all of the Sahara routes, it is estimated that the average death rate is still very high, over 20%.[106] But despite the immense losses of human life, this business was still considered to be very lucrative for the traders, which was evident in the profits drawn after they successfully crossed the Sahara. On both routes, enormous price increases for slaves were achieved. In 1850, between Kano and Murzuq, the price for young women increased by 160%, further along Tripoli it increased by 210%, and in Constantinople, the price was up to 300% in comparison to the starting price in Kano (Fig. 3.12).

What is mostly ignored is that the trans-Saharan slave trade had a strong impact on Europe, too. There, black slaves were quite common. Slaves were sold for agriculture and private household use, especially in Catalonia, Sicily, Naples, Venice, Genoa and parts of France. For example, documents from the Catalonian slaves indicate their Cyrenaican origin. A Sicilian census from 1565 alone confirms the presence of 117 black slaves around Palermo who were from the Bornu Empire.[107] It was not until the transatlantic slave trade gained in importance that this part of the European trans-Saharan slave trade diminished. Other European regions, which were then under the influence of the Ottoman Empire, were still receiving significant quantities of slaves. As a result, black slaves also found their way into Turkey (Izmir) to Cyprus, to the Balkans (Albania) and to the Aegean region (Crete).[108] The European powers also participated in the trans-Saharan slave trade in other ways. On the one hand, European traders acted as financiers. It was in this way that the English traveller James Richardson found out the following information during his stay in the Sahara in 1845/1846: "*Three-fourths of the slave-traffic of The Great Desert and Central Africa are supported by the money and goods of European merchants, resident in Tunis, Tripoli, Algiers and Egypt*".[109] On the other hand, the European powers offered themselves as transportation service providers for the transport of slaves between the North African coast and the Ottoman territories.[110] Up until and in the early nineteenth century, the Ottoman Empire was lacking in suitable transport vessels. European powers such as France, but also states of maritime towns such as Venice,

Location	"Good Woman"	"Good Man"
Kano	32	10 - 12
Murzuq	85	40
Tripoli	100	60 - 65
Constantinople	130	90 - 100

Fig. 3.12 Price development between Kano and Constantinople in 1850 (Richardson 1848; Tambo 1976)

Fig. 3.13 Caravan near Kufra in the 1930s

Genoa, Pisa and Amalfi, took over the transport in exchange for money. It was only in the middle of the nineteenth century that the central powers of the Ottoman Empire had their own appropriate merchant ships.[111]

3.10 Decline of the Trans-Saharan Trade

Jacqueline Passon

Since the beginning of the nineteenth century, the decline of the caravan trade was initiated by two fundamental changes in the general conditions. To begin with, the West African coast increasingly became the target of trade. Portuguese ships had already been travelling within these regions since the twelfth century. In addition, there was competition between the Spaniards, the French and the English to secure the trading posts on the Mauritanian and Senegal coasts as well as in the Gulf of Guinea. The land route that led there from southern Sudan was considerably shorter than the one on the Mediterranean coast, and so, particularly the slave and gold trade was relocated to this route in the direction of the Atlantic coast. In contrast, goods from North Sudan and the Sahel were still transported northwards through the Sahara.[112] In the 1850s, European colonial powers largely restricted the slave trade. These developments slowly contributed to the eventual offside position forced upon the trade centres of Ghadamis and Murzuq. This carried serious consequences for Murzuq because seven-eighths of the city was financed by the slave trade.[113]

Up until the beginning of the twentieth century, the trans-Sahara trade was a significant impetus for the Fezzan region (Fig. 3.14). The region formed an intersecting point between inner Africa and the Mediterranean and Europe.[114] Nevertheless, Libya was able to continue with transnational trade for much longer than was previously thought, until well into the twentieth century, despite the colonial demarcation, the change in transportation and the political upheavals in modern times. At the beginning of the 1960s, goods were still transported by caravan. Here, one speaks of transnational caravans. The camels were eventually replaced by the car during the 1960s of the previous century. In the 1970s, the restructuring of trade by the state brought the caravan trade to a complete standstill.

Fig. 3.14 Deep into the Sahara, in secret enclaves between rocky escarpments, are found acres of pastures. You only need to know where to go (Furlong 1909, 216)

3.11 Fezzan as a Transit Region in the Late Twentieth and at the Dawn of the Twenty-First Century

Jacqueline Passon and Sabrina Swifka

3.11.1 The Intertwining of the Fezzan Oases

The geographical position of the Fezzan between two such different physio-geographical regions as the Mediterranean Sea in the north and the Savannah region in the south determined its strategic importance for many centuries. Due to its location, it played a commercial intermediary role between the two regions, which ensured a solid livelihood for the Fezzan oases participating in the trade. It even brought prosperity to some of them.

Leaving the historical perspective and considering only the interrelationships of the Fezzan oases at the beginning of the twenty-first century, it can be said that the trade in goods has never come to a standstill. A strong revival of the Fezzan as a transit and trade area occurred at the end of 1980s until 2011. Since the southern borders of Libya have become more permeable for the transport of goods in this period of time, the region had once again developed into a trans-regional transport route and trading region. The result of these developments was a lively trade relationship with Chad, for example, which had been established from the 1990s until 2011. Libya supplied clothing, automotive spare parts, Libyan cigarettes, as well as food (oil, rice, flour) and petrol, while the goods being supplied from Chad to Libya included mainly camels or goats.[115] Due to the deteriorating security situation and the lack of adequate border control after the 2011 revolution, the Fezzan faces, once again, tremendous difficulties. Today, the Fezzan is linking southern Libya to the Sahel and sub-Saharan migrant routes to northern Libya and onto Europe. Incentives for smuggling of all types of people, oil, gold, weapons, drugs far outstrip those for making money through legal means.[116] Like in the nineteenth century, regional powers, different ethnic groups and transnational actors are joining local conflicts or using the south as a transit zone.

Generally, until 2011, there were two types of transport: trade that can be compared to the traditional caravan type with camels and trade on a motorised basis. Camel caravans crossed the Sahara in a south to north direction. An essential difference, compared to the traditional caravan trade, was that during this period the camels were not loaded with goods but were themselves the commodity. In the past, the Tuareg traded primarily in meat; several times per year, camel farmers from Niger or Chad crossed the Sahara with several hundreds or even thousands of camels to Murzuq or al Kufrah, in order to sell the animals on the local camel markets. The reason that these difficult trips were still worth making was, that in the sub-Saharan region, there are wide grassland areas covered with vegetation where the camels can be kept during the rainy season. The meat of animals coming from Niger or Chad continued to be more cost-effective than the meat from Libya, despite the long transport. At the same time, large motorised caravans, comprising 35 cars on average, carried goods and people to Libya (Fig. 3.15).[117]

If the transport patterns are considered, it has to be pointed out that from the densely interwoven web of trading routes, three former caravan trails were formed as main caravan routes: A section of the old Bornu trail connected the al Gatrun oasis with markets in the northern part of Niger, as well as in the northern part of Chad. Another, less frequented route started from Ghat and connected it with oases in Algeria and Niger. The third route in south-western Libya linked the al Kufrah oasis with the markets in eastern Chad.[118] The route sections of the more recent business activities do not exactly correspond to the flourishing trade routes of the previous centuries, but run more directly. In modern times, they do not necessarily need to run along the water holes which were once essential for survival in the desert. Technical innovations changed the logistics of the desert crossing at the beginning of the twenty-first century. Satellite communication made the use of "mobile oases" possible. The water, food and wood supply via trucks enabled herdsmen to bring their camels to Libya using more direct, shorter routes. In this way, the truck had become an important station during the crossing of the desert; it functioned as a "mobile oasis". As such, it had the same function and relevance as the natural oases for the caravans on their way across the desert.[119]

Camel herds move north on these routes between the months of September and May, whereby al Kufrah in particular is the most important trans-shipment point for camels in Libya. Until 2011, about 60,000 camels reached the southern Libyan oasis per year.[120] In contrast, two or three times a year, caravans from Niger or Chad, with up to 400 camels, reached the market in Fungul which is located 30 km to the east of Murzuq.

On the other hand, these routes are used by transport vehicles that deliver migrants, and naturally also smugglers. In the middle of the last decade, scientists from the Geographical Department of Sebha University tried to observe and quantify the traffic, which was not an easy task. According to a study conducted in this context, the number of trucks that left al Gatrun, in the direction of Niger, during an eleven-month period (reporting period: 11 January 2005 to 31 September 2006) was at 806. Exactly the same number of trucks has crossed the frontier in the opposite direction.

Fig. 3.15 Modern-day transportation of people and goods

The number of the counted passengers was at 4896, although it was difficult for the researchers to determine the exact number of passengers because many of them were crossing the border illegally. Therefore, it must be assumed that the actual number of the passengers was much higher. According to the interviewed shippers, the phasing-out of trade on this route was due to the overloading of Africa's Atlantic ports.[121]

3.11.2 Camel Markets in Fungul and al Kufrah

In the south of Libya, two camel markets were established to process trade in camels. One of them is situated in the south-west of Libya immediately adjacent to Murzuq. At the market in Fungul, animals brought from Niger and Chad to Libya were primarily traded. Here, they were sold to local farmers and butchers. How long the animals remained at the market before they were sold depended on the market situation, as well as on the supply and demand. Usually, the animals were kept here for several days up to one month. Ultimately, the traders remained at the market in Fungul until the last animal was sold.

The camel market in Fungul (Figs. 3.16 and 3.17) was small compared to the market in al Kufrah in the south-east of Libya, which is the largest in the country at 45 ha. The structure of the markets, however, is the same. The markets consist of a large open area. Most of the herds can move around freely, which means that the herdsmen have to keep their animals together by day. At night, the animals have their legs tied together, so that they are not able to run away. In several fenced corrals, a handful of camel herds are accommodated, where they can remain unattended. The herdsmen stay in small, simply built straw huts. In the area of the market, the camels are provided with sufficient straw and are regularly given water to drink.[122]

The camel trade in al Kufrah (Fig. 3.17) was one of the most important economic pillars in the economy of the oases and was dominated by the Zuwaya, a local tribe. The cattle market in al Kufrah lies on the southern edge of the settlement. The camels are either sold there or are traded on. Many of the animals continued their trip to the second largest camel market of the country on the Mediterranean coast. It is located in Misratah, a town about 200 km to the east of Tripoli. From Misratah, some of the animals were sent to Tripoli and to the Libyan centres of consumption. The

Fig. 3.16 Modern camel caravan: camel herds from Nigeria move north to Fungul between the months of September and May. At the camel market in Fungul, they are sold to local farmers and butchers

camels were also traded in Egypt, Tunisia or Italy. Usually, the camels were fattened in Misratah, which increased their value so that a price between €480 and €600 (800–1000 LYD) for a young camel and between €600 and €780 (1000–1300 LYD) for a fully grown camel had to be paid. While the Libyans prefer young and fat camels, the fully grown camels are often traded in Egypt, where they are preferred. Camel meat does not play an important role in the preparation of daily meals, but during religious celebrations and at weddings, camels are needed in considerable numbers. For example, at a wedding in al Kufrah, which continues for several days, the hosts need, between 40 and 50 camels for the catering of the guests.[123]

However, there are still camel caravans arriving in Libya today. Given the current uncertainties facing the Fezzan and its oases, however, it is very difficult to estimate the number.

3.11.3 Recurring Persistent Patterns in Trade

Libya's south-west has been a transit zone between sub-Saharan Africa and the Mediterranean coast over centuries. Cross-border trade remains prominent nowadays. However, smuggling has surpassed licit trade. Since 2011, smuggling has grown exponentially and dominates the local economy today. According to the International Crisis Group human trafficking, fuel and gold smuggling is widespread and highly visible, though drugs and weapons also pass through the region surreptitiously. The breakdown of government authority and sustained political and legal uncertainty led to competition over smuggling routes. At the same time, the easy access to the enormous stockpiles of weapons left in Qadhafi-era arms depots led to fighting conflicts among local groups.

The majority of the illegal migrants enter Libya from Niger or Chad; smaller numbers cross the border from Algeria. What is interesting to remark about the smuggling routes is that they are divided into segments controlled by different groups. As in the nineteenth century, the Tubu control the southernmost portion, used by the majority of migrants, from the border with Niger to Sebha. In contrast, the Tuareg dominate the route from the Algerian border to Sebha. The next station of the trip, from Sebha to Shweref, is in the hands of Magarha and the Warfalla traffickers. Beyond that, other tribes take over. As a synthesis of the existing material in this book shows, a long-term analysis reveals

Fig. 3.17 Photographs above show traders from Nigeria on their way to the camel market in Fungul (photographs below)

	Camel Market in Fungul	Camel Market in Kufra
Size of Caravans	300 – 400 Camels	2,000 – 3,000 Camels
Duration	25 – 30 Days between Marte and Fungul	14 Days between Tiné and Kufra
Travelling Companions	6 – 8 People	8 – 14 People
Trips per Year	3 – 4 Times	4 – 5 Times
Losses	4 – 5 Camels per Trip	5 – 10 Percent per Trip
Sales Price	Adult Camel with Young Camel: 1,300 LYD	Single Adult Camel: 700 – 900 LYD
	Single Young Camel: 500 LYD	Single Young Camel: 600 – 700 LYD

Fig. 3.18 Camel markets in Fungul (as of November 2006) and Kufra (as of November 2006). There is a price difference between the camels from Sudan and those from Chad. The Sudanese camels cost about €60 (100 LYD) more. In the ranking, the Sudanese (usually stronger) camels are at the top, followed by the camels from Tiné and those from the pastoralists in the southern regions of Chad

recurring persistent patterns. Already in the nineteenth century, trade, particularly in slaves, proved to be highly lucrative. As in the past, trade generates high profits and many people benefit from it. The tribes involved are unlikely to give it up. As a result, many young Tubu students, for example, give up their studies, because the promise of immediate cash is so attractive. Lack of education, fluid and shifting loyalties within tribes and ethnic groups contribute to a partially fragmented society (Fig. 3.18).[124]

The Fezzan has preserved its long traditions that are not only reflected by its trading practices but also by the mentality of the local people.

Notes

1. Wright (1989, 34).
2. Passon (2011, 169).
3. Lupton (1979, 25).
4. Hornemann (1802), Passon (2011, 170).
5. Lyon (1821), Passon (2011, 170f).
6. Denham et al. (1828), Eydoux (1938), Passon (2011, 171).
7. Passon (2011, 198).
8. Barth (1857), Richardson (1848).
9. Barth (1857), Eydoux (1938).
10. Duveyrier (1863, 1864), Krause (1882).
11. von Beurmann in PM (1862/63).
12. Rohlfs (1881, 1874).
13. Nachtigal (1870).
14. Krause (1878, 1882).
15. Weis (1996), Schulz (1996).
16. Liverani (2000, 496–520).
17. Wright (1989, 32–39).
18. Wright (1989, 44).
19. Savage (1992, 363).
20. al Yaqubi in Hopkins and Levtzion (1981, 22).
21. Hopkins and Levtzion (1981, 378).
22. Savage (1992, 363).
23. Thiry (1995, 458).
24. Wright (1989, 38–44).
25. Wright (1989, 40).
26. Wright (2007, 24–32).
27. Wright (2007, 35, 37).
28. Harmann (1998, 9–94), Baier (1980, 57–78), Wright (2007, 37); This concept was put forward by Lloyd Fallers and elaborated by Abner Cohen and Philip Curtin. Baier (1980, 58), Fallers (1967), Cohen (1971, 266–281), Curtin (1975, 59ff); See also Avner Greif and his case study of Maghribi Jewish traders in the medieval period. He discussed how partnerships and other trade relations were most efficient between members of a commercial coalition or trade network. Greif (1993, 525–548).
29. Duveyrier (1864, 256ff).
30. al Hachaichi (1912, 220ff). They were living in Wadai, Kanem, Bornu, Adamawa, Nupe, in the Hausa cities of Kano, Zaria and Sokoto; at Ghat, Ghadamis, Murzuq, in Zinder and at Djadjidoune, a small village four days north of Zinder. Ghadamis merchants lived in Tripoli and Tunis and traded between these two cities by sea. al Hachaichi (1912, 220–222).
31. Baier (1980, 58).
32. Barth (1857, 178).
33. Baier (1980, 57–78).
34. Quoted in Wright (2008, 131).
35. Baier (1980, 57–78).
36. Laing to Horton, 16 October 1825 quoted in Boahen (1964).
37. Despois (2012).
38. Richardson (1849).
39. Richardson (1849, Vol. II, 115ff).
40. al Anies (2009).
41. Interviews with elders from Ghat, April 2006; Despois (2012).
42. al Anies (2009).
43. Barth (1965 Vol I, 156), Baier (1980, 62), see also Boahen (1964, 122ff).
44. Baier (1980, 57–78).
45. Barth, 35.
46. Ghalbun (1930–31 and 1936); Hornemann, 100; Ahmida (1994, 19–41), Wright (2007, 69–88).
47. Interviews with elders in Ghat, April 2006.

48. Hornemann, 64ff.
49. Hornemann, 68.
50. Hornemann, 65.
51. Ciammaichella (1987, 36).
52. Oehms (2005, 3 and 13), Anderson (1984, 325–348).
53. Cordell (1977a, 31), Anderson (1984, 325–348).
54. Cordell (1977a, 31), Anderson (1984, 325–348).
55. Wright (2007, 127–136).
56. Wright (2007, 119ff); FO 160/65 Crowe to Palmerston, 9th November 1847.
57. Wright (2007, 120); FO 84/737, Crowe to Palmerston, 23rd June 1848.
58. Wright (2007, 127–136).
59. Khaldun translated by Rosenthal (1967, Vol. 10, 337f).
60. Khaldun translated by Rosenthal (1967, Vol. 13, 342).
61. Ackerman-Lieberman (2012), Lydon (2012).
62. Lydon (2012, 222).
63. Lydon (2012, 207).
64. Lydon (2012, 3).
65. Interviews with elders in al Qatrun in November 2006; Attir (2005).
66. Collection of source material compiled by Mohammed Abdel Aziz (Dougal), collected 22.11.2006; Collection of source material compiled by al Haj Muktar Achmed Taher (al Bacchi), collected 19.11.2006.
67. Lydon (2008, 89–113).
68. Trivellato (2007, 81).
69. Collection of source material compiled by Mohammed Abdel Aziz (Dougal), collected 22.11.2006; Collection of source material compiled by al Haj Muktar Achmed Taher (al Bacchi), collected 19.11.2006; Interviews with elders in al Qatrun in November 2006.
70. Girsh or Qirsh, Gersh, Kurus are all names for currency denominations in and around the territories part of the Ottoman Empire. Piastre is another name for Kurus or Girsh. The value of the Ottoman piastre by the late nineteenth century was worth about two pence sterling, Krause (2001).
71. Collection of source material compiled by Mohammed Abdel Aziz (Dougal), collected 22.11.2006; Collection of source material compiled by al Haj Muktar Achmed Taher (al Bacchi), collected 19.11.2006.
72. In 1858 the Ottoman code of land law was promulgated, and over the course of the century private ownership and registration were introduced in the settled areas of Tripolitania and urban areas of Cyrenaica, supplanting collective tribal landholding. Particularly along the Tripolitanian coast and in the Jabal al-Gharb, tribal lands were divided and ownership assigned to individuals who paid a small fee for a certificate of registration, Anderson (1984, 332).
73. Mohsin Khan: "*O you who believe! When you contract a debt for a fixed period, write it down. Let a scribe write it down in justice between you. Let not the scribe refuse to write as Allah has taught him, so let him write. Let him (the debtor) who incurs the liability dictate, and he must fear Allah, his Lord, and diminish not anything of what he owes. But if the debtor is of poor understanding, or weak, or is unable himself to dictate, then let his guardian dictate in justice. And get two witnesses out of your own men. And if there are not two men (available), then a man and two women, such as you agree for witnesses, so that if one of them (two women) errs, the other can remind her. And the witnesses should not refuse when they are called on (for evidence). You should not become weary to write it (your contract), whether it be small or big, for its fixed term, that is more just with Allah; more solid as evidence, and more convenient to prevent doubts among yourselves, save when it is a present trade which you carry out on the spot among yourselves, then there is no sin on you if you do not write it down. But take witnesses whenever you make a commercial contract. Let neither scribe nor witness suffer any harm, but if you do (such harm), it would be wickedness in you. So be afraid of Allah; and Allah teaches you. And Allah is the All-Knower of each and everything.*" Quran 2:282 (surat l-baqarah).
74. Collection of source material compiled by Mohammed Abdel Aziz (Dougal), collected 22.11.2006; Collection of source material compiled by al Haj Muktar Achmed Taher (al Bacchi), collected 19.11.2006.
75. Barnes (1987, 67).
76. Interviews with historians from the CLAR.HS in Tripoli in November 2014. Collection of source material compiled by Mohammed Abdel Aziz (Dougal), collected 22.11.2006; Collection of source material compiled by al Haj Muktar Achmed Taher (al Bacchi), collected 19.11.2006.
77. Collection of source material compiled by Mohammed Abdel Aziz (Dougal), collected 22.11.2006; Collection of source material compiled by al Haj Muktar Achmed Taher (al Bacchi), collected 19.11.2006.
78. Kieckhefer (1995).
79. Collection of source material compiled by Mohammed Abdel Aziz (Dougal), collected 22.11.2006; Collection of source material compiled by al Haj Muktar Achmed Taher (al Bacchi), collected 19.11.2006; Interviews with elders in al Qatrun in November 2006.
80. Laabdallaoui and Rüschoff (2005, 36–37).
81. Evans-Pritchard (1963, 16).
82. Ahmida, 97; Lydon (2012, 391ff).
83. Lydon (2012, 242f).
84. Lydon (2012, 206–230), Lydon (2009, 25–60), Meerpohl (2009, 157–183).
85. Lydon (2009, 25–60).

86. Khaldun (1892), Hopkins and Levtzion (1981, 339).
87. Thiry (1995).
88. Wright (1989, 41).
89. Nachtigal (1967), Krause (1882).
90. Lovejoy (1984, 85–116).
91. Richardson (1857–59), Nachtigal (1967), Krause (1882).
92. Liverani (2000, 496–520).
93. Lovejoy (1983, 66ff).
94. Wright (1989, 41).
95. Hunwick (1992, 5–7), Wright (1989, 26–54), Wright (2007, 1–8).
96. Wright (2007, 23).
97. al Yaqubi in Hopkins and Levtzion (1981, 22), al Bakri in Hopkins and Levtzion (1981, 63f), Savage (1992, 351–368); Wright (2007, 24–27), Martin (1969, 15–27).
98. Wright (1989, 28).
99. Savage (1992, 351–368).
100. Wright (2007, 1–8).
101. Wright (1989, 42).
102. Wright (1989, 43).
103. Wright (1989, 43–48), see also Lovejoy (1983) and Renault (1982).
104. Tambo (1976, 187–217), Austen (1975, 36–76).
105. Austen (1975, 36–76).
106. Mahadi (1992, 125), Schiffers (1962, 94).
107. Verlinden (1980, 74–78).
108. Wright (1992, 180), (2007, 34).
109. Richardson (1848, 478).
110. Renault (1982, 181).
111. Wright (2007, 128).
112. Wright (2007, 41–57).
113. Weis (1996).
114. Herzog (1961, 255–262).
115. Kohl (2004).
116. International Crisis Group 2017.
117. Interviews with merchants from Niger at the camel market in Fungul, November 2006.
118. Meerpohl (2009).
119. Meerpohl (2009, 204–210).
120. Meerpohl (2009, 245).
121. Ahmed Sassi und Mansur Hammadi, 2006.
122. Interviews with merchants from Niger at the camel market in Fungul, November 2006.
123. Meehrpohl (2009, 248–250).
124. International Crisis Group 2017. The group points out, that the tribes and ethnic groups of the Fezzan are not monolithic entities.

4 Navigating and Mapping the "White Spot"

Jacqueline Passon and Klaus Braun

4.1 Navigating the Desert—Concepts of Spatial Orientation

Jacqueline Passon and Klaus Braun

Searching for "Timbuktu": Does the map show the way?

For a very long time now, man has been trying to create an image of the surface of the earth. The expansion of man's horizons has always been, and continues to be, connected with the making of new maps. Maps, like no other object, document man's urge to discover the unknown and to attain knowledge. For a long time, knowledge about the structure and appearance of the newly claimed lands brought political, economic and strategic advantages. Geographic knowledge meant power. For this reason, maps were also a well-kept secret.

The portrayal of knowledge on a map reflects, to a large extent, the view of the world by the map's author, which makes the map a powerful instrument from another point of view. Ultimately, the map-maker decides what the recipient may know. He can also knowingly mislead the map reader. In this way, he has the power of interpretation over the knowledge that he would like to transmit to the recipient. Does he actually possess the sovereignty in matters of interpretation about the knowledge displayed on the map? Too little attention is often paid to the actual process of knowledge generation and to the preparation of the maps related to it. Critical cartography must question the data that are included on a map. What is the origin of the data and whose knowledge or perception do they represent? In doing so, from the perspective of a critical cartography not only one should ask questions about the handed down Eurocentric rules, on the basis of which the knowledge was obtained, formed and transmitted, but one must also take into consideration the strategies and practices in the process of generating and transmitting the knowledge both on the European and—with regard to the production of the North African maps of the nineteenth century—on the Arabic/African side. This means that, on the one hand, light needs to be shed on the type of appropriation of space of all persons involved. On the other hand, all the multifaceted negotiation processes in the generation of knowledge must be examined very carefully. As regards the production of the North African maps of the nineteenth century, it could be demonstrated that these data present not only the collected knowledge of the travelling researchers but also the knowledge which was, and is, deeply connected with the local knowledge of the indigenous population.

As Europeans, we are calibrated to the cultural technology of deciphering a given map. It is difficult for us to imagine how we could orient ourselves without this auxiliary tool in a foreign environment. However, if we would like to gain access to the maps of the nineteenth century and to the trade related to them in the Sahara and the surrounding areas, we must detach ourselves from the idea that the map is the only possibility, in respect of our concepts for spatial orientation, to orient ourselves in a given space. Even today, the Sahara-knowledgeable guides need neither a map nor a GPS. Instead, they rely on two systems of navigation that have been used for thousands of years.[1] The most prevalent of global navigation methods has been the demarcation of paths by stone cairns and upright stone slabs (Arabic *alamat*; sing. *alam*) which are even visible from afar. This so-called *alamat* orientation system is a universal system known from many roads and paths around the globe. As Förster points out, the earliest evidence of such road markers comes from roads established and used by Pharaonic expeditions during the Old Kingdom around 2600/2500 BC. Late prehistoric travellers had clearly only been able to move by developing mental maps of landscape features. "The *alamat* system modified the former landscape-based orientation in a way that was no longer

J. Passon · K. Braun (✉)
Department of Physical Geography, University of Freiburg, Freiburg, Baden-Württemberg, Germany
e-mail: klaus.braun@geographie.uni-freiburg.de

J. Passon
e-mail: jacqueline.passon@geographie.uni-freiburg.de

K. Braun and J. Passon (eds.), *Across the Sahara*,
https://doi.org/10.1007/978-3-030-00145-2_4

(a) (b) (c) (d) (e) (f)

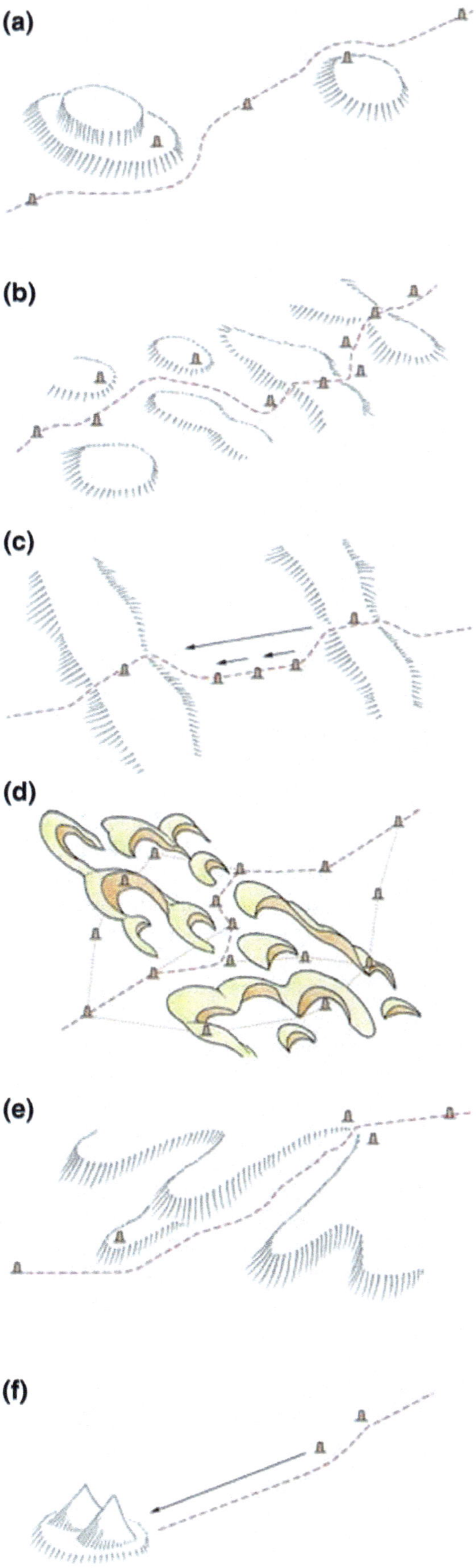

◄ **Fig. 4.1** Schematic illustrations of navigational strategies to cope with topographical obstacles as deduced from the survey along the Abu Ballas Trail. **a** Road signs are set at larger distances when the landscape is plain and open, either on the ground or on small isolated hills. **b** In undulating or hilly terrain road signs are set at short distances, either on the ground or on top of a hill overlooking the terrain. **c** Ridges and deep valleys are crossed with the help of a hierarchical *alamat* system: road signs are set on top of the ridges where they are visible at a certain distance while a line of signs leads through the valley. **d** In large dune fields where dune sand is constantly drifted by the wind, road signs indicate an array of alternative routes and bypaths. **e** Across wadi systems the road follows natural ascents or descents that is marked at its entrance and exit by landmarks. **f** Natural landmarks such as cones (e.g. the double cone of the so called "Two Tits" situated at the Abu Ballas Trail in Egypt) or other terrain structures are forming additional elements of the navigational system of the caravan routes. Text modified by the author (see Förster et al. 2010, 68)

restricted to developing mental sequences of the landscape configuration of individual routes, but allowed the following of the artificial lines of cairns" (Figs. 4.1, 4.3 and 4.4).[2]

"*When you lose your way, you learn it*". This illustrates an old proverb of the Tuaregs. What meaning can be extracted from this? Could a deviation from the route be related to the chance to collect important experience? Maybe one goes back along the route once more for a while and tries then to find the correct route by paying more attention. In any case, one learns to look more carefully this way. Just this quality, i.e. looking more carefully, is of central importance for the appropriation of a given space, and it is absolutely vital for anyone who participates in the navigation of the caravan. For a journey across the Sahara to be completed successfully, the caravan guide (*saykh*), along with his helpers (*khabir/taksif*), had to have, in addition to precise insights into the political and cultural structures of the societies of the central and sub-Saharan areas, extensive knowledge, in particular, about the section through which the caravan was to be guided. The basic prerequisites for this were that the *khabir/taksif*:

- knew the sections of the route and their condition precisely, whereby he must have already travelled these routes fairly often
- knew the landscape characteristics (e.g. mountain ranges, wadi courses, vegetation sections, individual trees, prominent rocks (Fig. 4.2) or other landmarks and prominent points of reference; this also included the artificially created direction signs in the form of the so-called stone manikins or stone cairns).

As a rule, a *khabir/taksif* had already worked as a helper or herdsman with his father or other relatives who were working as guides. These apprenticeship years permitted him to memorise the sections of the route, as well as the respective distances and the corresponding durations. The memorised landmarks served him as a support for orientating on the individual route segments. By carrying out these

Fig. 4.2 After leaving the sand sea of Awbari, the wayfarer passes several tumuli of stones, "more than eight feet high, evidently placed to direct the caravans over the trackless portions of Sahara" as Richardson chronicled. He continues and explains that on the way to Ghat the traveller has to cross the region of the Genii, "the land of mystery and disembodied spirits. The whole country is intersected and bounded on every side with the ranges of black, gloomy, and fantastically shaped mountains (photo), distinguishing the country of the Ghat Touaricks, where their friends and confederates, the Jenoun or Genii, dwell with them in the most harmonious friendship" (Richardson 1848, 436f)

activities, the *khabir/taksif* gained a better perception of the region, its physical characteristics and the distances between the different points of reference. Thanks to his mental mapping, he had a good understanding of the exact location of the points of reference, the distance between them and how much time the caravan needed to cover the route.

In regions where there is no noticeable point of reference on the horizon to indicate the direction, the *khabir/taksif* needed additional knowledge of orientation. This included, among other things, knowledge about the position of the sun. Using this, the *khabir/taksif* could estimate the approximate time. By observing the five prayer times, he was accustomed to constantly paying attention to the position of the sun. One could estimate at any time where north was and in what direction one had to travel. In addition, one knew approximately how many hours the caravan had travelled since the last prayer.

At night, there was no problem maintaining direction when one had exact knowledge about the firmament, its constellations and its changes during the seasons. At night, the stars served as orientation for the *khabir/taksif*. The polar star, in particular, helped to maintain direction. He was able to determine the time by the changes in the constellations and knew how many hours the caravan had already travelled in the darkness or when it was time to set up camp for the night. In addition, he knew how many hours remained for him and the caravan until the Morning Prayer.

Along with the knowledge of the route, the points of reference and the celestial bodies, the *khabir/taksif* had to have knowledge about the pastures and the availability of water. These were important points of reference within the mental map of every caravan guide, whereby the water points were of central importance:

Fig. 4.3 Across wadi systems, the road follows natural ascents or descents that is particularly marked at its entrance and exit by landmarks (see above). Road signs are set at larger distances when the landscape is plain and open, either on the ground or on small isolated hills

- The course of the route did not always correspond to the shortest distance between the start and end points, but ran from water source to water source.
- The very long segments without water sources, and thus particularly difficult and hazardous, were—when possible—shortened by the construction of wells.[3]

Until the last century, the caravan guides used the oases, wells and water points on the route, in order to provide the animals and themselves with sufficient water. The trans-Sahara routes led from water source to water source (Fig. 4.5). This allowed for filling up the water supplies at regular intervals.[4]

The firewood supplies were important for survival to the same extent as the supplies of water and food. Since there was no tree or bush vegetation on many sections of the route, the *khabir/taksif* had to assume that no wood could be found. He had to know in which wadi or valley there was vegetation, in order to be able to stock up or increase the firewood supplies. This meant that he had to have an idea about how much firewood was necessary for the preparation of a meal, how many meals had to be prepared per day and for how many days the firewood would last. For this purpose, firewood was already collected in the start area and transported along from there. If the areas with wood resources were directly along the route, he would lead the caravan through them; if they were located off the route, he would send assistants of the caravan with their pack camels to collect firewood and store it. They then had to compensate for the delay by a faster tempo. Since wood was an important and scarce resource, it had to be used extremely sparingly, just as the water supplies which were not unnecessarily consumed.[5]

In reality, the *khabir/taksif* did not exactly follow the tracks of the most recent caravan; rather, he would follow the general direction and take his own route. The offset between the used routes could vary between several metres up to a kilometre. In order to be able to maintain the direction, the *khabir/taksif* would search for a conspicuous

Fig. 4.4 Back then, as today, road signs need to be checked and repaired. Taking care of the maintenance of the landmarks (alamat) along the trade routes was an important task of the caravans

point of reference far away and then proceed to follow it without losing sight of it. Depending on whether the route was running westwards or eastwards, the passages varied or side valleys were used. These data are based on the scientific research of the caravan routes in western Libya and in the centre of the country. In travels along different caravan routes which, among other things, were accompanied by travel guides whose families were traditionally closely connected with the caravan trade, the corresponding data were collected by means of observations and inquiries.[6] Along with that, the findings correspond to the results of Meerpohl, i.e. the informative data about the historical, as well as recent caravan trade, which Meerpohl had been able to collect during an expedition through parts of the eastern Sahara.[7] The work of Laydon on the Western Sahara also substantiates this approach to the current practice of navigation in the Sahara: caravan itineraries were not exactly predictable or direct, for they were largely dictated by the environment. Based on interviews with caravaners in 1880, the American consul, Mathews, also acknowledged this when he wrote: "*[...] the akabahrs do not proceed in a direct line across the trackless desert to their destination, but turn occasionally eastward or westward, according to the situation of certain fertile, inhabited and cultivated spots, interspersed in various parts of the Sahara like islands in the ocean*".[8]

In two field experiments further conclusions regarding navigation in the desert areas were made. The results obtained also correspond with Meerpohl's findings. The field experiments attempted to obtain conclusions about navigation by comparing data from the mental map of the guide with landscape features from high-resolution satellite photographs. For example, the guide described the topographical features that are important to him on the following segments along the route from al Fugah to al Qatrun (see *From Murzuq to Tajahri*).[9] It was interesting to see that these features could be traced on the high-resolution satellite map. This means that the caravan's *khabir/taksif* was able to create a mental logbook of positions: prominent points of reference, which, for example, designated a certain descent in a wadi system or the ascent from there, respectively (Fig. 4.7), were saved in his mental map by means of certain points, such as stone manikins, places of prayer or other characteristic landscape features. As already mentioned, the *khabir/taksif*, besides

Fig. 4.5 Wells or other characteristic objects are the prominent landmarks in each mental map. For "path finding" in the Sahara, central significance is given to countless caravan tracks, way markers (alamat) and above all to watering holes and wells. The water points were the

◀ prominent landmarks in each mental map. The images and map portray the situation around the Wadi Tabunia on the caravan route from Tripoli to Murzuq (see From Tripoli to Mizdah). The Bir Tabunia Ⓒ is still important and kept in working order today, as it is a crossroads of old and new ways through the desert. Along the caravan route, there were further wells (see map: Abyar at Tabuniyah). On the ground, "single-track" paths Ⓑ are visible, which can be interpreted as caravan tracks in the terrain (see QuickBird scene from February 2003 and photograph on the right side)

knowing and recognising the prominent points, also had to remember the approximate duration of the individual segments of the route. In order to determine all that, the guide was asked to specify the days he needed for the next segments of the route. With some help when reading the satellite map, he was able to specify the days required for the next segments of the route. Furthermore, he was always able to indicate the direction to Mecca during the journey. Based on this knowledge, he was able to determine the cardinal points without any problems. In this way, he had a precise spatial perception which allowed him to show the directions in which the next settlements were located, even when they were hundreds of kilometres away. Without even thinking, he was able to show the correct direction. For the guiding of caravans, this means that the *khabir/taksif* always logs the positions in his memory and compares them with the landscape features. And finally, overall, it is to assume that the *khabir/taksif* embodied the knowledge of spatial orientation.

European explorers, such as Georg Schweinfurth, tried to attribute an insufficient geographic understanding to the Africans. They used it as an explanation to misunderstandings and errors in their own designations and maps on the basis of the different concepts of space.[10] However, it was not the different concepts of space that hindered the generation of knowledge but, partially, the arrogant self-posturing of the European travellers who had attributed an inadequate concept to the Africans.

During their crossing of the Sahara, traders, pilgrims, scientists or adventurers (Fig. 4.6) have all relied on local guides with certain knowledge and skills in orientation. The trading stations in the south, or the ports in the north, could only be accessed with a caravan that could, in addition to safety, offer orientation in one of the most inhospitable areas on earth. Even if the European traveller had had detailed maps, he would have had little chance of success when venturing out on his own. On the search for Timbuktu, the

Fig. 4.6 A homeward-bound kafla suddenly loomed up before the American explorer, writer and photographer Charles Furlong. He was the first American to explore parts of the Sahara. This experience led to his writing of "The Gateway to the Sahara" in 1909 (Furlong 1909, 198)

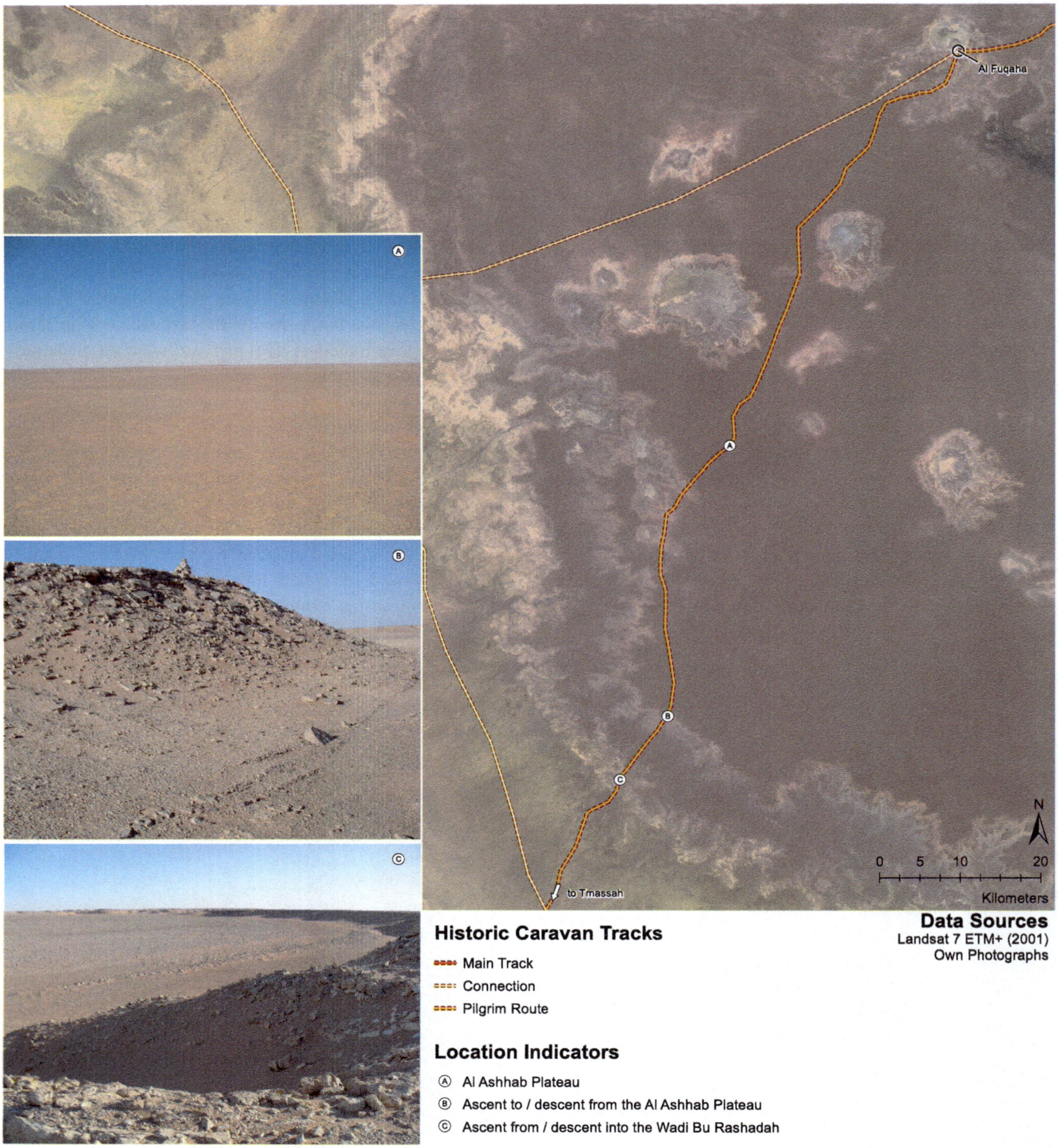

Fig. 4.7 Where is the way to Tmassah? To maintain the direction across the endlessly wide, stone-strewn, water and vegetation-free plateau (al Ashhab Plateau Ⓐ), a maximum amount of ability and skill is needed. After the descent from this high plain, which stretches over two altitude levels Ⓑ, Ⓒ, there is a supply station for caravans, a "caravan resting place" providing rest and supplies for the livestock (see From Zilla to Murzuq). The road follows natural ascents or descents that is particularly marked at its entrance and exit by artificially created direction signs in the form of so-called stone manikins or stone cairns Ⓑ

map shows something, but not the route. Certainly, Europeans in the nineteenth century believed in the power of maps. Detailed, as never before, they showed the topographical features; however, in the promising search for the treasures of the black continent, this power had reached its limits. The success of a given expedition depended essentially on the capabilities of the caravan guide and his assistants and was tied to the multifaceted negotiation processes with different groups and stakeholders.

4.2 From Mental Maps to European Knowledge

Jacqueline Passon

4.2.1 Justus Perthes Publishing House—Motor of Map Production of the Nineteenth Century

In the nineteenth century, map production is closely associated with the Gotha publishing house of Justus Perthes. The magazine Petermanns Geographische Mitteilungen (Petermann's Geographic Notices), which was first published in 1855, soon became the undisputed media of the time. August Petermann, Cartographer who enjoyed high respect internationally and whose name is closely associated with the publishing house Justus Perthes, had a keen sense for expeditions and ensured that cartographically useful material was collected during the expeditions in Africa, and that data transmitted to Gotha were as accurate as possible. These data were then used to draw precise maps. He established a scientific practice of mapping, by giving instructions to the explorers such as to keep itinerary records. His main care was to train the travellers in the proper use of instruments. Initially, the cartographic translation of the collected data required a check for plausibility of the material delivered by the travellers. By means of meticulous comparative analyses, contradictory presumptions were deleted, secured knowledge separated from yet unsecured knowledge, and the newly obtained knowledge was classified in the current state of research. Between 1855 and 1878, Petermann advanced to become a dominant figure. His approach allowed him to publish first-hand information about regions unknown to Europeans. In this way, more than a hundred maps about this region appeared in Petermanns Geographischen Mitteilungen (Petermann's Geographic Notices).[11]

At this point, we have to ask how the process of knowledge generation has taken place on the part of the European explorers and the concomitant development of maps about this area in detail. Petermann and other cartographers proposed the most promising routes to the explorers. How did the travellers handle these proposals in action and how much leeway did the Europeans have when following a route?

4.2.2 Negotiation Processes

Tripoli, the capital of present-day Libya, seemed to be the most appropriate gateway to the Sahara due to its topographical situation. Another factor was that Tripolitania and parts of the Fezzan were considered to provide extremely safe passage. Once arrived here, the travellers first had to familiarise themselves with the Ottoman and European officials and to gain their support for the respective projects. In addition, a decision had to be taken in what manner the journey to the south was to be carried out. In order to reach the first important station of a journey to the south, to the caravan cities Ghadames or Murzuq, one either joined a caravan or, if the necessary money was available, one organised his own small *kafla*. Apart from the negotiations with the Ottoman and European officials in Tripoli and the oases controlled by them, another step was to find local guides and loyal companions as well as to establish contacts with various ethnic groups the territories of whose were on the route.

A journey to Central Africa consisted of different stations where the journey always had to be replanned and new negotiations had to take place. This meant that one had to get involved repeatedly with new political officials, tribal leaders but also companions. The following journey of the British and German explorers Richardson, Barth and Overweg from Tripoli into the Aïr Mountains is an example of how the negotiation processes took place at the various stages.

The fact that the travellers had to rely on local guides was decisive for a journey through the desert. As described in a previous chapter, these guides moved along paths familiar to them and stored in their mental map. They adhered to the already available paths, primarily used by traders and caravans, but also by pilgrims and pastoralists, which, however, did not mean that the explorers could not leave the caravan for smaller explorations. After the participating parties put together a small *kafla* in Tripoli, the expedition around Richardson, Barth and Overweg reached at first Gharian located in the Jabal al Nafusah. From there, their way led them to Mizdah and to the Tabunia wadi. Here, they set up camp at a so-called caravan meeting point. Such resting places had an enormously important, highly practical function and were located at places with strategic importance for travel. Where the topography changes in the terrain, i.e. sections without water and vegetation or sections difficult to pass because of the composition of the ground (gravel and sand), there the traveller would find facilities that can be described as caravan meeting places. In the Tabunia wadi, caravans coming from the north or south met. Depending on from which direction one came, the Hamadah al Hamra, the red rock desert, was either ahead or behind. Prior to ascending into this hostile plateau, a caravan coming from the north had to make vitally important preparations for the onward journey, for the crossing of this section lasted about 6 days. Water was available again only at Bir al-Hassi (*see From Tripoli to Mizdah*). Even the small *kafla* of the European explorers had to give water and food to the camels and replenish water supplies prior to the upcoming difficult march. Depending on the size of the caravan, this could be accomplished in one to three days.[12]

Such opportunities allowed the explorers to carry out their own explorations of the surrounding area of the respective location. Barth used the rest day for a field trip to a nearby Roman tomb as well as for a small outing to Qaryah al Gharbiyah, a former Roman military camp and outpost along the Limes Tripolitanus. Barth's travelogue revealed numerous references to the desire the German researcher dedicated to the places of antiquity. However, the description of the burial monument near Bir Tabunia clearly shows the interpretative framework in which the researcher was operating: *like a solitary beacon of civilization, the monument rises over this sea-like level of desolation, which, stretching out to an immense distance south and west, appears not to have appalled the conquerors of the ancient world, who even here have left behind them, in lithographed proof, "a reminiscence of a more elevated order of life than exists at present in these regions*".[13] Inspired by the antique architectural monuments, the explorers measure the contemporary Arab culture and way of life against antiquity which was a model to them. For instance, Rohlfs concludes 17 years later: "*In view of the most glorious architectural monuments and derelict buildings, which often required only few repairs, [...] these people lead their lives in deplorable conditions. They never ask: Why cannot we construct such buildings and live in them? Never does a stone dam, which suggests the perimeter of a former water reservoir or the arch of an antique aqueduct make them think about the current poverty of the country*".[14]

After 39 days, on 6 May 1850, the small travel party finally reached the first big caravan station on their journey, the mercantile city of Murzuq. Murzuq, however, was not only an important hub for various goods that were traded here from north and south but also a place where negotiations about travel itineraries and travel conditions had to be carried out. The European travellers also were requested to organise their onward journey into the mountain massive of the Aïr. Unknown to them, it is situated about one thousand kilometres as the crow flies to the south of Ghat. Upon arrival in Murzuq, they did not waste any time and immediately started to prepare for their journey to Ghat or Aïr. The negotiation processes regarding the route ahead, however, proved to be extremely difficult, as they soon had to realise: "*Unfortunately our stay in Murzuk seemed likely to become a very long one, as the chiefs from Ghat, who were to take us under their protection, were not yet sent for. (...) No doubt, in order to visit Air, a country never before trodden by European foot, with any degree of safety, we wanted some powerful protection*"[15] (Fig. 4.8).

It was clear to the travellers that without the protection or the friendship of a powerful leader or another respected person of the Tuareg, who could provide safe passage, continuing their journey through the Aïr Mountains would not be possible. Each crossing of the desert meant attacks on travellers by hostile clans and tribes. For this reason, not always the shortest route between two stations was taken, but the route that offered the highest level of security. Choosing an *amidi*, a friend and protector, was an important matter for European travellers, but also for traders. The "friends and protectors" were generously rewarded for this service: "*The northern Tuarek, when they occupied the country round Ghat, established a sort of tribute, or gherama, to be paid by merchants passing through their territory, and on payment of which the trader should be no further molested, but enjoy full protection*".[16] This tribute was already mentioned by Leo Africanus, who travelled through North Africa at the beginning of the sixteenth century.[17]

The toleration of Europeans in the desert depended on the benevolence of the local rulers. Adolf Krause, another German–African explorer of the nineteenth century, talks about a kind of "historical right" that the Tuareg assumed in the assignment of the competence when it came to accompanying travellers.[18] How did this come about? The Sahara desert has always been a much contested space between the rulers in the north and the ones in the south. The desert, however, always developed its own dynamics. The invasion of the Bedouin tribes Hilal and Sulaim in the Maghreb in the eleventh century also changed the life in the western and south-eastern Sahara. Arab nomads advanced into the south-eastern Sahara from the Nile in Sudan in the twelfth century. In the fourteenth century, individual groups moved to the Western Sahara, where they were able to gain control of the local tribes. The regions in the centre of the desert and its southern borders between Timbuktu and Kuwar remained, however, under the control of the Berber-speaking Tuareg, who founded their own sultanate Agadez in the Aïr plateau in the sixteenth century.[19] Anyone who had the intention to cross this area had to require the consent of the Tuareg. In this context, the desert is always stylised as a lawless space, where there was no law and order and where travellers had to bow to anarchist vagabonds. Much too often, the European understanding is taken as a reference benchmark, as it was also done by many travellers of the nineteenth century. The relationships and structures of African societies, however, are not related to categories of European legal understanding. The spread of Islam had an important influence on the political and economic conditions within the Sahara and also in the areas to the south.

Fig. 4.8 Adiri, a typical oasis in the Wadi ash-Shati, as seen by Barth in the nineteenth century. In the background, one can see a so-called qser. Barth noticed that this desert castle is just one of many in southern Fezzan. He furthermore stresses that this building might be associated with the presence of the Garamantes in this area. He also mentions tombs built by the Garamantes, which are situated nearby (Barth 1977, 197)

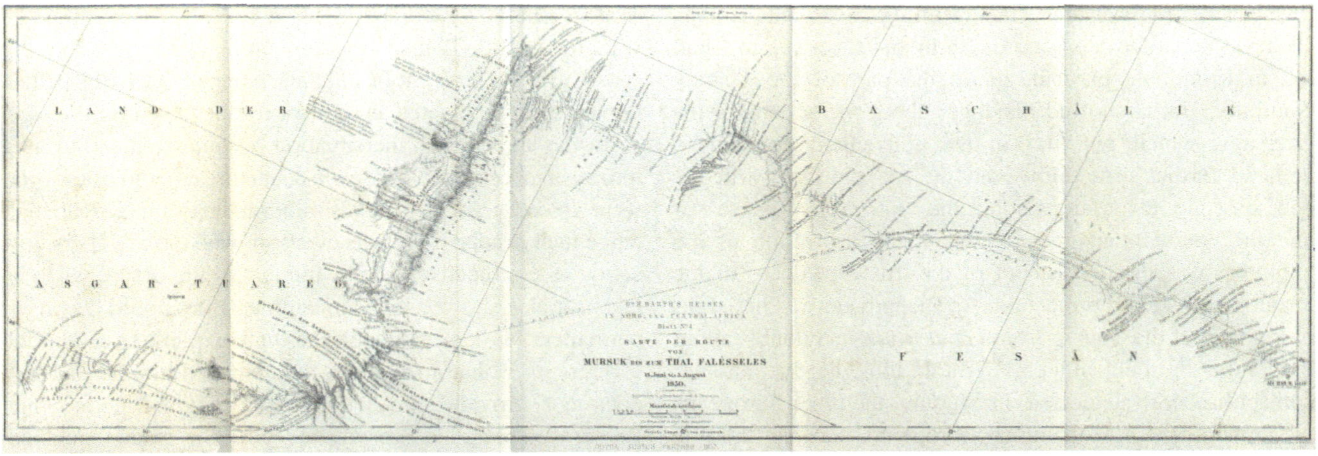

Fig. 4.9 What is in the map? We have to ask how the process of knowledge generation has taken place on the part of the European explorers. This map was drawn by A. Petermann (Justus Perthes 1857; Barth 1857)

Commerce relied more and more on the Sharia, the Islamic law, and Islamic scholars were used as judges or mediators in case of disputes.[20] This, however, should not obscure the fact that political crises, caused by recurring shift of power and religious disputes, destabilised these areas time and time again.

To be able to continue the expedition, Richardson, Barth and Overweg had to negotiate with the Tuareg, who were in control of the area of which the path led. The negotiations naturally included among others many material concessions, as Hadj Lameen, brother of the governor of Ghat Richardson reminded: "*He does not forget to remind us that the Tuaricks expect many presents. I have no doubt they do. He says we must be generous at all the following places:—Ghat, Aheer, Agadez, Damerghou, Zinder, Minyou, Tesaouah, Kashna, Kanou, Sakkatou, Bornou, Begharmi, Mandara, and to the Tibboos of Bilma; not to mention the intermediate towns and villages. However, if the presents be valuable, we may expect in some places rations of food in return*".[21] In accordance with the usual procedure, the director of the expedition, Richardson, was to negotiate with the Tuareg chiefs and possibly conclude a contract. It was beneficial that Richardson had already met Hatita, a leader of the Tuareg from Ghat, during his last journey in this area: "*During my former visit to Ghat, when I travelled as a private individual, known as Yakob, I made acquaintance with Hateetah, a Tuarick Sheikh, who had assumed the title of Consul of the English. It is the custom in that country for every stranger on his arrival to put himself under the protection of one of the head men, to whom alone he makes presents, and who answers for his safety*".[22] While Hatita together with an escort was on his way to Murzuq, Barth and Overweg, however, no longer had the patience to wait for Hatita's arrival. The two Germans meanwhile preferred to continue their way to Ghat without Richardson and with a different Tuareg caravan. One could now only speculate about the motives of the Germans' behaviour. Halfway through the route to Ghat, Richardson and his companions finally caught up with them again.

The analysis of the travel reports shows that the different assessments of the situation between Richardson and the two Germans further complicated the negotiations with the various actors. In particular, the resulting divergent actions, such as the earlier departure of Barth and Overweg to Ghat, led to irritations, particularly on the part of the Tuareg. Naturally, Hatita showed his anger about the actions of the Germans, which put him in the convenient position to demand further gifts. However, the main difficulty in the course of the negotiations about the route from Ghat to Air and the concomitant question about the protection of the explorers was the assessment of the trustworthiness of the leaders who offered protection. As Richardson was in favour of the help of the Tuareg from Ghat, Barth had doubts about the range of their influence. Barth thought that another contact had been neglected. In Murzuq, the three Europeans met a man named Mohammed Boro, who, carrying the title *Serki-n-turawa*, "*Lord of the wise men*", lived both in Agadez and in Sokoto and who was, according to the former governor of the Fezzan, Hassan Pasha, a man of great influence, even if currently he was not holding public office.[23] While Richardson did not think much about him, Barth pointed out his advantages: "*Nevertheless, Mohammed Boro was at present a man of great influence and with very important connections and could be of great benefit to us, yet in the opposite situation he could become highly dangerous to us. It is extremely regrettable that Mr. Gagliuffi, for reasons unknown to me, underestimated the importance of this man with respect to the success of this expedition and treated him accordingly. I believe that Hadj el Amin was the one who deliberately spoke disparagingly of Mohammed Boro's character, fearing that we would attach less importance to the connection with the chiefs from Rhat if we had an influential man from Agadez with us. Therefore he depicted him as conniving and told that he had dealt too much with the Turkish and wanted to use their power to obtain again his previous rank and position. He assumed much more importance than he actually had. In short, he was a man, whose friendship was not worth the trouble soliciting, particularly if it requires a slight sacrifice*".[24]

Yet, the story shows that many different interests had to be balanced and jealousies were the order of the day. It is interesting in this context that Mr. Galiuffi, who held the position of British Vice Consul in Murzuq and in addition was an influential dealer with many contacts, obviously pursued here very much his own interests, as even his compatriot Richardson noted critically in his records. The researchers had to note that the desire for personal gain of their negotiating partners was greater than the one for constructive cooperation, and not only on the part of the Tuareg.[25] However, the story also shows the constraints the travellers were subject to. Another notable fact of this event is that the above-stated excerpt from Barth's German diary cannot be found in the English version.[26] There might be many explanatory possibilities for this, and they might be surely linked with the fact that Barth wanted to present himself as prudent mediator and strategist in his travel report. However, on no account did he want to snub the British Government by publicly expressing criticism about their public officials. After all, it was the British Government who acted as sponsor for the expedition.

In the later stages, the situation becomes critical, turning into an open conflict where even death threats from Boro's part were voiced. Even if this was rather meant as a tactical challenge than dead serious, this event showed the travellers how serious the protective accompaniment in the desert has to be taken. It also showed that Mohammed Boro could become a permanent burden. Although the three explorers secured the services of Mohammed Boro, they angered him as Gagliuffi does not provide him with sufficient gifts. Only when Richardson provided him with new gifts, the situation relaxed.

Overall, it can be concluded that negotiations with the Tuareg were very sluggish. The chiefs finally delayed the onward journey of Richardson, Barth and Overweg until July 1850. On the one hand, this might be due to the difficult situation in the target region; however, these delaying tactics can mainly be explained with the fact that, from the perspective of the Tuareg, the gifts that the travellers had given them were considered insufficient. Only four months after the departure from Tripoli did the explorers reach Ghat and continued their dangerous way into the mountain massif of the Aïr.[27]

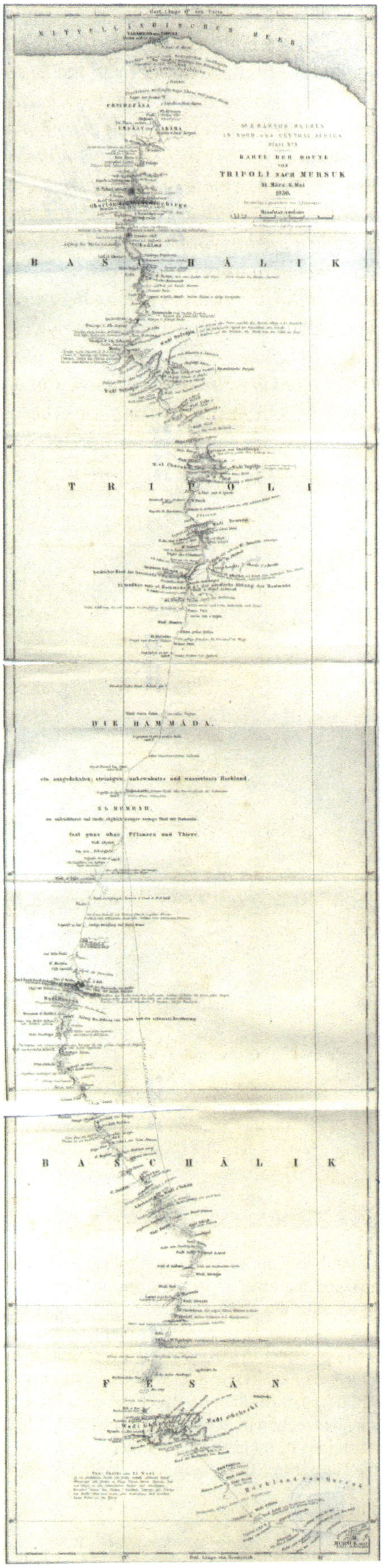

◀ **Fig. 4.10** What is in the map? The fact that the travellers could obtain reliable geographic data only in close cooperation with the local population is decisive for the process of the generation of knowledge and the production of maps. This map was drawn by A. Petermann (Justus Perthes 1857; Barth 1857)

4.2.3 Summary

The fact that the travellers could obtain reliable geographic data only in close cooperation with the local population is decisive for the process of the generation of knowledge and the production of maps. Even though Petermann and other cartographers proposed the most promising routes to the explorers, they could not really freely decide on location which route to take. If the first stage starting in Tripoli, which initially led to the first major stopover either to Ghadamis or Murzuq, was rather easy to manage by joining a caravan or by the travellers organising a small caravan themselves, in order to travel on the established commercial or pilgrim routes, the onward journey to Central Africa turned out to be much harder. A journey through the desert included lengthy negotiation processes, which, the further one penetrated to the south, became more difficult. This was due to the unclear political balance of power. In order to find and travel on a suitable route, negotiation processes had to be carried out with very different personalities. This included European, Ottoman or Sudanese officials, guides and companions of caravans or representatives of different ethnic groups of the desert and the sub-Saharan regions. In order to make such negotiation processes successful, the travellers had to have not only diplomatic skills and patience, but their success was mainly dependent on their historical understanding as well as their knowledge of the political conditions of the desert and Sudan, and their intercultural empathy. The successful development of negotiation processes was the basis to generate knowledge. The most important abilities were summarised well by Barth as he carried out negotiations with the Tuareg in the region of Ghat: "*It was a serious undertaking to enter into direct negotiation with these Tuarek chiefs, the absolute masters of several of the most important routes to Central Africa. It required great skill, entire confidence, and no inconsiderable amount of means, of which we were extremely deficient*". And if the negotiations stagnated, even science could make no progress. Barth grants us insight into his state of mind: "*The way in which negotiations took place spoiled almost our whole stay at this place and denied us the opportunity to grain greater insight in the exploration of the interesting and almost unknown ethnological relationships of these areas*".[28]

One could state many more examples that show that the travellers were always forced to refrain from carrying out their plans. For clarification purposes, another scene is to be stated here. The group around Richardson, Barth and Overweg was on the way to Ghat with Hatita, as a decision already taken before had to be changed much to the detriment of Barth: "*There had been much talk for some days to the effect that we travellers, together with Hatita, should take the nearer but more difficult road to Ghat across the range, while our luggage should go by the longer but smoother road round the mountains; but it was at length decided that we should all go by the longer road, and none but the Sfaksi, who was anxious to overtake the caravan as soon as possible, took the more difficult path, which, for geological observations, might have proved the more interesting*".[29] The way decisions were made often was not understandable for Europeans and would also appear irrational to them. They only had limited access to the logic of the activities of the indigenous people. One is easily tempted to speak of irrationality and anarchy. However, this explanatory approach would be inadequate.

For the creation of his work, apart from the ancient authors and the available treatises of Arab scholars of the Middle Ages and current contributions of European explorers of the eighteenth and nineteenth centuries, Barth mainly used indigenous sources. In part, he named these sources. For example, he used a Bornu chronicle: "*(...) a chronicle (divan), or rather the dry and sterile abridgement of a chronicle, comprising the whole history of Bornu, from the earliest time down to Ibrahim, the last unfortunate offspring of the royal family who had just ascended the crumbling throne of the Bornu empire when the last English expedition arrived in that country; two other still shorter lists of the Bornu kings; a detailed history of the first twelve years of the reign of the king Edris Alawoma, consisting of two parts, in my copy one of 77 and the other of 145 pages, and written by a contemporary of the above mentioned king, the Imam Ahmed, son of Sofiya and a short document containing information about embassies*

Fig. 4.11 Back then, as today, passing through the territories of a tribe in the Sahara is bound to a permission given by the council of elders. The figure above shows a meeting between the German–Libyan members of the joint venture and the elderly people from al Qatrun. After an initial welcome, a common prayer was spoken. Subsequently, issues of the visit were discussed with the committee, to which ample time should be devoted. Negotiations about where to go and what to do in the area were held. For this, one needs to part with plans already carried out or "western" solutions and keep up with the reality of a tribal community. Back then, for the pioneers of Africa-related research, this was a major challenge: communication with different stakeholders in the Sahara required a strong intuitive understanding as well as familiarity with the local "political and cultural landscape". As shown, negotiations with different stakeholders were decisive for the process of the generation of knowledge and also for map production. When selecting the routes, the explorers could not always decide freely which way was to be taken. On the one hand, this was dependent on the safety prevailing on the distance to be covered and on the sensitivities of the respective protective powers. On the other hand, it was dependent on which guides were accompanying the travellers. Ultimately, the European explorers moved along the paths stored in the mental map of their guides and companions

Fig. 4.12 A debate of the elderly people from al Qatrun

sent to Tripoli by some Bornu kings, and published in the Bulletin de la Societe Geographique de Paris 1849".[30] Furthermore, Barth replied on informants or rapporteurs. In his work, there are numerous references to the fact that the detailed geographic descriptions were based on "information by rapporteurs", who the German scientist occasionally questioned, but otherwise very much appreciates their accuracy and value. Barth made map sketches based on his travels and the geographic data based on indigenous sources. Together with detailed written descriptions about places he visited himself or he had heard about, he sent the map sketches to Dr. August Petermann who then started the scientific process of map drawing (Figs. 4.9 and 4.10).

Negotiations with different stakeholders (Figs. 4.11 and 4.12) were decisive for the process of the generation of knowledge and therefore also for map production. When selecting the routes, the explorers could not always decide freely which way was to be taken. On the one hand, this was dependent on the safety prevailing on the distance to be covered and on the sensitivities of the respective protective powers. On the other hand, it was dependent on which guides were accompanying the travellers. Ultimately, the European explorers moved along the paths stored in the mental map of their guides and companions. The travellers could obtain reliable historical and geographic data only in cooperation with the local population. The knowledge of the European explorers laid down in writing and encrypted in maps was essentially based on local knowledge. In contrast to Barth, other researchers such as Georg Schweinfurth tried to attribute inadequate geographic understanding to the Africans and explain misunderstandings only due to the different concepts of space. However, not only the different concepts of space were an obstacle during the process of generating knowledge, but the successful or less successful execution of the negotiation processes.

Points of reference in the landscape are stored in the mental map of a desert guide and can be traceable on a satellite map. It can be assumed that a desert guide embodies all the relevant facts, information and skills for navigating in the desert, which is far more than only storing knowledge in a mental map. Apart from knowing and recognising points of reference, a desert guide can remember the approximate duration of the individual stretches. On a satellite map, he is able to state the number of days he will require for particular stretches of the road.

Notes

1. Interviews with desert guides collected in April and November 2006, March 2008, October 2009.
2. Förster et al. (2010, 54); The way the ancient Egyptians navigated by means of *alamat* has been studied along the Abu Ballas Trail where lines of hundreds of stone cairns have been recorded (see Förster et al. 2010, 49–75).
3. Tilho (1920, 94).
4. Ciammaichella (1987, 32).
5. Meerpohl (2009).
6. Observations and interviews with desert guides collected in April and November 2006, in March 2008 and in October 2009.
7. Meerpohl (2009).
8. Laydon (2012, 221).
9. Field experiments with desert guides conducted in November 2006; Meerphol (2009).
10. Fritsch (2009, 87–101), see also Fritsch and Voigt (2008, 1–10); For transcultural communication and negotiation processes in Eastern Africa, see Voigt (2012, 27–38).
11. Schelhaas and Wardenga (2011, 89–107).
12. Barth (1857), Richardson (1853).
13. Barth (1857, 133).
14. Rohlfs (1874, 21).
15. Barth (1857, 224).
16. Barth (1857, 83).
17. Barth (1857, 258).
18. Krause (1882, 266–356), see mainly 328–335.
19. Austen, 80f.
20. See also Lydon (2012).
21. Richardson (1853, 104f).
22. Richardson (1853, 120f).
23. Barth (1857), Richardson (1853).
24. Barth (1857, 174).
25. Barth (1857), Richardson (1853).
26. Barth (1857–1858).
27. Barth (1857), Richardson (1853).
28. Barth (1857, 262).
29. Barth (1857, 96).
30. Barth (1857, 253f).

On Major Trans-Saharan Trails

5

Jacqueline Passon and Meike Meerpohl

5.1 A Dense Network of Routes and Oases

Jacqueline Passon

The map "Important Caravan Tracks and Oases" offers an overview of the progression of the most significant caravan and pilgrim routes and local trading routes. It reveals the location of the most important oases in present-day Libya.

Because of its geographical location, at the point where the Mediterranean reaches its furthest southern point, Libya has been a transit country for goods and travellers between the Mediterranean region and the countries south of the Sahara since time immemorial. Adapting to the climatic and environmental conditions, the transport of goods and travellers proceeded by caravans right up into the twentieth century.

Since ancient times, caravan transport has played a significant role for the economy of northern and inner-African countries. In addition to the starting and finishing points at Tripoli, the transit hub of trade on the way to the south was formed by the oases of Ghadamis as well as Zawilah and later Murzuq. From Tripoli, Murzuq could be reached via many routes: the most important caravan and postal route led to Murzuq via the Bani Walid and the Zillah oases. Two routes, that ran further southwards, led to the

Fig. 5.1 Numerous caravan tracks running straight across Libya. They describe the traditional paths that were handed down since primeval times which merchants, herdsmen and pilgrims were supposed to follow. These trade and pilgrim routes connected Europe, the Middle East and the African mainland

J. Passon (✉) · M. Meerpohl
Department of Physical Geography, University of Freiburg, Freiburg, Baden-Württemberg, Germany
e-mail: jacqueline.passon@geographie.uni-freiburg.de

K. Braun and J. Passon (eds.), *Across the Sahara*,
https://doi.org/10.1007/978-3-030-00145-2_5

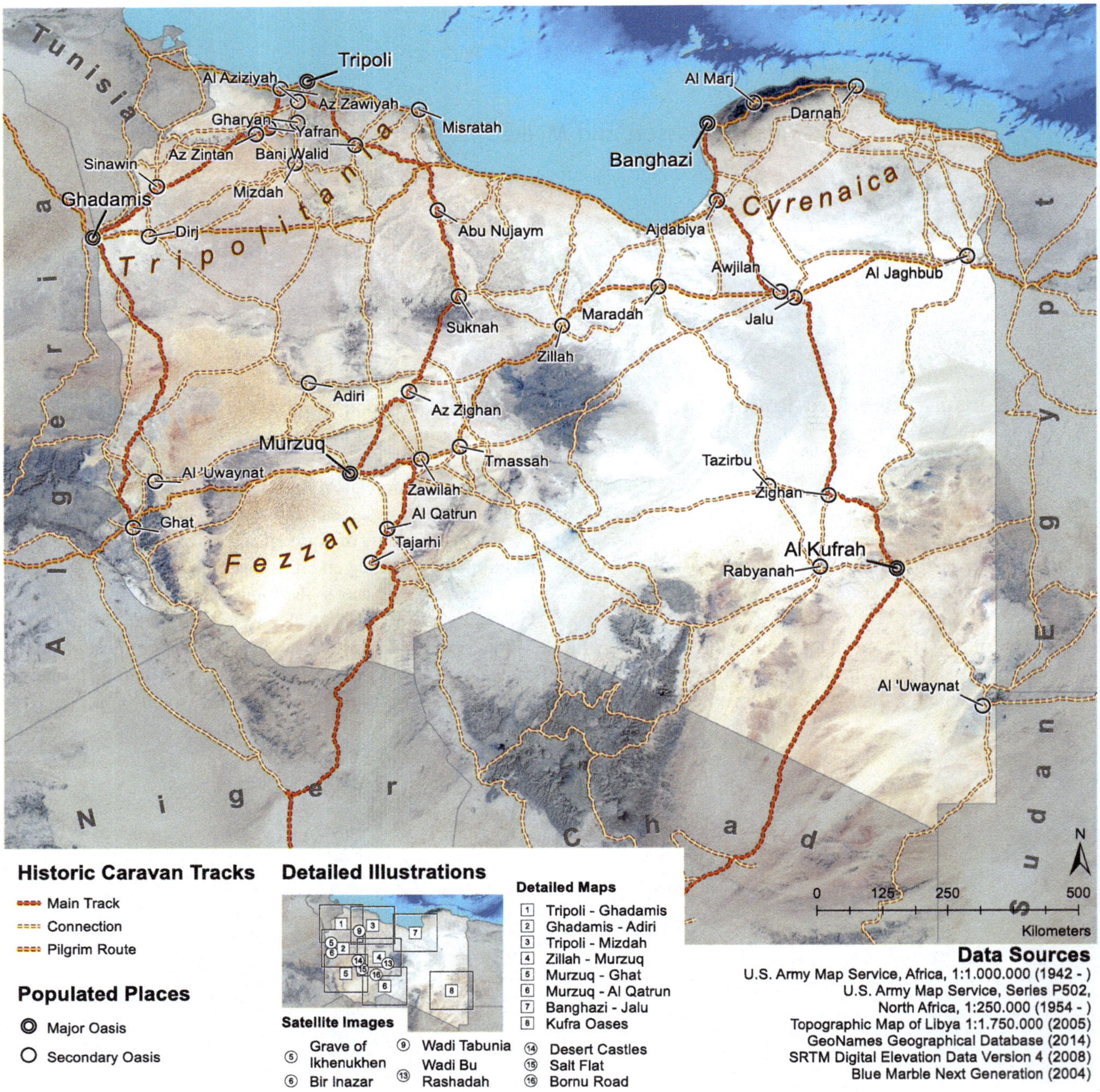

Fig. 5.2 Map: Important Caravan Tracks and Oases

trading metropolis of Fezzan via Gharyan, Mizdah and the Wadi ash-Shati. These various routes eventually diverged in Murzuq. The northern routes met on the ancient north–south route, a significant pilgrimage route between Cairo and Timbuktu, which established the connection to Ghat. The continuation of the northern routes to the south formed the famous Bornu route, which led, on the one hand, to the Sokoto Caliphate and the Bornu Empire, both of which lie in present-day northern Nigeria and, on the other hand, to Kanem and to the Sultanate of Wadai in present-day Chad. Ghadamis could also be reached via numerous routes. The main road led to Ghadamis via the mountain villages of Yafran, Az Zintan and the Sinawin Oasis. Routes also spread out from the larger settlements of the Jabal al Nefusah and met near Sinawin, where they then followed the main road. A further junction before Ghadamis was formed by the eastern oasis of Dirj, where two roads from Mizdah met with a road coming out of

Sinawin. One of these roads joined the pilgrimage route about 90 km east of Dirj and ran between Fez (Morocco), Ghadamis and Cairo (Egypt). From Ghadamis, one could reach Ghat and from there Agadez (Niger) and Timbuktu (Mali).

Another significant route marked the way from Benghazi to the Wadai Empire (Chad) and the Sultanate of Fache (Sudan) via Awjila and al Kufra. Due to the difficult political situation in the region, which affected this route, the caravans there were often exposed to unpredictable risks, which allowed this route to be used only at times. The road became particularly important during the second half of the nineteenth century. With regard to the value of the goods, the significance of the east–west routes cannot be compared to the north–south connections. The coastal road, which stretched out along the Gulf of Sidra, served as a pilgrimage and postal route. The exchange of goods between Europe and Africa was conducted via the north–south routes. Europe supplied mainly industrial products such as weapons, dishes, materials, tea, spices and paper. Africa exported gold, leather and furs, cotton products, indigo, ivory and ostrich feathers, but most importantly slaves (Figs. 5.1 and 5.2).

5.2 From Tripoli to Ghadamis

Jacqueline Passon

The map "Tripoli–Ghadamis" shows the various routes that led from the north to the significant trade metropolis of Ghadamis. An important and regularly frequented caravan route ran via the Jabal Nafusah. Two other routes went via the Mizdah Oasis.

From Tripoli, Ghadamis could be reached by various roads. The shortest and fastest way led over the mountain villages of Yafran, As Zintan and the Sinawin Oasis to Ghadamis. On his way to Ghat, the British traveller Richardson went along this regularly used route. There he collected much valuable information about the Tuareg and established important contacts with the leaders of this large Berber ethnic confederation. From Tripoli, the route went first towards West to Az-Zawiya, where it branched off to the south and led through the Jefara Plain in the direction of Wadi al Athel at the foot of the Jabal al Nefusah. From here, it followed the difficult climb into the mountain range to Yafran, which at this time was home to an Osman garrison. The mountain range is a hilly limestone massif, which is called after the largest Berber tribe in this region, the Tamazight Nefusah. It is a mountainous desert plateau interspersed with deep valleys and fertile oases. Because of its rough topography, governments located in the coastal areas historically found the Jabal al Nefusah hard to control. People often fled there to escape government rule. Therefore, the region has historically been associated with groups seeking refuge. Due to its rough terrain and a large number of secluded valleys, the Jabal used to be ideal for retreat. It was used as a hiding place by the original population, the Berbers, when the Arabs attempted to settle in Jabal al Nefusah in the wake of the Muslim invasion in the middle of the seventh century. It remains one of the few areas in Libya, where Berber culture still thrives. Beyond that, it was home to the oldest Jewish settlements in Libya. Many Jews escaped from Tripoli during the incursion of the Spanish in the sixteenth century.

Yafran (Fig. 5.3), hub of the mountain region, is composed of several villages spread over the mountain slopes overlooking the Jefara Plain. It was known in Tripolitania for its renowned synagogues. The Jewish presence in Yafran was the oldest in Tripolitania. Until around 1945, there had been a synagogue on this site for almost 2000 years. The remains can be seen today.[3] Economically, the region prospered from agriculture and, of course, from the trans-Saharan trade (Figs. 5.4, 5.5, 5.6 and 5.7).

The next stops on the way to Ghadamis were Jadu and Nalut. Caravans usually stopped within reach of an oasis. Like in Jadu, there was always an arrival or a departure point for caravans, where the caravans rested (Fig. 5.8). From here, the way continued to an important crossroads of the caravan trade, the Sinawin Oasis. According to Richardson, it lays a four days' march from the mountains. The road was very good to travel. As Richardson reports, the greater part was a "*beautiful broad carriage-road*".[4] Sinawin itself in the middle of the nineteenth century drew a bleaker picture, as we can gather from James Richardson's report, which says: "*Seenawan is but a handful of date-trees, thrown upon the wide waste of the Sahara, with one or two pools of sluggish running water, sheltering beneath its palms thirty or forty inhabitants. There are four or five spots of vegetation, gems of emerald on the rugged brow of the Desert. The houses, if such they are, consist of half a dozen or more of mud hovels huddled together, here and there a little stone stuck in the walls, and some dark passages running beneath them. One or two had a couple of stories and a stone wall round them. Yet, within, they are cool, and have dark rooms to protect the inhabitants from both heat and cold. There are also two or three mud and stone burges, or round towers, to protect the few dates and spots of green. Nevertheless, in this pretence of existence, surrounded by the frightful sterility of the Desert, glowed the warmth of true hospitality.*[5]

In Sinawin, construction relics such as the well, and two water basins from the traditional settlement mentioned by Richardson in his account, still persist to this day (Figs. 5.9 and 5.10). The old town of Sinawin, however, has left its golden age far behind (Fig. 5.10). Today, the old settlement

Fig. 5.3 During the nineteenth century when the British traveller Richardson visited the Castle (kser) of Yafran, Turkish soldiers were lounging here. He describes it "as a most formidable thing to look at from a distance, but a wretched mud-built place in reality. To the Arabs, however, it is a terrible bulwark of strength, and for them impregnable".[1] Today, it houses a hotel with a great panoramic view

is abandoned, and the wells are filled with waste. Nevertheless, the various routes, which radiated outwards from the larger settlement centres of the Jabal al Nefusah, come together at this oasis as they did in the old days. The Dirj Oasis to the east is a further crossroads before Ghadamis. Here, two routes from Mizdah came together: a road met about 90 km east of Dirj with the pilgrimage road that ran between Fez, Ghadamis and Cairo. The others led directly from Mizdah to Ghadamis. Both came together in Dirj with a route from Sinawin.

The caravan routes departing from Mizdah to Ghadamis separated at the nearby Fassanu Oasis. The southward route was

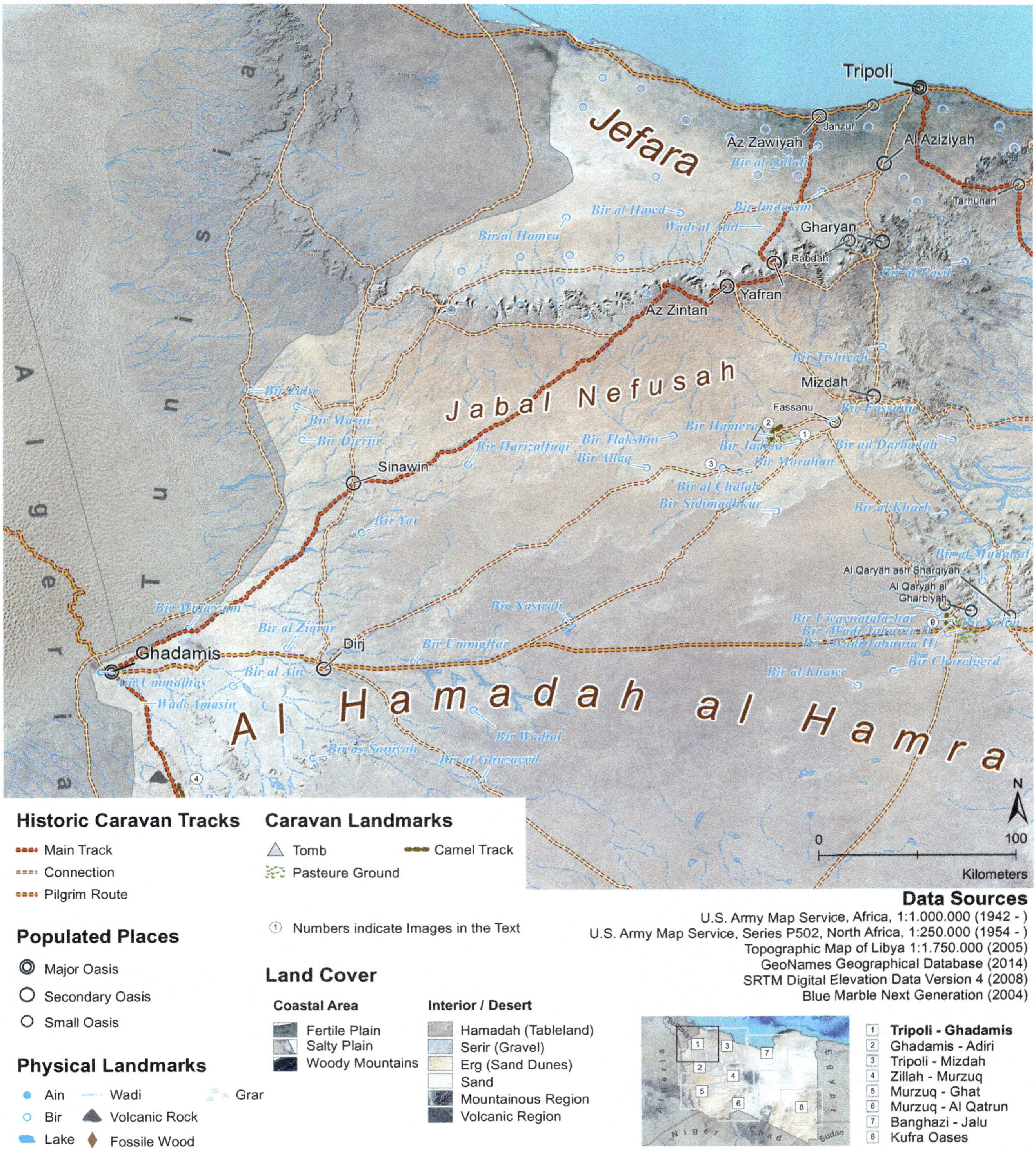

Fig. 5.4 Map: Tripoli–Ghadamis

Fig. 5.5 Yafran overlooks the Jefara Plain

Fig. 5.6 Jabal al Nefusah is also known as al Jabal al Gharbi (Western Mountain) or Adrar n Infusen, as the Berbers called it. The region extends some 170 km south-west of Tripoli to the Tunisian border and includes the areas around Yafran, Gharyan and Nalut. Yafran was situated on the ancient caravan road which linked Tripoli and Ghadamis with the Sudanic inland ports. The drawings above from de Mathuisieulx's publication "Attraverso la Libia" published in 1912 give a strong impression of the village and the Ottoman castle at the beginning of the twentieth century. As part of their effort to improve security, the Ottomans established garrisons throughout the Jabal like in Yafran or Nalut. For many centuries, Berbers were living together with Jews in the same neighbourhoods, as was customary elsewhere in Jabal al Nefusah. Both communities promoted trade and the exchange of ideas across cultural and religious divisions. The long-standing relations between Jews and Berber tribes guaranteed the safety of the trade routes.[2] The last picture shows a very small caravan on the route to Ghadamis

Fig. 5.7 An abandoned old Berber village which is situated 5 km east of Yafran in the Nefusah mountain region. Two main routes connected Tripoli and the Sudanic empires by crossing the Jabal al Nefusah: The Tripoli–Ghadamis–Ghat route to Hausaland and Air and the Tripoli–Fezzan–Kuwar route to Bornu and Lake Chad. The Berbers supplied the needs of passing caravans

Fig. 5.8 An arrival or a departure point for caravans from the oasis of Jadu

Fig. 5.9 An Italian postcard from around 1920 shows the spring of Sinawin

Fig. 5.10 Dams and canals were constructed in order to allow for an all-season irrigation of the gardens of Sinawin. On the right: remains of the old town of Sinawin a formerly important crossroads of the caravan trade

once travelled and described by Rohlfs. He characterises the region as "*appallingly desolate and uniform*" and goes on: "*The wide valley in which Mizdah lies divides into two arms above the place; the main direction goes from West to East. Far and wide there is not a tree to be seen. Some herbs grow, though they are kept down by the strong sand storms, on the Southern bank.*"[6]

The more northward route follows the watering holes of Bir Fassanu, Bir Jadida, Bir al Hamera, where caravan tracks can be still traced today (Fig. 5.11), and Bir al Chalab. These wells are said to have been used since antiquity. The wells were built by the Italians during the colonisation of the country in the 1930s, which is testified, for example, by a plaque installed at Bir

Fig. 5.11 For "path finding" in the Sahara, central significance is given to countless caravan tracks as one can find, for example, near Bir al Hamera (see map ②)

Jadida by the Italians (Fig. 5.12). Today, the wells are maintained by the Libyan Government. The wells are equipped with troughs for cattle (Fig. 5.13), and they are regularly sought out by the herdsmen or the camel herders.

Rohlfs, who needed eight days altogether for his trip from Mizdah to Ghadamis, also speaks in his memories of one of the above-mentioned wells, the Bir al Chalab, and provides an insight into the functioning of a caravan. He reports that the caravan, with which he travelled, had set up their tents at the Aghadir-el-Cheil (horse watering hole), a place that is located in a straight direction about 10 km from the Bir al Chalab, according to Rohlfs. There, it emerged that the convoy was indeed equipped with enough feed for the animals, but that the water supplies would only last two more days. This posed a problem since the travellers were faced with at least five more days' march. In desert areas, where water was scarce, it was critical for the survival of the caravan to employ paid scouts called *taksif* to find water. The scout, who knew the area very well, was then sent with other camel drivers and a few servants as well as all of the camels to Bir al Chalab, so that the animals could be watered and the water holes could be freshly filled.[7] Today, Bir al Chalab (Fig. 5.14) is a small modern high-tech oasis, which is among other things, equipped with solar panels, in the middle of the inhospitable semi-desert area, and serves as a shelter for herdsmen. A well supervisor (Fig. 5.16), who is in service for a number of weeks, oversees the facilities. Regularly, water trucks arrive (Fig. 5.15), which transport the valuable resource to the individual, dispersed cattle troughs. The water is provided free of charge by the Libyan Government.

The two routes that led from Mizdah to Dirj in their onward course also involved the difficult climb into the Hamadah al Hamra, a plateau of rocky desert, that was difficult to cross for the long processions of heavily loaded camels (see *From Ghadamis to Bir Inazar and Adiri*).

Fig. 5.12 Bir Jadida (see map ①) is a well that was cemented by the Italians

Fig. 5.13 Troughs for cattle. Today the desert is still used by herdsmen and their cattle

Fig. 5.14 Today, Bir al Chalab is a small modern high-tech oasis in the middle of the semi-arid Sahara (see map ③)

Fig. 5.15 Water trucks arrive at Bir al Chalab in order to transport the valuable resource to the individual, dispersed cattle troughs

Fig. 5.16 The Oasis is organised by a well supervisor

5.3 From Ghadamis to Bir Inazar and Adiri

Jacqueline Passon

The map "Ghadamis–Bir Inazar–Adiri" shows the course of several caravan routes in the direction of Ghat to Adiri. This region is characterised by difficult terrains, which is devoid of any fixed settlement.

The Ghadamis–Air–Kano route was, commercially speaking, one of the important routes especially in the nineteenth century. Due to the successful establishment of the Sokoto Caliphate, political and commercial live was concentrated in Sokoto and Kano. This part of the Sahara was under the control of two powerful branches of the Tuareg of the Sahara, who ensured the security of the routes.[8]

The caravan routes, which started from the trading metropolis of Ghadamis and led to Ghat, are characterised by the fact that there are no fixed settlements along the entire route that can offer rest and supplies. Because of this, the caravans made a habit of creating a store at the halfway point of the route, where they laid down provisions for the return journey. According to Duveyrier, who had seen the sacks left behind by other caravans on the way to Ghat, it was no

Fig. 5.17 A caravan track in the barren rocky desert, the Hamadah al Hamra (see map ④). Richardson rightly pointed out: "All is either earth, sometimes as hard-baked as stone, or large blocks of stone, but chiefly very small chips of stone covering the entire surface".[10] The photograph at the bottom right show the course of Wadi Amsin where sparse vegetation is found

idle fear that other travellers might seize them. On the return journey, the caravans gathered up the provisions that they had deposited. The French explorer added that in the Sahara there were routes, looked over by the population. For their use, low duties had to be paid by the caravans in order to be guaranteed safe passage. These routes are examples for the safe passage that could be found even in the desert. As a result of this system along the Ghadamis–Air–Kano route, not only large annual caravans but also small caravans were to be seen on it throughout the year. Large caravans usually had an armed escort. Nevertheless, there were other routes that crossed territories where anarchy still reigned. Large armed caravans could use these but only as long as they had a means of defence.[9]

The way from Ghadamis to Ghat is long and arduous: after leaving the Ghadamis Basin, there was the climb into, and subsequent crossing of the Hamadah al Hamra (Fig. 5.17), a barren desert full of rocks. The surface of this territory is almost always rocky and totally devoid of vegetation, except in small basins where the limestone has dissolved. Richardson characterises the way to Bir Amsin and further on to Bir Inazar as follows: "*As far as the eye can stretch on every side is one vast, solitary, lifeless, treeless expanse of desert earth! (...) Ground strewn over with small flints and other sharp chips of stone. Saw nothing alive in The Desert but one solitary bird ... Arrived at the well of Maseen* [Bir Amsin], *at 4 p.m. Much the same scenery as yesterday. The road good, not quite so stony as yesterday, and scattered over with pieces of very fine quartz and shining felspar. No sand in quantity, and a little herbage for camels*" (Fig. 5.18).[12]

Following this were the Hamadat Tinghert and the western offshoots of the sand sea of Awbari (Fig. 5.19) that had to be crossed. This was a very difficult route for the camels that frequently upset their loads in mounting or descending the groups of hills (Fig. 5.20).

For "path finding" in the Sahara, central significance is given to countless caravan tracks, way markers and above all watering holes and wells, as caravans must orientate themselves to these. The photographs on the following pages show, for example, the situation in the region of Bir Inazar (Fig. 5.22). The Bir Inazar, an important well even today, is a crossing point of old and new ways across the desert. "Single tracks", recognised in the terrain as paths, prove that this place was a central meeting point for countless caravans. This is backed up by the fact that in immediate proximity the grave of Sidi Mohammed ag Ikhenukhen (Fig. 5.21) can be found. He was one of the greatest chiefs of the northern Tuareg in the nineteenth century and assistant to European travellers and Sahara explorers like Henri Duveyrier and Alexandrine Tinné.

The caravan routes in the western part of modern-day Libya were primarily controlled by the Tuareg in the nineteenth century. They consisted of a number of large ethnic groups, who lived in the desert in the region that had to be traversed in

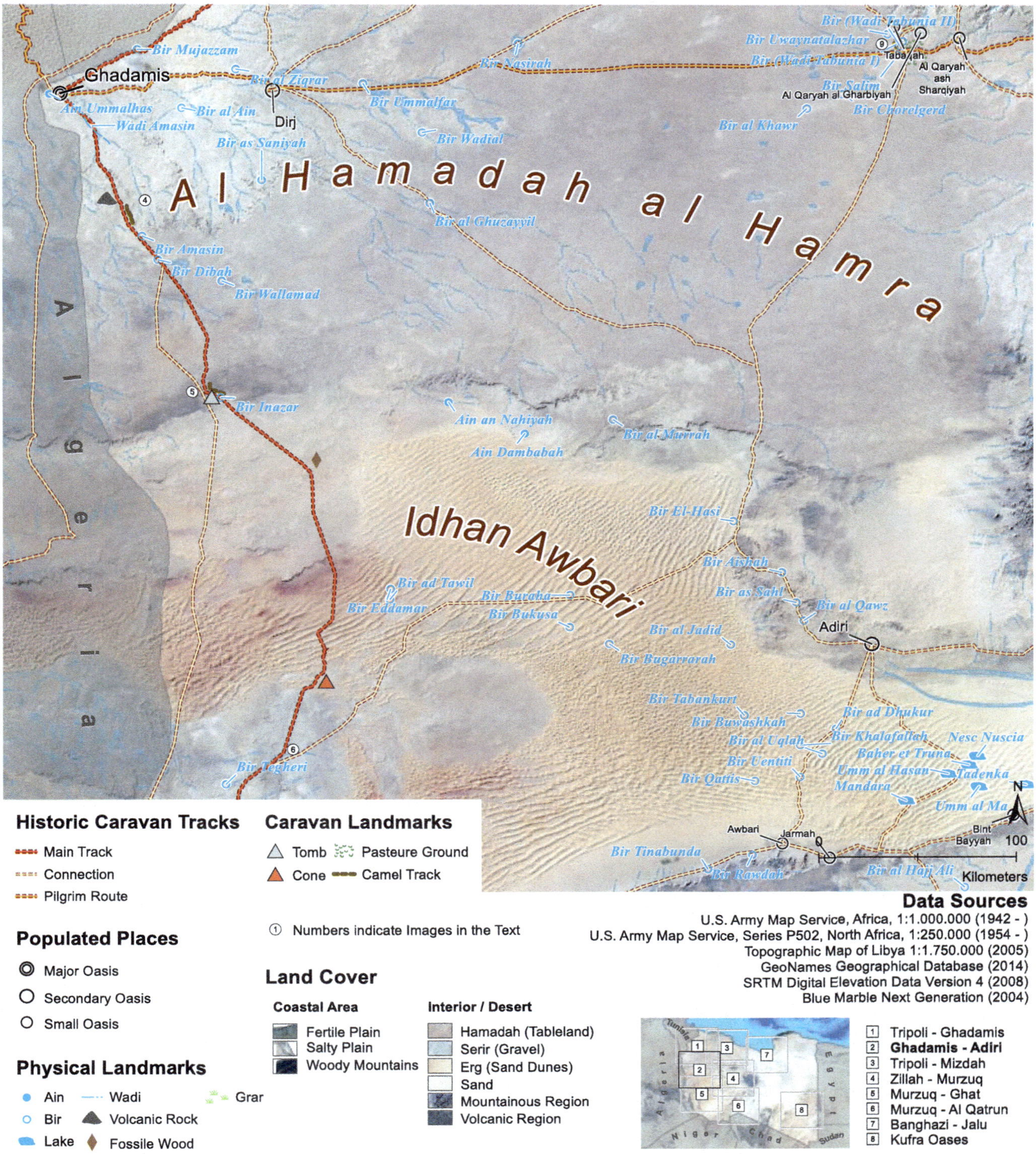

Fig. 5.18 Map: Ghadamis—Bir Inazar—Adiri

Fig. 5.19 Crossing the Awbari sand sea was a great challenge both for man and camels. Richardson reports: "The Arabs smooth the abrupt ascents, forming an inclined plane of sand, and then, in the descents, pull back the camels, swinging with all their might on the tails of the animals. No herbage—no stone—no earthy ground—all, everything one wide waste of sand, shining under the fervid sun as bright as the light, dazzling and blinding the eyes[11]"

order to reach the southern edge of the Sahara. Richardson and Rohlfs also met with them in order to negotiate safe passage on their way from Ghadamis to Ghat. Without the protection or friendship of one of the powerful leaders or another person respected by the Tuareg, a journey across the desert was more than just a bit risky. For European travellers, and also for traders, the choice of an *amidi*, a friend and protector, was an important affair at that time.

Duveyrier, who was travelling in North Africa on assignment for the government of Napoleon III, met with Ikhenukhen for the first time in 1860. As he was the political leader of the northern Tuareg, he guaranteed Duveyrier protection and friendship on his travels from then on. For example, he accompanied him on his journey from Ghadamis to Ghat. Ikhenukhen held the position of a king, *amanokal*, of the entire collection of the Adjer-Tuareg. Among the tribes of the Adjer, the Oraghen and Imanghassaten were the most powerful and important. The leader of the Oraghen was simultaneously king of the entire collection of the Adjer. In December 1862, a treaty was signed in Ghadamis between France and the political leader of the Tuareg of the Adjer—by name Ikhenukhen in order to kick-start a lively flow of trade, which has, however, never materialised. Through this contract with France, Ikhenukhen was also obliged to protect all French people who travelled across the Adjer region. The Tuareg leaders were looking for allies against a threatened occupation of their land by the Ottomans, who had advanced into the Sahara from Tripoli and had already occupied Murzuq, the capital of Fezzan. The English travellers in turn found their protection in this region

Fig. 5.20 Vast, undulating mounds of sand lying in the hot desert sun but they do not—as many would think—make up the majority of the Sahara desert. Most of it consists of stony plains (Serir) or rocky areas (Hamadahs)

Fig. 5.21 Grave of Sidi Mohamed ag Ikhenukhen (see map ⑤) is situated on a ridge on the caravan route from Ghadamis to Ghat. The topographic position emphasises his importance as a chief of the northern Tuareg in the nineteenth century. Along the Ghadamis–Air–Kano route the northern Tuareg acted as transporters and ensured the safety of the caravans. The grave with its engraved stone is one of the few visible reminders of this powerful man of the northern Tuareg: "Mohammed known as Ikhenukhen, son of Osman, died in 1876"

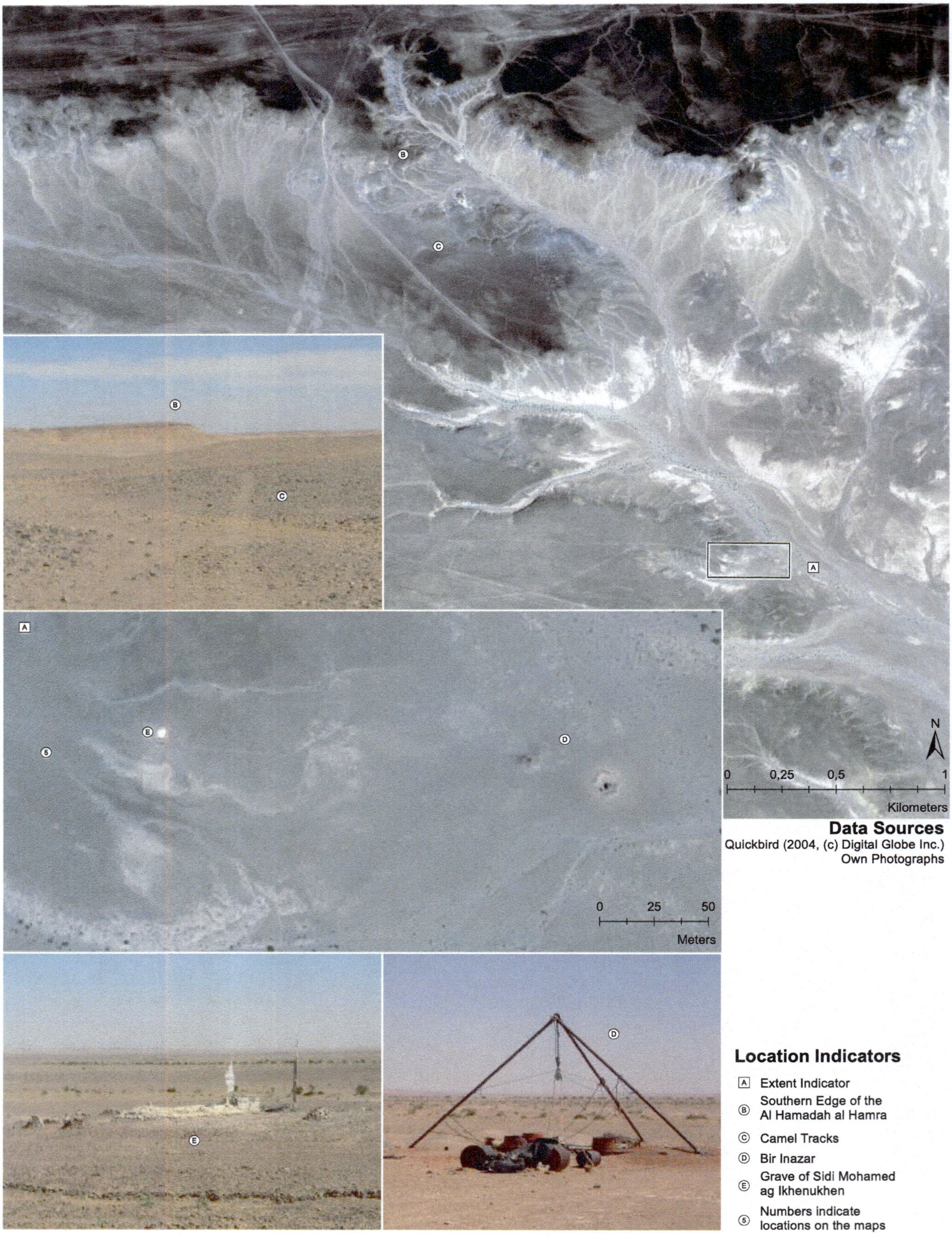

Fig. 5.22 Bir Inazar: The images portray the situation around the Bir Inazar in the Wadi Inazar on the caravan route from Ghadamis to Ghat. The Quickbird image from September 2004 gives an impression of the position of Bir Inazar and the grave of Sidi Mohamed ag Ikhenukhen

◄ (see map ⑤ and ⑥). It also shows the course of caravan tracks on the right bank of Bir Inazar where sparse vegetation has grown. On the ground "single-track" paths are visible which can be interpreted as caravan tracks in the terrain Ⓒ. Numerous tracks prove that this place was a focal point for countless caravans. The Bir Inazar Ⓓ, which is still important and kept in working order today, is a crossroads of old and new roads through the desert. This is backed up by the fact that in immediate proximity the grave of Sidi Mohamed ag Ikhenukhen Ⓔ can be found. Ikhenukhen was one of the greatest chiefs of the northern Tuareg in the nineteenth century and assistant to European travellers and Sahara explorers like Henri Duveyrier and Alexandrine Tinné as well as to traders on the route between Ghadamis, Ghat and further on to the south

Fig. 5.23 In Wadi Wani Miskiki groundwater can be found close to the surface (see map ⑥). The khabir/taksif of a caravan had to have knowledge about the pastures and the availability of water. These were important points of reference within the mental map of every caravan guide, whereby the water points were of central importance

from the Imanghassaten. That is why Richardson, who travelled the route from Ghadamis to Ghat in 1845, found his *amidi* in Hatita, a leader of this tribe. Krause, a German traveller and historian, spoke of a kind of "historical law" that the Tuareg saw evolve for the dispersal of responsibility in the accompaniment of travellers. The travellers coming from the north (Algeria) belonged to the Oraghen, those from the east (Tripolitania, Fezzan) to the Imanghassaten.[13]

After leaving the sand dunes of Awbari, the road stretches quite good to travel ahead. Most of the stones scattered en route to Ghat turn into black shingles, and all the regions have a volcanic look. The whole region is intersected and bounded on every side with the ranges of black, gloomy, and uniquely shaped mountains (see *rom Murzuq to Ghat*). It is, therefore, no surprise that many mysterious stories arose against the backdrop of this rugged scenery (Figs. 5.23 and 5.24).

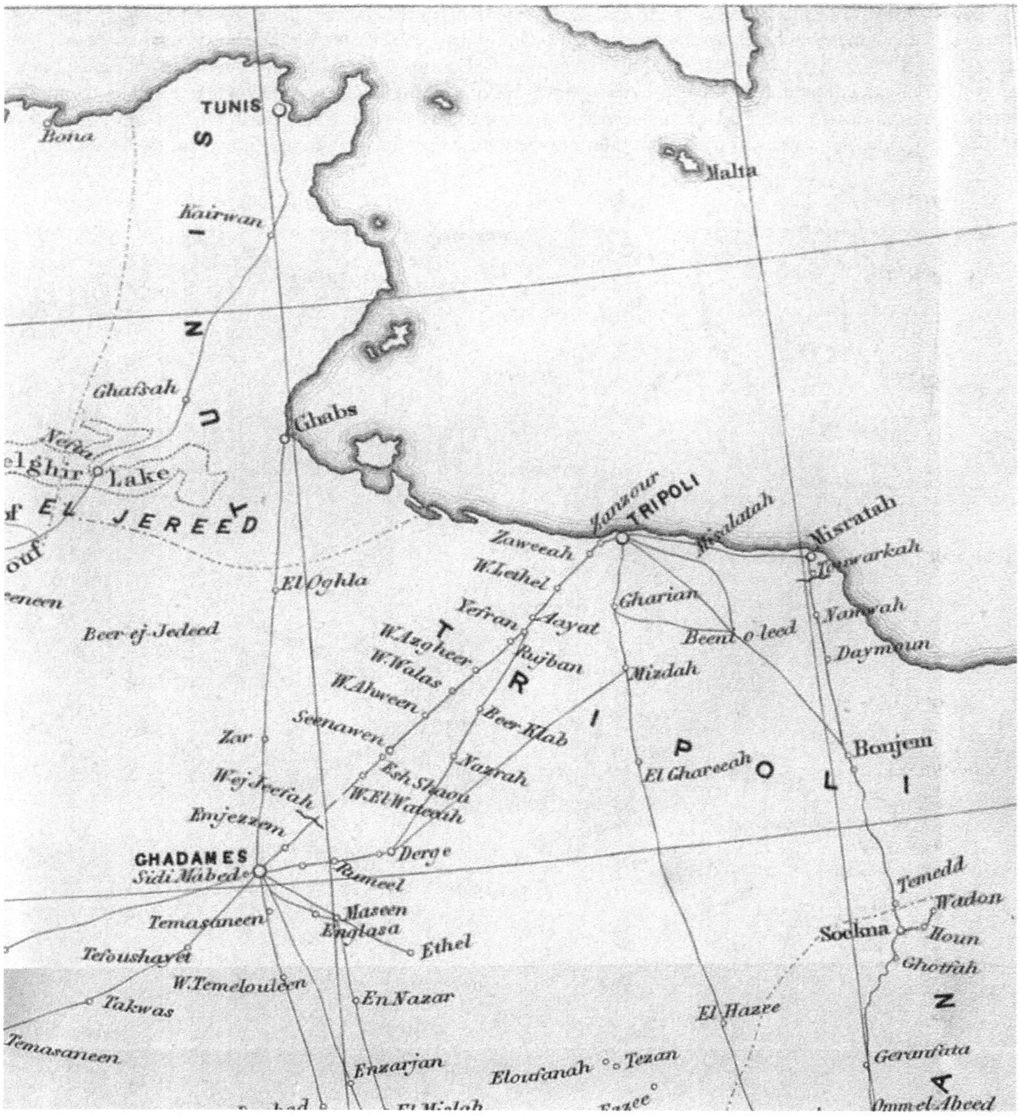

Fig. 5.24 This map—according to Richardson (1845)—reflects the different paths that led to Ghadamis

5.4 Excursus

5.4.1 Ikhenukhen—the Leader of the Tuareg from Adjer

5.4.1.1 Ikhenukhen Seen Through the Eyes of Henri Duveyrier 1864

Ikhenukhen belonged to the tribe of the Oraghen, which was counted among the Adjer. According to tradition, this tribe had its roots in the surroundings of Sokna. They inhabited Fezzan, the land of Ghat and Ahaouagh, a territory found on the left bank of the Niger, east of Timbuktu. At this last station, the tribe divided; one part remained in the Ghat area, and the other larger part moved towards Ahaouagh. In Ghat, the Oraghen seized power, which they had held before the nineteenth century.

Ikhenukhen was the son of Osman. After his death, he was, by custom of the Tuareg (Fig. 5.25), heir of the title *amchar*, a Berber term used historically to designate a lay tribal chief. However, he passed this up in favour of his cousin, Mohammed-eg-Khatita, who was the husband of his sister Zarah. He did not want to put himself under the constraint of being settled, as befit an *amchar* of the Adjer. Ikhenukhen held the position of *amanokal*, an overlord of the entire collective of the Adjer. He had three sons and three daughters. According to rules deriving from the matriarchal regime, political succession was, in principle, transmitted to the eldest brother of the preceding *amanokal*, to the eldest son of his maternal aunt or to the eldest son of his eldest

Fig. 5.25 People of the Tuareg as drawn by Duveyrier (1864)

sister. Thus, the heir of Ikhenukhens power was the son of his sister Outiti.

Ikhenukhen and his brothers are praised by a native poet. The following lines make the high opinion the poet had of them clear:

> *"The sons of Osman are stronger and braver men, who do not dirty the blood of their parents and who measure the corn generously. If a man goes to seek them, he can only try to fight with them. Their thoroughbred camel mares come neither from Adjer nor from Air, neither from the place of the Arabs, who pay taxes; and if one of them is mistaken, do not believe that she does it to escape and return to her home country. Their pack camels have feet as large as drums and the burdens they bear are like the peaks of mountains. They have mares with beautiful manes, which are saddled day and night. In them God has united the abilities and necessary qualities for the path of the journey. It is not only today that the sons of Osman shine in this splendour; all of Ahaggar and Adjer know it."*

This excerpt not only tells something of the history of the Tuareg of the north and Ikhenukhen, but it also provides an interesting glimpse into the standards used in their culture to measure recognition, wealth and social status. In the song by the unknown poet, strength and bravery are praised as the highest-valued personal qualities. The poet lists the most important possessions as camel mares of noble origin, able-bodied pack camels and noble horses.

According to the French African explorer, the reason why Ikhenukhen took such a high position within the Tuareg is that he is said to have also been the most skilled among the Tuareg with the broadsword and shield. Duveyrier gives the following judgement of the then already 76-year-old leading personality of the Tuareg:

> *"After almost seven months of living together with Ikhenukhen and careful observation, I am of the opinion and convinced that the qualities of his heart and his spirit, his generosity and integrity of character contribute to his outstanding skill in the handling of the weapon. (...) he bears the stresses of nomad life like the youngest of his sons. Everything, his behaviour, his voice, his methods of leadership, unveils a man of a yet uncivilised society; but within all the inherited mistakes of his race, one cannot help but recognise a man with great commitment to his principles, a boundless devotion to what he knows to be his duties."*

Regarding the fighting abilities of Ikhenukhen and the nobles of his family, Duveyrier writes that these had 100 soldiers available mounted on camels. In total, the most influential of the Adjer had over 200 warriors available. By European standards, 200 warriors might have represented quite a weak contingent. If one regards the conditions in the desert, however, the number of 200 warriors was sufficient, because there were few wells from which 200 camels could drink quickly. Yet, a further reason was that between one stage and the next there were sometimes distances of 200–300 km. With regard to the wealth of Ikhenukhen, the French explorer mentions that he numbers among the richest of the Adjer. His wealth consisted principally of camels, of which he is said to have owned around 60. Because of the power of the Oraghen, many tribes had to pay tribute, in order not to fear an attack from them.[14]

5.5 From Tripoli to Mizdah

Jacqueline Passon

The map "Tripoli–Mizdah" shows the course of two important caravan routes which led from Tripoli to Murzuq: a westward route via the small oasis of Mizdah and an eastward route via Bani Walid. This last one has evolved into a proper caravan and post road due to its more favourable natural conditions.

Two important caravan routes led from Tripoli to Murzuq, the capital of Fezzan. The shorter ran via Gharyan and Mizdah and divided there again into a westward route and an eastward route. The stretch with a westward direction was traversed and described by the British explorer and traveller Richardson as well as by his companions, the German African explorers Barth and Overweg. The eastward route was well known to Rohlfs. The other, longer route, which Duveyrier, Lyon, Vogel and Nachtigal followed, deviated considerably to the east from the first route. In spite of the insignificant detour, this path was much more frequented and evolved into a proper caravan and post road. The reasons for this lay in the rest and supply opportunities. On the one hand, this stretch had watering holes at regular intervals. On the other hand, with its regular

Fig. 5.26 Wadi Sofedjin is one of the most fertile valleys in Tripolitania. The area around Bani Walid has been inhabited since ancient times, as can be seen from Roman remains (photograph on the right). The Limes Tripolitanus marks the most southerly spot in Roman Libya. The drawing shows the north gate of the Roman fort of Abu Nujaym as seen by Lyon in 1819

settlements, it offered the desired stop-off points and guaranteed a greater degree of security for the caravans. This route would usually take around thirty days, while the westward route would only require about twenty days.

The former main caravan route led first through the Wadi Melgha via the Tarhuna Plateau to Bani Walid. The oasis, which consisted of a row of individual villages on both sides of the Wadi, was considered to be the most fertile of the African interior. Already in ancient times, the region was richly populated. At the end of the nineteenth century, an estimated 3000–5000 people still lived here. There is much archaeological evidence attesting to the propensity of the area at the museum of Bani Walid. From there, the route led to Abu Nujaym (Fig. 5.26), an important outpost of the Limes Tripolitanus which sunk into complete insignificance following constant raiding by steppe nomads. According to Nachtigal, in the nineteenth century, this place was half covered with sand and featured a derelict castle, several huts, a few palms and barely 200 inhabitants, who were engaged in barter trading with the passing caravans. Barter trading is a system of exchange, where goods or services are directly exchanged for other goods or services without using a medium of exchange, such as money. The wells of Abu Nujaym are the only watering holes in a wide area. Only in Suknah (see From Zillah to Murzuq), where the road bends towards Zillah and the oases of Awjila and Jalu, the life-giving wetness is again to be found. From there, a chain of oases can be reached via the Black Mountains (Jabal as Sawda), which stretches from Az Zighan to Murzuq.[15] Today, the stretch outlined has been overlaid in large parts by the modern road system, so few relics can be traced (Fig. 5.27).

The westward route, which in Roman times formed the main road in Fezzan, was, however, easy to follow from Mizdah. In this respect, it should be regarded more closely. In 1850, Richardson, Barth and Overweg decided to investigate this route and therefore to cross the Hamadah al Hamra, a barren rocky desert. Four years earlier, Richardson had chosen the easier route via Suknah on his return to Tripoli. Thus, they were the first Europeans to pursue this route. From Tripoli, the stretch led first across the Jefara Plain in the direction of al Aziziya and beyond to the foot of Jabal Nefusah (Fig. 5.28) and to Rabdah. The place offered favourable conditions for rest before beginning the climb on the steep path over the Jabal (Fig. 5.29).

Rabdah was divided into a west and an east village (Fig. 5.30). The date palm groves of the two village communities were only separated from each other by a small distance. The topographical element dominating the scenery is Jabal Manterus, which is volcanic in origin, and as a result, its colour stands out well from the settlement surrounding it. Barth and his companion Overweg climbed the interesting bicorn of the dark-coloured Jabal Manterus, where on the eastern and higher summit lies the tomb of a Marabout, a holy shepherd called Sidi Bu-Maza. The westward village of Rabdah, like the eastward village, was fed by a copious spring. In the nineteenth century, for example, the water in the westward village was collected in a basin of about 15 m in length and 9 m in width. Barth was quite

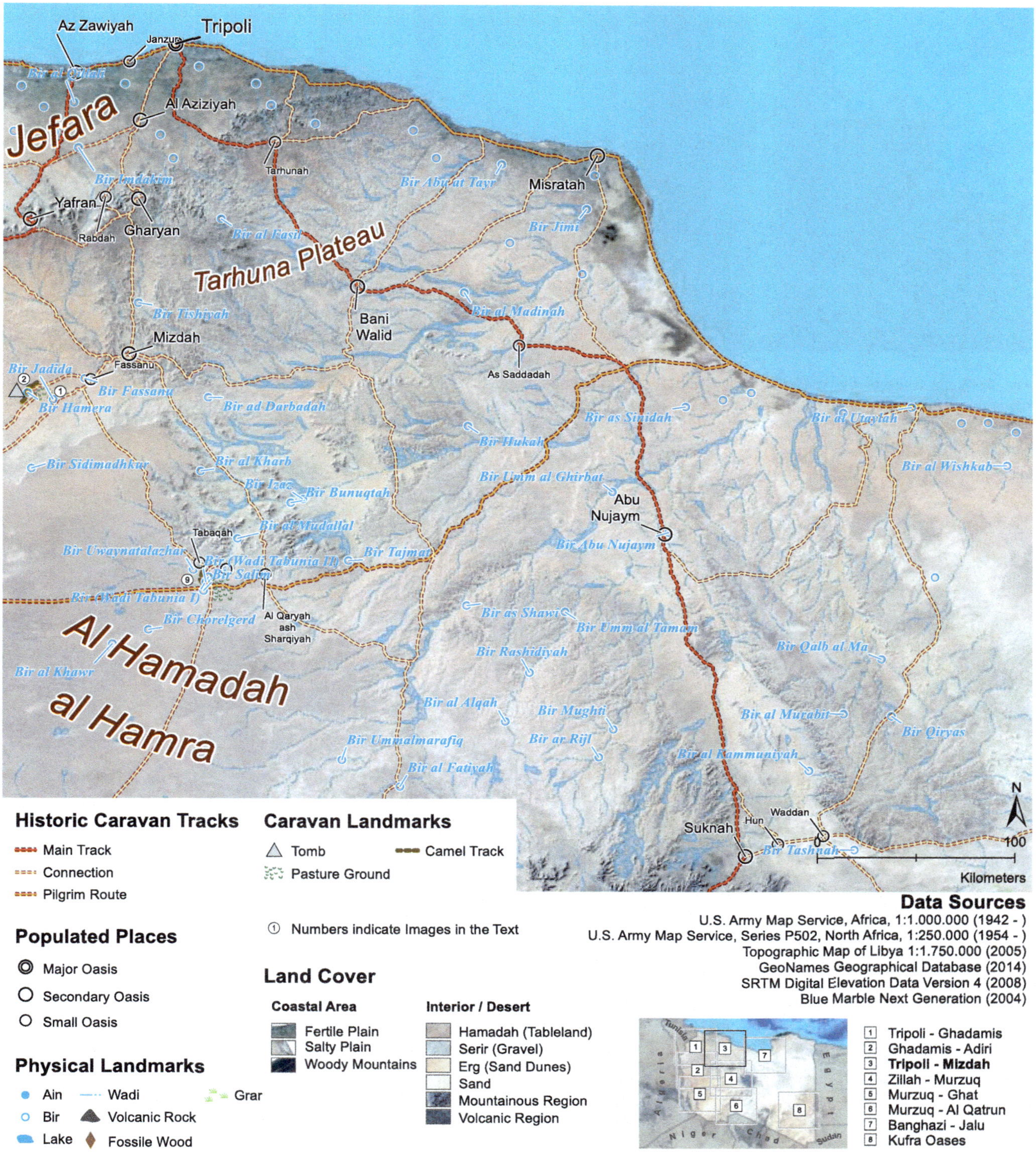

Fig. 5.27 Map: Tripoli–Mizdah

surprised by it as he recounts: "*It* [Rabda] *is fed by a copious spring, which arrested our attention. Following it up to trace its source, we were greatly surprised to find, in the heart of some date-trees, a basin fifty feet in length, and about thirty in breadth, in which the water was continually bubbling up and sending forth a considerable stream to spread life and cheerfulness around.*"[16] In Italian times, the basin was renovated and cemented (Fig. 5.32).

Fig. 5.28 En route to the south, the Jabal Nefusah, a limestone escarpment running parallel to the Libyan Mediterranean coast, had to be overcome. This was a difficult march over steep paths and through gorges of the Jabal Nefusah. On the right-hand edge of the picture, the town of Qawasin can be seen, which is situated on the foot of the Jabal Tekut and is just like Jabal Manterus of volcanic origin. On the top are the ruins of a chapel of Sidi Ramadan

Fig. 5.29 Routes to Sudan climbs into the Nefusah mountain range and traverses the Hamadah al Hamra

Fig. 5.30 Holy scenery, the holy chapels of Rabdah (above). Remains of the abandoned old town of Rabdah (below)

From an agricultural perspective, Rabdah was characterised by the cultivation of dates and a large quantity of onions. For a long time, the westward village functioned also as the residence of Hamid, a powerful Arab chieftain, who ruled the entire mountain region in the first half of the nineteenth century. He must, however, have been defeated by the Ottomans and must have lived in Barth's time in Bani Walid. On the northern side of the westward village, there are also seven holy chapels, which were known as "al Hhararat" (Fig. 5.31).[17]

From Rabdah, the difficult march continued over steep paths and across gorges of the Jabal Nefusah to Gharyan (Fig. 5.33), which in those days were controlled by the Ottomans. The Ottomans established a fort in Gharyan, too, in order to guarantee the safety of the region as well as of the trade routes. Gharyan has been a major settlement in the mountains. Today, it even has a university. Lyon and later Barth mention the underground apartments, typical for this landscape. Traditionally, the families lived in caves or in underground residences (Figs. 5.34 and 5.36) as much as 10 m deep, where, they were safe from marauders, and were they conducted their daily life, worked and raised their families.[18] Those residences are still abound today and are maintained by their owners and used for tourism.[19]

Fig. 5.31 Holy chapels of Rabdah, which were known as "al Hhararat"

Fig. 5.32 Water is a valuable commodity, which was collected by the people of Rabdah in a water basin that was cemented by the Italians

Fig. 5.33 Gharyan, which is about 80 km south of Tripoli, has been a major settlement in the mountains. It is considered to constitute a unit with the nearby towns of Tighrinna and Banu Abbas. As indicated above, the Jabal Nefusah used to be ideal for retreat and was used as hiding places by various population groups. Governments historically found the mountainous region hard to control. The Ottomans built a castle in order to guarantee the safety of the region. But this place has only been successfully controlled by the Italian colonists. Economically, the region prospered from agriculture.[20] Barth gives a brief account of the situation in the middle of the nineteenth century: "The villages, at least those above the ground, are generally in a wretched condition and half deserted; still the country is in a tolerable state of cultivation, saffron and olive-trees being the two staple articles of industry".[21] Another branch was the trans-Saharan trade, several of whose routes passed through Gharyan to Tripoli. The market was held on a normal weekday on the open ground as the photograph above from Laronde's publication "La Libye a travers les cartes postales 1900–1940" shows. Once Italian control was fully established, the modern town of Gharyan came into being with buildings above ground (photograph below, Laronde 1997). The Italian settlers expanded the range of agricultural products to tree crops, including figs, apricots, almonds and olives. Tobacco was also grown in the Gharyan area and kitchen garden crops were widespread[22]

Fig. 5.34 The photograph above from Laronde's publication (1997) shows the so-called troglodyte house which is situated in Gharyan. Lyon vividly reports: "As the natives live, as I have observed, underground, a person unacquainted with the circumstance might cross the mountain without once suspecting that it was inhabited. All the dwelling-places being formed in the same manner, (…). The upper soil is sandy earth, of about four feet in depth; under this sand, and in some places limestone, a large hole is dug, to the depth of twenty-five or thirty feet, and its breadth in every direction is about the same, being as nearly as can be made, a perfect square. The rock is then smoothed so as to form perpendicular sides to this space, in which doors are cut through, and arched chambers excavated, so as to receive their light from the doors. These rooms are sometimes three or four of a side, in others a whole side composes one; the arrangements, depending on the number of the inhabitants. In the open court is generally a well, water being found at about ten or twelve feet below the base of the square. The entrance to the house is at about thirty-six yards from the pit, and opens above ground. It is arched overhead; is generally cut in a winding direction, and is perfectly dark. Some of these passages are sufficiently large to admit a loaded camel. The entrance has a strong wall built over it, something resembling an ice-house. This is covered overhead, and has a very strong heavy door, which is shut at night, or in cases of danger. At about ten yards from the bottom is another door, equally strong, so that it is almost impossible to enter these houses, should the inhabitants determine to resist"[23]

Fig. 5.35 Gharyan is famous for its functional tableware which is made by local craftsmen. Even the people from Tripoli come to Gharyan to buy dishes for daily use. And of course, tourists used to have a little stop there

Fig. 5.36 Pictures taken from de Mathuisieulx's book "Attraverso La Libia" (1912) showing a troglodyte house and the castle of Gharyan

Fig. 5.37 Remains of the Roman and later Muslim settlement of al Qaryah al Gharbiya

Fig. 5.38 Drawing taken from Barth shows that the Roman gate of al Qaryah al Gharbiyah had not been collapsed at that time. Barth reports: " (...) the building at Gharyan consists of three archways, flanked by towers with receding walls. The two smaller gateways have been almost entirely filled with rubbish; the upper layer likewise is gone, and only those stones which form the arch itself are preserved"[26]

Fig. 5.39 Traces of a mosque in the old town of Qaryah al Gharbiyah

Fig. 5.40 In Wadi Tabunia, water can be found close to the surface. It was a favourable terrain for resting

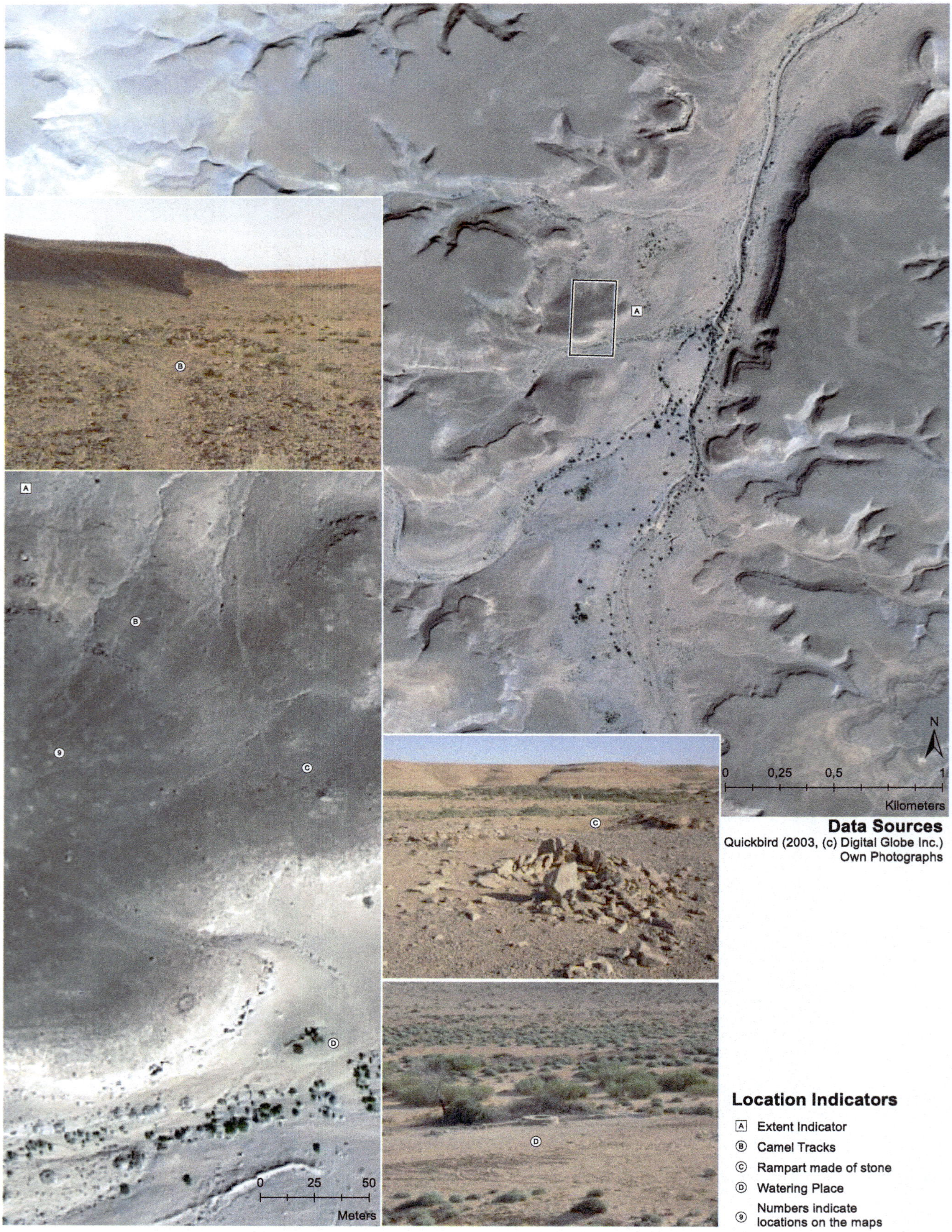

Fig. 5.41 Wadi Tabunia: The images portray the situation in the Wadi Tabunia on the caravan route from Tripoli to Mizdah and further on to Murzuq. The Quickbird image from February 2003 gives an impression of the position of the remains of a caravan-resting place in Wadi Tabunia

◄ (see map ⑨). Features include ramparts made of stone, behind which tents could be set up Ⓒ, and a well Ⓓ, which is still important and kept in working order today. Numerous tracks Ⓑ prove that this place was a focal point for countless caravans

From here, the way led further into the Mizdah Oasis (see Mizdah—important crossroads in western Libya), which enjoyed great significance despite its small dimensions. As it is here that the two important caravan roads from Tripoli to Murzuq and from Tripoli to Ghadamis come together.

The route then followed the edge of the Dahar Plateau towards the south and led through barren and stony territory until it entered the Wadi Tabunia, which cuts into the northern edge of the Hamadah al Hamra. The caravans traversed the edge of the valley. Barth speaks of a well of Tabunia, at which the caravans he accompanied made their camp. As the map indicates, there are in fact several wells in that area. From the information recorded by Barth, it is today no longer possible to locate the place he called Bir Tabunia. Relevant preparations had to be made before the imminent difficult march across the Hamadah al Hamra could really shart. Usually, a whole day was needed for the camels to drink at the wells, to graze and rest and to refill the water stores. Barth used this day to undertake an excursion to the nearby oasis al Qaryah al Gharbiyah, a former Roman military camp at the Limes Tripolitanus (Figs. 5.37, 5.38 and 5.39).

At the fault edge of the Hamadah, at the central plateau in the mountains of Qaryah, there are two villages that have this fault area to thank for their names: Al Qaryah al Gharbiyah (the western Qaryah) and al Qaryah ash Sharqiyah (the eastern Qaryah). Both places were presumably devastated by robberies at that time. According to Barth's descriptions, the old town of the western village (al Qaryah al Gharbiyah) had already fallen into disrepair. In earlier times, it had stood in high regard because the chapel of a saint was said to have been there. Around 30 inhabitants, fit to bear arms, lived in near-derelict dwellings, the ruins of which can still be visited today.[24] Besides the remains of the Islamic settlement (Figs. 5.37 and 5.39), there are further remains of the Roman camp (Fig. 5.38), which have been nowadays partially reconstructed by archaeologists. The Roman site was appreciated accordingly by Barth and was presented as a contrast to the Islamic culture. Capturing the Zeitgeist of the nineteenth century, Barth has a tendency towards evaluating the Roman achievements higher than the way Arab people live and work: "*We crossed the ravine, leaving the grove on our left, and ascended the opposite cliffs towards the ruined cluster of miserable cottages* [al Qaryah al Gharbiya], *when, having traversed the desolate streets, we encamped outside the Roman gate, the massive and regular architecture of which formed a remarkable contrast to the frail and half-ruined structures of the village. We were greatly astonished to find such a work here*."[25]

For Barth, as well as for most of the nineteenth-century travellers, the monuments created by the Romans in this vast area rose "*like a solitary beacon of civilisation over this sea-like level of desolation, which, stretching out to an immense distance south and west, appears not to have appalled the conquerors of the ancient world, who even here have left behind them, in "lithographed proof," a reminiscence of a more elevated order of life than exists at present in these regions*".[27]

Underneath the place, there was a palm grove with about 350 date palms, which still exists today. The springs of the small oasis were already collected in a basin in the nineteenth century. To a certain extent, dates, wheat and barley were cultivated. According to Barth, the place was avoided by the caravans as they regarded the water as harmful.[28] Since the caravans camped outside the settlement anyway and entered the oasis when necessary,[29] several things indicate that they halted in Wadi Tabunia. Presumably, the climb to the plateau was also easier to accomplish from there. The remains of a supply station for caravans, a caravan-resting place (see From Zillah to Murzuq), which was arranged here above the Wadi, also support these theories. Along the caravan route, there were further wells (Abyar at Tabuniyah) and stone shelters in close proximity to this site, similar to that found at the caravan-resting place. The satellite image (Fig. 5.41) shows the location of the caravan-resting place. As equipment for this place, there were ramparts made of stone, behind which tents could be erected (Fig. 5.41), and in the near vicinity piles of stones used as loading stations for camels. A well also belonged to this place, which was located in the immediate vicinity underneath the plateau in the Wadi (Fig. 5.40). Furthermore, spacious pastures offered camels the possibility to graze.

It is interesting to note that Barth remarks that mosque attendance decreased greatly in this area in the mid-nineteenth century. One possible explanation is that he did not mention the place Tabaqah, a central hub of communication of theological knowledge. The town was located on a pilgrim road, which led from Fez via Ghadamis and Tabaqah to Cairo and finally to Mecca (see map). The place housed a large *zawiya*, primarily a meeting place for spiritual pursuits and religious instruction (monastery), with a *madrasa*, an institution of learning where the Islamic sciences are taught, that had been laid out on the plateau above the Wadi (Fig. 5.43). The size of the site alone proves its great significance. The first monastery sites in North Africa are documented from the thirteenth century. They are comparable with hermitages. The *zawiya* has its origins in the kubba shrine, a tomb surmounted by a dome, as well as in the *rabita*, a hermitage. In the course of the dissemination of the Sufi movement across North Africa, it was accompanied by an enormously quick expansion. It evolved finally into

Fig. 5.42 New town of Tabaqah is situated amidst the Wadi

Fig. 5.43 Old town of Tabaqah. The place became a hub of theological knowledge

Fig. 5.44 Bir al-Hasi: An important well at the southern edge of the Hamadah al Hamra as seen by Barth (1977, 193)

centres of religion and political power. A rural *zawiya* like Tabaqah also provides lodging facilities for pilgrims (funduq) and contains libraries, schools, mosques and workshops. In addition, a *zawiya* may function as an intellectual centre, a sanctuary-offering asylum and a political focus. In the past, monasteries often played an important commercial role by protecting trade routes and creating networks of exchange among its members.[30] The important structures of a medieval *zawiya* have lasted until today: they include a place for prayer, a shrine, a madrasa and accommodation for students, pilgrims and travellers.

The old town of Tabaqah is laid out on the terraces of the Wadi, whereas the new town was built below the monastery (Fig. 5.42). The grouping of theological knowledge is still a characteristic of the place today. Tabaqah houses a theological faculty as a branch of the University of Gharyan, which is equipped with a comprehensive archive. "Religion" generates a persistent structure of the place (Fig. 5.43).

After leaving the Wadi Tabunia, the difficult march across the Hamadah al Hamra began. Water was next to be found only at Bir al-Hassi (Fig. 5.44). The importance of the well is underlined convincingly by Barth: "*No name could be more appropriate to this place than el Hasi (the well). There is no need of any discriminating surname; it is "the Well"—the well where the traveller who has successfully crossed the Hamadah may be sure to quench his own thirst and that of his animals. But it is not a cheerful resting place, though it is the great watering place on this desert road, as he has to cross the fearful "burning plain" of the Hamadah before he reaches the spot.*"[31] It took about six days to cross the highland of the Hamadah. The caravans progressed only very slowly in this sector of the route, while the camels made a habit of walking alongside each other when the territory allowed for it. Adiri, situated in the Wadi ash-Shati, one of the important lifelines of the Fezzan oases, was then reached via the waterless rocky desert. Finally, the way continued through the sand dune landscape of Idhan Awbari to Murzuq.

5.6 From Zillah to Murzuq

Jacqueline Passon

The map "Zillah–Murzuq" shows the southern progress of the caravan routes from Tripoli to Murzuq and depicts the course of the trade routes from Zillah to Murzuq as well as a part of the pilgrim road from Cairo to Timbuktu.

In early 1862, the German African explorer Karl Moritz von Beurmann undertook an attempt to clarify the fate of his compatriot Eduard Vogel, who had died in Wadai in the first half of 1856. His original plan intended to advance from Benghazi on the direct route via the al Kufra oases group (see *From Benghazi to al Kufra* via *Jalu*). The port city was the only place on the Mediterranean coast with which the Kingdom of Wadai had a trading relationship at that time. However, the direct route to Wadai was not traversable due to feuds and wars between various local groups living in the south of Libya. Therefore, Beurmann was forced to travel to the African interior via Murzuq. Already at the end of the eighteenth century, Hornemann, who with his travels had heralded the epoch of the European scientific explorations of Africa, had traversed the ancient north–south way, an important pilgrim road between Cairo and Timbuktu. In particular, the route he had taken passed over the famous oases of Siwa and Awjila through the basalt volcanic region of Harudj al Aswad and Tmissah to the trade metropolis of Murzuq.[32] As recent research indicates, the Cairo–Timbuktu route already has been travelled since the sixth or fifth century BC. This route seems to refer to Herodotus' list of the Libyan desert people. In fact, it might have been a straight caravan road linking the lower Nile valley to the Niger bend.[33] Beurmann chose a more northerly route. This route was yet new and unknown in Europe, and it led him from Benghazi via Ajdabiya, Awjila, Maradah and then to Zillah (see map Benghazi–Jalu) (Fig. 5.46).

The oasis of Zillah (see *Zillah-Oasis of the Palm Trees*) was only seldom visited by Europeans. It was not until 1878 that the German geographer, explorer and adventurer Gerhard Rohlfs first visited the place on his way from Tripoli to al Kufra. It counted, however, among the richest in the eastern Sahara because of the existence of over 100,000 date palms. Furthermore, the inhabitants offered significant camel breeding, which could not be found in any other oasis. The place is enclosed on all sides by steep sloping mountains belonging to the Jabal as Sawda. In the Middle Ages, it enjoyed great significance as a center of trade and marked the northern boundary of the Kanem and Bornu Empires. Al Idrisi reports in his book *The pleasure of him who longs to cross the horizons*, completed in 1154, about Zillah: "*From Awjila to the town of Zala is ten stages in a westerly direction. This is a small town with a lively market, full of commercial activity. There live a people mixed of Hawwara Berbers among whom one finds protection and generosity. From Zala, too, one may go to the land of the Sudan.*"[35] Nothing remained of this splendour when von Beurmann visited the oasis in 1862. According to his information, around 500 people lived then in the oasis. He further reported that the inhabitants sustained themselves on the cultivation of date and grain but did not engage much in trade.[36]

From Zillah, the stretch led first along the Wadi al Jarad in the direction of al Fuqaha. The caravans kept mostly to the banks of the Wadi al Jarad (Fig. 5.45) as dangers loomed in the Wadi itself. During rainstorms, even when the rain did not fall locally, the Wadi swelled to ranging torrents, which represented a danger for man and beast. The path markers (Fig. 5.45), still in use today, show where the Wadi can be crossed. Not far from these markings there is a prayer site, which is part of a campsite. The short part of the marking

Fig. 5.45 Course of the caravan route along the Wadi al Jarad (large photograph, see map ⑩). The "tumulus of stone" or path marker shows where the Wadi can be crossed (small photograph at the bottom). Those tumuli of stones, set by the people to mark the route, are called alam.[34] Within the same area, there is also a praying site (small photograph at the top), as a sign pointing to Mecca

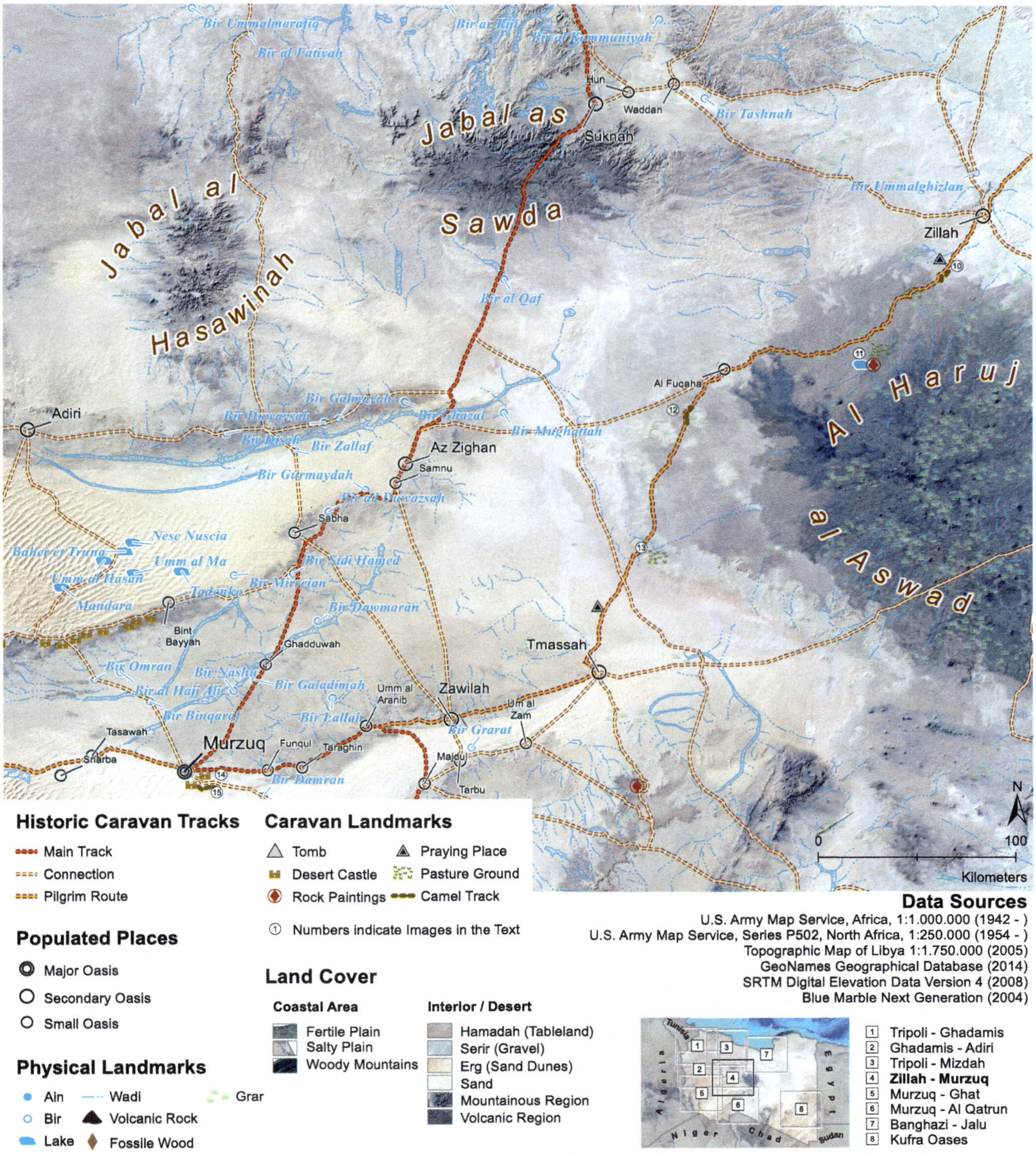

Fig. 5.46 Map: Zillah–Murzuq

points to the north-east and shows the direction of Mecca (Fig. 5.45).

The way leads further across stony territory, the Harudj al Aswad in the eastern Jabal as Sawda (Figs. 5.47 and 5.49). This area consists of boulders of up to one metre in diameter. Between the sometimes densely, sometimes more widespread angular and often very sharp-edged rocks, there is gravel or sand. The relief in this section of the route only

Fig. 5.47 For the caravans, it was difficult to cross the stony Harudj al Aswad

Fig. 5.48 A "grar" (pasture ground) at Harudj al Aswad

Fig. 5.49 Rocky Harudj al Aswad

Fig. 5.50 Caravan tracks at the endlessly wide, stone-strewn, water and vegetation-free plateau (see map ⑫). This area is called al Ashhab

Fig. 5.51 Watering hole at Harudj al Aswad: as the shadows of night fell fast, the caravans used to encamp. But there was neither the time nor the mood for campfire romanticism. James Richardson reports that when a caravan passed a well, there was the greatest confusion to get all the camels to drink. The people quarrelled and fought about this, as well as for their turn to fill their gurbahs and skin bags for water[37]

allowed the caravans to progress very slowly. Not far from the caravan route, there is a watering hole (Fig. 5.51), next to which there is a pasture (Fig. 5.48). On the overhangs of the rocky walls, the careful observer can discover rock art (Fig. 5.52), which could be evidence that this place has been a sheltering place for travellers and inhabitants of the desert since time immemorial. From here, the route continued across rocky terrain (Serir) to the nearby oasis al Fuqaha (see *Al Fuqaha-Isolated Oasis in the Libyan Desert*), which von Beurmann was the first European to visit after a six-day march in April 1862.

The old town, which lies in an undrained hollow surrounded by steep limestone walls, was inhabited in the mid-nineteenth century by 300–400 inhabitants, who lived on date and grain cultivation.[38] Like most other traditional settlements in Libya, the old town was abandoned in the 1970s in favour of a modern settlement. The elders of the place report that until the early 1960s, goods were still brought in by caravan. Local traders passed by, for example, from al Fuqaha via Zillah, Maradah and Ajdabiya to Benghazi and back. Such a journey lasted around 15 days. On 15–30 camels, dates were transported to the Cyrenaican metropolis. They primarily brought back wheat and barley on their return. An older inhabitant of the oasis describes further that caravans from Murzuq also came to al Fuqaha. In this context, these caravans can be described as transnational caravans. The camels were finally replaced by cars in the course of the 1960s. In the 1970s, according to the local Libyan elders, trade was restructured by the state and the caravan trade was completely disrupted.[39]

When leaving al Fuqaha in the direction of Tmissah, travellers were offered only an endlessly wide, stone-strewn, water and vegetation-free plateau (Serir) for the next one to one and a half day. This area is called al Ashhab. Dirt trails still cross the area today. After the descent from this high plain, which stretches over two altitude levels, there is a supply station for caravans. It is a "caravan-resting area" located there for this purpose, and its layout and facilities have a counterpart in Wadi Tabunia (see map Tripoli-Mizdah). The spot offers places for prayer, countless piles of stones used as loading stations for camels (Fig. 5.53), and a rampart made of stone behind which tents could be erected (Fig. 5.54). The satellite image (Fig. 5.54) shows the location of the caravan-resting area, which is directly next to a wadi that is called Wadi Bu Rashadah. It is said that many watering holes have been there that are, however, untraceable today. With the plants growing there, the wadi also offered sufficient grazing opportunities for the sustenance of the camels. Furthermore, barley was said to have been cultivated in this area. This underlines again the significance of

Fig. 5.52 Rock art at Harudj al Aswad is an eloquent testimony which indicates that this spot has been a sheltering place for travellers and inhabitants of the desert since long ago (see map ⑪)

Fig. 5.53 Piles of stones used as loading stations for camels at the caravan-resting area (see map ⑬): the camels knelt down. Then being only half as large, each camel could be loaded much easier. James Richardson reported: "They invariably grumble and growl, sometimes most piteously, when they are being loaded, as if deprecating the heavy burden about to be placed upon them and appealing to the mercy of their masters"[40]

this place, which also lies one to one and a half-day march away from Tmissah, the next southward oasis. The shelter was of vital importance for caravans coming from both north and south. Depending on the direction, the difficult passage across the water and vegetation-free high plateau lay before or behind the travellers. The place provided rest and supplies for the livestock.

Tmissah (Fig. 5.55) was described by von Hornemann at the end of the eighteenth century as an insignificant village. The indigenous population, which according to Beurmann numbered 300–400 inhabitants in the mid-nineteenth century, constructed their living quarters between the rubble of the older houses. The place with its derelict *qser* was further surrounded by a wall, which, however, provided little protection because in many places it had fallen down. Tmissah was surrounded by date forests, which provided the staple food. In contrast to the town, von Beurmann found the gardens and fields in good condition: in summer, figs, pomegranates and melons could be found in abundance. Furthermore, the inhabitants also owned large herds of sheep and goats.[41] In the eighteenth and nineteenth centuries, Tmissah apparently did not benefit from the trans-Saharan trade anymore. The situation was different in medieval times. The Arab geographer al Bakri mentions Tmissah in the eleventh century as an important commercial point with various markets (*bazaars*) that after 918–19 assumedly

Fig. 5.54 Ashhab Plateau and Wadi Bu Rashadah: the images portray the situation at the ascent or descent © from the Ashab Plateau, which stretches over two altitude levels on the caravan route from Zillah to Murzuq. After the descent from this high plain—when travelling southwards—there is a supply station for caravans, a "caravan-resting place", providing rest and supplies for the livestock. The Quickbird image

◄ from January 2007 gives an impression of the position of the remains of this caravan-resting place. Features include a rampart made of stone behind which tents could be set up Ⓔ, and numerous piles of stones used for loading and unloading baggage Ⓓ. The satellite image shows also the course of the Wadi Bu Rashadah. With the plants growing there, the wadi offered sufficient grazing opportunities for the camels Ⓑ

Fig. 5.55 Abandoned old town of Tmassa

participated in the Kanem trade.[42] The fate of the town's economy may have been the result of the decline of Zawilah, the nearby mediaeval political and economical heart of the Fezzan.

According to al Bakri, the traveller passed Zawilah, a historically significant landmark, two days ahead. Once the prosperous capital of the region and a flourishing caravan centre, it had then declined into a village.[43] Zawilah was the mediaeval Islamic capital of the Fezzan and since 918–19 dominated by Hawwara Berbers, predominantly Ibadis. However, the origins of Zawilah are much older. British archaeologists suspect the establishment of the oasis settlement to have been in the late first or second centuries AD. The presence of substantial Garamantian activity in the Zawilah region is evident from the numerous cemeteries and the presence of ceramics from the first to fourth centuries AD. In contrast to previous assumptions, that means, that trans-Saharan trade and Zawilah both existed prior to Islam.[44]

Under the suzerainty of the Banu Khattab dynasty, the town then developed both into a bustling economic centre of the trans-Saharan trade and a hub of contact between the expanding Islam and central Sudan (Fig. 5.57). It was a main point of departure for caravans going through Kuwar to the Lake Chad region via the so-called Bornu Road (see *from Murzuq to al Qatrun*), which was the continuation of the main north–south road coming from Tripoli. Furthermore, a subordinate trade and a pilgrim route from Cairo to Timbuktu passed through Zawilah. This east–west route reached Cairo, where it entered the Egyptian metropolis by the Bab Zawilah, the Zawilah gate. From the eighth to the twelfth century, the traders of Zawilah controlled commerce with the new political and economic centres east of Lake Chad. They were also able to dominate the roads northwards to Tripolitania and north-eastwards to Benghazi or Cairo. The trade in black slaves brought considerable prosperity to the central Saharan trade dominated by Zawilah. The demand of slaves in the new Islamic areas of North Africa, Egypt and primarily in the Middle East presented great opportunities. Thus, it is not surprising that slaves became the most important export items. Zawilah's slave traders, many of them were Berbers of the Hawwara tribe, were famous throughout the Islamic world for the number of captives they brought from central Sudan. By the tenth century, most of the black slaves sold in Muslim countries were imported via Zawilah. Its traders established a network of strong trading links from Sudan across Fezzan to the North African coast. The Kuwar oases lay astride the Bornu Road and developed into a major hub for goods dominated by the Ibadi Berbers. They were also closely related to the Jabal Nefusah. The Berbers of the mountainous region and Kanem in turn were closely connected through the exchange of goods for a long time. Slaves were sold by local Sudanic rulers to traders from Zawilah. According to al Bakri, these traders brought them by caravan on a journey that lasted forty days from Kanem to Zawilah, from which they were further transported to their final destinations.[45]

In the Middle Ages, the area around Zawilah was irrigated by wells and *foggaras*. Dates and vegetables were cultivated in large quantities.[46] However, in the last third of the twelfth century due to al Idrisi "*[the] Arabs roam the country causing as much trouble to the people [of Zawilah] as they can.*"[47] Sharaf al-Din Karakush, an Armenian *mamluk*, who had the support of the Arab tribes of Sulaym, occupied parts of the Fezzan in 1176–77. The term *mamluk* is most commonly used to refer to Muslim slave soldiers and Muslim rulers of slave origin. The quiet times were now a thing of the past. Since the Hilal and Sulaym Arabs had invaded Libya in the eleventh century, the political, economic and religious situation in the Maghreb changed constantly. The power vacuum created in the Fezzan region enabled raiders from the Sudanic state of Kanem to appear at the frontier post of the Maghreb, and by the end of the twelfth century, the Negro kings of Kanem had taken control of the Fezzan. Historically, there is much doubt about the political and commercial power Kanem wielded in Fezzan. But the representatives of the Kanimi, known as the Banu Nasr, wanted to keep the trade routes to the north

Fig. 5.56 Remains of old Taraghin, where the Kanem rulers established their capital by the end of the twelfth century

Fig. 5.57 Most important remaining landmarks at Zawilah: The tombs of the Banu Khattab (left) and the Al-Fath mosque (right)

functioning and tried to strengthen their diplomatic relationship with the Hafsids, a dynasty ruling Ifriqiya from 1229 to 1574. Their territories were stretched from the east of modern Algeria to the west of modern Libya. Ibn Khaldun mentions that Kanem maintained friendly relations with the Hafsids of Tunis. Kanem probably controlled the Fezzan through a subordinate king, whose own capital was newly established in Taraghin (Fig. 5.56), 70 km west of Zawilah.

Between 1300 and 1500, a number of political conflicts and hostilities took place in Kanem and Bornu. In the early fifteenth century, the struggle for political supremacy between Kanem and Bornu led to a reduction of the trans-Saharan trade that Zawilah still lived on. At the same time, the capital was probably reassigned to the town. The Khurman, a Fezzan group from the Wadi al Ajal, probably made Zawilah the capital of Fezzan once again for a short time. However, political and economic decline had already begun. The town definitely had lost its rank that of a capital and trading centre in the first half of the sixteenth century when the Awlad Muhammad dynasty took political power in Fezzan. Murzuq became the new capital of the country, remaining so until the twentieth century. Trade now took place there, and Murzuq became the vital caravan centre of the Fezzan and a stopping place for pilgrims from the west on their way to Mecca.[48]

What is important to note is that much of the social hierarchy was still intact in Zawilah, even in the Ottoman period. The inhabitants still had a big social standing within Fezzan because of the *sharif* status of some inhabitants, which is mentioned by Hornemann. According to the German traveller, Zawilah was known as the *"town of sharifs."* Many leading and wealthy men and relations of the Sultan of Fezzan were said to live in Zawilah.[49]

The remainder of the route follows an almost continuous chain of oases into the trade metropolis of Murzuq. Von Beurmann left Murzuq in the summer of 1862 with a trade caravan to continue his journey to the African interior on the famous Bornu Road (see *from Murzuq to al Qatrun*).

5.7 Excursus

5.7.1 Ibadi Berbers of Zawilah—Masters of Trade

Zawilah lay astride several caravan routes and developed into a major hub for goods, especially slaves, dominated by the Ibadi Berbers who became the masters of trade in the Fezzan region during the Middle Ages. Among others, the Arab historians al Yaqubi, al Istakri or al Bakri provide a clear insight into a bustling economic centre of the trans-Saharan trade and its inhabitants. The documents below are taken from the writings of al Yaqubi, al Istakri and an anonymous writer.

Al Yaqubi: Kitab al buldan (889–90)

"Beyond Waddan to the south is the town (balad) of Zawila. Its people are Muslims, all of them Ibadiyya, and go on pilgrimage to Mecca. ... They export black slaves from among the Miriyyun, the Zaghawiyyun, the Marwiyyun and from other peoples of the Sudan, because they live close to Zawila, whose people capture them. I have been informed that the kings of the Sudan sell their people without any pretext or war. The skins known as al zawiliyya come from Zawila. It is a land of date-palms, where sorghum (dhura) and other [grains] are sown. Various people live there from Kurasan, al Basra and al Kufa. Fifteen day's journey beyond Zawila is a town (madina) called Kuwar, inhabited by Muslims from various tribes, most of them Berbers. It is they who bring in the Sudan [slaves]. Between Zawila and the town of Kawar, and adjoining Zawila as far as the route to Awjila and Ajdabiya live people called Lamta, who greatly resemble the Berbers."[50]

Al Istakhri: Kitab Masalik al mamalik (first half of the tenth century)

"Zawila is on the frontier of the Maghrib, a town of middle size with an extensive district bordering on the land of Sudan. ... The black slaves who are sold in the Islamic countries are taken from among them. These slaves are not Nuba, Zanj, Habasha, or Buja. They are a race apart, deeper and purer black than all others. ... Most of those black slaves converge on Zawila."[51]

Anonymous: Kitab al Istibsar fi aja ib al amsar (1191)

"Zawila is a great and very ancient city in the desert. it is near the land of Kanim, who are the Sudan. ... It is the place of assembly for caravans and slaves are brought to it. It is the point of departure for Ifriqiya and other countries."[52]

5.7.2 Zawilah—Heart of the Medieval Fezzan

Al Bakri, who is one of the most important sources for the history of the Western Sudan, provided in his *Book of routes and realms* (*Kitab al masalik wa l-mamalik*) also a vivid account of Zawilah during one of the more crucial periods of its history. For the eleventh century, al Bakri notes that Zawilah was like the town of Ajdabiya which lies fourteen stages (this corresponds to fourteen halting places) ahead. He described Zawilah as a town without walls in the midst of the desert, which had a cathedral mosque, a bath and markets. Traders crowded the *bazaars* along the streets as caravans met there from all directions. From there, the ways of those setting out radiate.

By al Bakris words, Zawilah lay between the *maghrib* and the *qibla* from Atrabulus. The *maghrib* stands for the west, and the *qibla* from Tripoli is about south-east. From an agricultural perspective, palm groves could be found as well as cultivated areas which were irrigated by means of camel. However, trade was the prime concern of the inhabitants of Zawilah. From there, slaves were exported to Ifriqiya and other neighbouring regions (Fig. 5.58).[53]

Fig. 5.58 Most important remaining landmarks at Zawilah: The tombs of the Banu Khattab

5.8 From Murzuq to Ghat

Jacqueline Passon

The map "Murzuq–Ghat" shows the southern connection of the caravan routes from Ghadamis to Ghat as well as their further course. The cross-link roads to Murzuq are also shown.

Ghat (see *Ghat—A Picturesque Qsar on the Frontier between Fezzan and the Algerian Sahara*), which is situated in the south-west of Fezzan, grew in the latter third of the nineteenth century to be the most important trans-shipment point for slaves. The Ottomans, who under pressure from the Europeans had banned the slave trade, did not manage to make their influence felt in this area—as opposed to in Murzuq—so that the slave trade from the African interior flourished.[54] In Ghat today, very little of its former glory is to be seen—above all, the old town is in a particularly bad condition. As in Ghadamis, there were attempts to establish tourism in Ghat as well. The nearby Akakus Mountains and the Ghat Festival, which took place annually in December, drew people to the most south-westerly town of Libya.[55]

There are two cross-link roads that ran between the two most important trading posts of the Fezzan, Murzuq and Ghat (Fig. 5.60). The stretch across the Wadi al Ajal (Wadi al Hayat) offered better rest and provision possibilities than the other road which ran parallel to the Wadi al Ajal, further to the south. Because of the natural conditions, watering holes can be found here at regular intervals. Furthermore, in Wadi al Ajal there is a series of settlements and desert castles that served as stopovers and to a large extent guaranteed safety for the caravans. However, there was a tragic accident on the southern route in the nineteenth century. Travelling women from Europe were, at that time, an exception in the desert. But Alexandrine Tinné was one of them. The intrepid Dutch woman wanted to be the first European woman to cross the Sahara. She had already reached Murzuq in 1868, where she met with the German doctor and explorer Gustav Nachtigal, who provided her with important contacts for the authorities and the Tuareg. The negotiations with the Tuareg lasted for half a year, before they made nine protection guards available for the expedition. Tinné, in making her way to Ghat, followed in the footsteps of Barth. After Barth had left Murzuq in 1850, he went to the oasis of Tasawah

Fig. 5.59 Caravan Route in Wadi Barjuj. The pilgrimage route followed the Wadi Barjuj too. The photographs on the right hand show the Akakus mountains

and then to the ancient city of Sharba (see *Sharba—Disappeared Under the Sand of the Sahara*) where he continued on his way to Ghat across the Wadi Barjuj (Fig. 5.59). The caravan road led further across different wadis. The traveller passed among others the Wadi Mathendous and the Wadi In Habeter, the middle course of Wadi Mathendous. This wadi is one of the main dry river beds on the southern edge of the Masak Plateau. It defines a wide area which includes the Wadi In Habeter and tributaries as the Wadi Telsaren. The valley and its tributaries are famous for their thousands of prehistoric rock art engravings along with numerous figures of more modern periods as horses and camels. Beds of dried rivers and water holes (Fig. 5.63) offered necessary possibilities to supply man and cattle. A number of landmarks bear witness to the human presence in this part of the caravan world. That includes caravan trails in Tajamoah Attorok (Fig. 5.61) that wind through the scenery like a motorway or various burial places that can be found in Wadi Telsaren (Fig. 5.62).

Tinné was probably taking the same path as Barth. But just a few days after she had set off, there was a tragic occurrence. After the travelling party had gotten involved in a quarrel, which Tinné sought to reconcile, she received a blow of a sword to the shoulder, broke down, was robbed of all of her clothes and possessions and bled to death in the desert. Since then, there has been much speculation about the background to this event: while Nachtigal cited greed as the sole motive for the attack, other travellers see the cause as having originated in tribal disputes. The reputation of the Tuareg leader Ikhenukhen is said to have been severely affected. Younger tribal leaders vied for power. The German historians Gootlob Adolph Krause and Erwin von Bary infer from this that the female explorer and her European companions were victims of political intrigue. The attack on the expedition was rumoured to have been planned in order to demonstrate that Ikhenukhen was no longer in a position to safely lead his wards through the territory of the Adjer Tuareg. Whatever the reason for this attack was, it is clear that travelling involved enormous risks at that time. Of the 200 travellers during the period between 1820 and 1870, 165 died because of illness or murder.[56]

The area between Murzuq and Ghat has already been settled for a long time. Testament to this is the thousands of rock drawings. The Masak Mastafet offers a unique universe of rock drawings, which first achieved scientific renown through Barth in 1850. As one of the first authors, he attested, at a time when research on prehistoric and early history was still in its infancy, to a very old age for the rock drawings and identified them as a historic resource and witness to the early ways of life of the inhabitants of the Sahara. He also saw in the rock drawings evidence of the climate change in the Sahara region during the last millennia. The rock drawings of Masak Mastafet were included in the list of UNESCO World Heritage Sites in 1985 (Figs. 5.64, 5.65 and 5.66).

The earliest drawings are commonly dated to the Late Palaeolithic Age with its large wild savannah fauna and to the

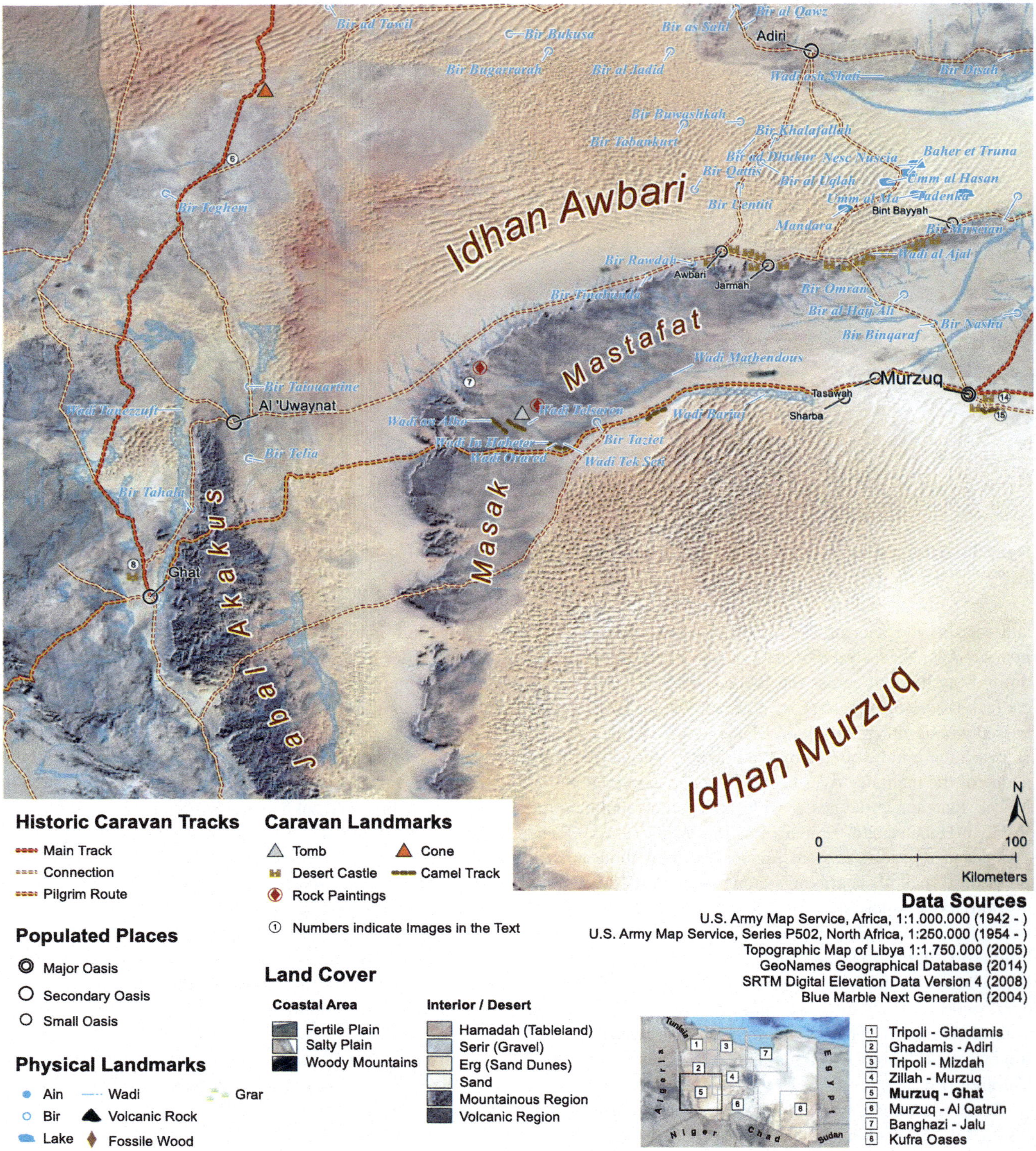

Fig. 5.60 Map: Murzuq—Ghat

Neolithic cattle herdsman culture. Today, it is assumed that the oldest drawings can be traced back to hunters and gatherers, who settled here in the early Neolithic Age. The rock drawings were drawn on to overhangs, in caves (Fig. 5.64) or on other surfaces that were protected from rainfall and sandstorms. The sites of the rock drawings are living spaces or cult sites. The drawings relate to the details of the lives of their creators, such as hunting experiences or camp life. The value of the rock

Fig. 5.61 Caravan route in Tajamoah Attorok: Hundreds of paths arranged in horizontal stripes run parallel to one another

drawings lies chiefly in the fact that, apart from the remaining settlement artefacts, they are the only evidence of the way of life of the people in this period. They tell of abilities, organisation and the societal relationships. They also provide clues as to climatic changes. Around 10,000 BC, the climatic and ecological conditions in this area improved again to the extent that from all sides, people could penetrate into the interior of the Sahara. Today, it is assumed that during the Holocene, there were extended pluvial periods. Related to this, the vegetation must have been thicker; furthermore, animals must have made their homes here, which then soon disappeared. The early rock drawings show large animals such as the wild buffalo (*Bubalus antiquus*). Around 7000 BC, there was a great cultural movement in the Sahara, which was dubbed the "herding phase" and relates to the onset of an arid phase. During this time, cattle raising was dominant. It is estimated that the period of "herding art" lasted more than 3000 years.

Towards the end of the second century BC, people of Mediterranean origin with horses and chariots moved from the Libyan Mediterranean coast towards the central Sahara. They expelled the Negroid population who had settled there. The Garamantes, who were first described as such by Herodotus in the fifth century BC, are also referred to by rock drawings, which show horses and chariots. For the Garamantes, who built sophisticated irrigation systems, the Wadi al Ajal, with the capital of Garama (Jarmah), the remains of which can still be visited today (Fig. 5.67), became a political and economic centre. They were described as farmers and cattle raisers, but were known as traders, too. As such, they controlled the trade routes between the Mediterranean and the sub-Saharan districts of Africa.

The last phase of the rock drawing art, which in some areas lasts until today, marked the so-called camel phase. This art is commonly associated with the Tuareg. The

Fig. 5.62 Old grave in Wadi Telsaren (Masak Mastafet)

Fig. 5.63 Water hole et al. Giltha (Masak Mastafet)

Fig. 5.64 Rock drawings were found, for example, in caves (see map ⑧). Rock art in Wadi an Albo

Fig. 5.65 Graffiti representing hippos and cows in Wadi Tek Seti (Masak Mastafet). The graffiti are located on the canyon's wall

drawings are testament to the activity of people in an extremely arid living environment.[57]

The route along the Wadi al Ajal is also lined with the so-called *qsur* (desert castles) whose structures are built with walls and towers. Similar to the constructions along the Zawilah–Murzuq–Barjuj depression and along the Bornu Road in the area between al Qatrun and Tajarhi, the conception and the facilities are reminiscent of antique limes period forts.

More information on history and architecture of these desert castles is given in the following chapter (see *From Murzuq to al Qatrun*).

Fig. 5.66 Malika (Queen)—rock drawing in Wadi In Habeter (Masak Mastafet/see map ⑦)

Fig. 5.67 Remains of Garama (Jarmah), the ancient capital of the lost civilisation of the Garamantes

5.9 Excursus

5.9.1 The Garamantes of Garama—Early Masters of Trade

The Garamantes were Saharan people, who founded a prosperous kingdom in the Fezzan area of modern-day Libya. They were a local power in the Sahara between 500 BC and 700 AD. The British archaeologist David Mattingly argues that the Garamantes are not just a vanished civilisation. In his opinion, they are a much maligned, misunderstood African people. To strengthen their own sense of importance, the Romans depicted other people as barbarians. It is not uncommon that ancient writers depicted the Garamantes as barbarians who threatened the Mediterranean world from their desert strongholds.[58] The first-century AD Roman historian Tacitus, for instance, described them as an "*ungovernable tribe ... always engaged in practicing brigandage on their neighbours*".[59]

Herodotus, writing in the fifth pre-Christian century, had a different picture of the Garamantes in mind. He records they were *"a very great nation"* who herded cattle, farmed dates and hunted the *"Ethiopian Troglodytes"* or *"cave dwellers"* who lived in the desert, from four-horse chariots.[60] According to Pliny the Elder, the Romans seem to have grown weary of Garamantian raiding. In 19 BC Lucius Cornelius Balbus marched against the Saharan people and captured 15 of their settlements.[61] The Garamantes, who used an elaborate underground irrigation system, controlled chains of wadis and oases on the desert caravan routes in Fezzan. Much of their wealth came from the control of the caravan trade. They brought the products of Saharan and sub-Saharan Africa to the Roman cities of the Mediterranean —salt, gold, semi-precious stones, ivory, wild animals for the arena and natron (a naturally occurring alkali used in embalming and glassmaking).[62]

5.9.2 Garama—Heart of Ancient Fezzan

Archaeologists excavated parts of the Garamantes' capital Garama (Jarmah), about 150 km west of modern-day Sabha. The ruins have been found beneath a medieval caravan city. Current research indicates that the Garamantes had a large

Fig. 5.68 Remains of a Garamantian cemetery near Garama. While most people were buried in shaft graves, some had elaborate monuments

number of other settlements. Four thousand people may have lived in Garama itself, with another 6000 in suburban satellite villages close by, and perhaps as many as 50,000 in the Wadi al Ajal area. The full extent of the Garamantian territory is still unclear. Scientists believe that Garama and the Wadi al Ajal were at the centre of a wider Garamantian Empire.

The Garamantes were farmers and merchants. Their diet consisted of grapes, figs, barley and wheat. They traded wheat, salt and slaves in exchange for imported wine and olive oil, oil lamps and Roman tableware. According to the Roman historians Strabo and Pliny, the Garamantes quarried amazonite in the Tibesti Mountains.[63]

In 2011, more than 100 fortified farms and villages with castle-like structures and several towns were discovered by British archaeologists, most of them dating back to the years between AD 1 and 500. These "lost cities" were built by the Garamantes, whose lifestyle and culture were far more advanced and historically significant than the ancient sources suggested.[64] The British archaeologist Mattingly states: "*In fact, they were highly civilised, living in large-scale fortified settlements, predominantly as oasis farmers. It was an organised state with towns and villages, a written language and state of the art technologies. The Garamantes were pioneers in establishing oases and opening up Trans-Saharan trade*" (Fig. 5.68).[65]

5.10 From Murzuq to al Qatrun

Jacqueline Passon

The map "Murzuq–al Qatrun" shows the route of the famous Bornu Road, which led from Murzuq into the African interior. In the map excerpt, the southern continuation of the caravan routes from Tripoli to Murzuq is also shown, as well as a part of the course of the trade route from Zillah to Murzuq and the pilgrimage road from Cairo to Timbuktu.

Fig. 5.69 A modern caravan from Nigeria heading north to Funqul, Libya

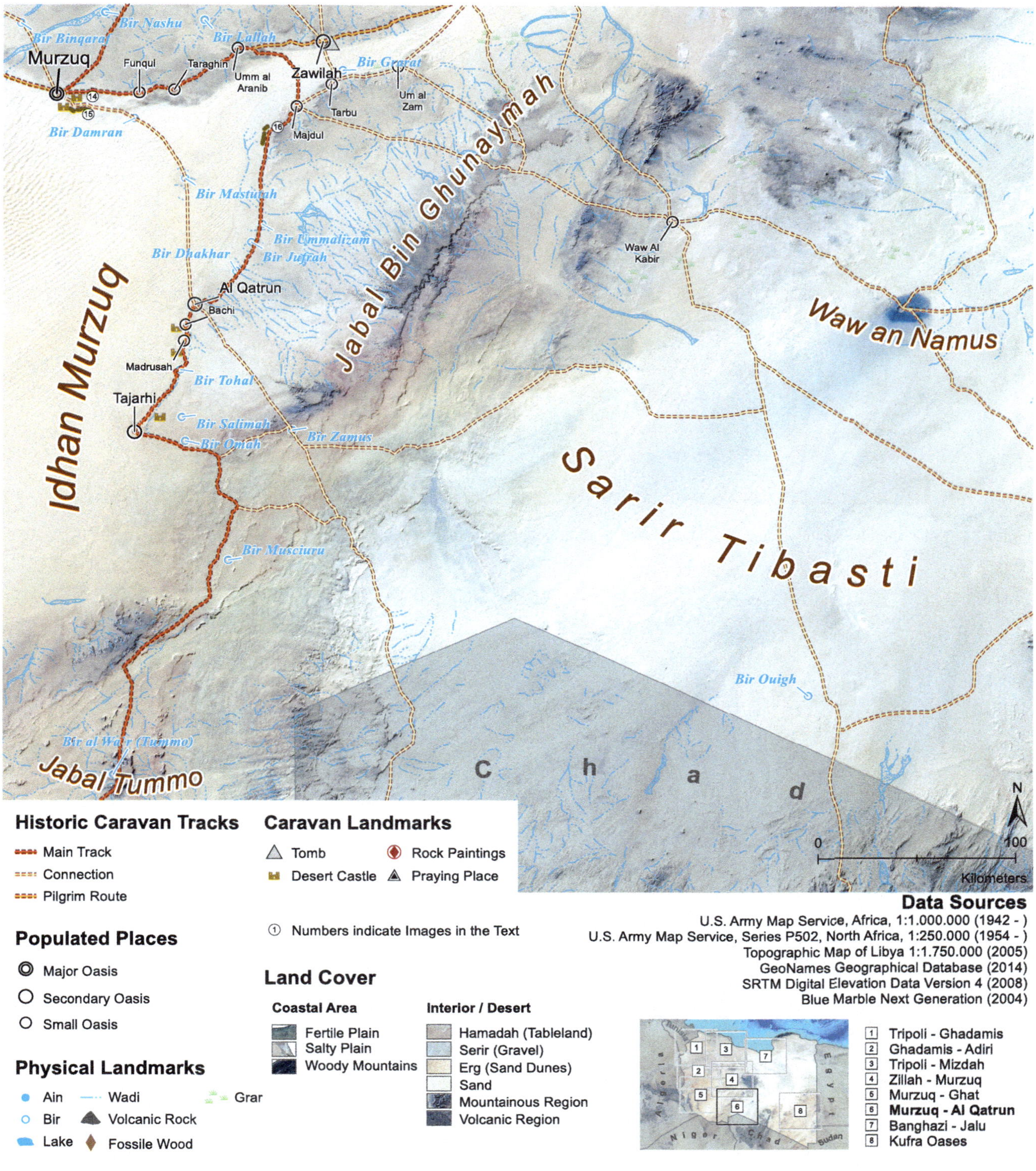

Fig. 5.70 Map: Murzuq–al Qatrun

Murzuq served as a hub of the caravan trade in the Fezzan since the first half of the sixteenth century. In particular, the slave trade flourished here until the late nineteenth century. Thousands of caravans must have come through the oasis during the course of history. So it is no surprise that even today there is evidence of numerous one-lane paths in the immediate surroundings. A few kilometres to the south-east of the town, one will find such tracks: in the satellite image

Fig. 5.71 Kuka, the nineteenth-century capital of Bornu, was an important market. The residence of the sultan as seen by Barth (1977, 205)

Fig. 5.72 Traces along the Bornu Road

(Fig. 5.74), a series of paths can be recognised, which may be identified on-site as the tracks of caravans. The tracks are part of the connecting route that led from Murzuq to al Qatrun. Due to the more adverse natural conditions in this section, the route went across the north-eastern extension of the dune landscape Idhan Murzuq, a dune landscape which is part of

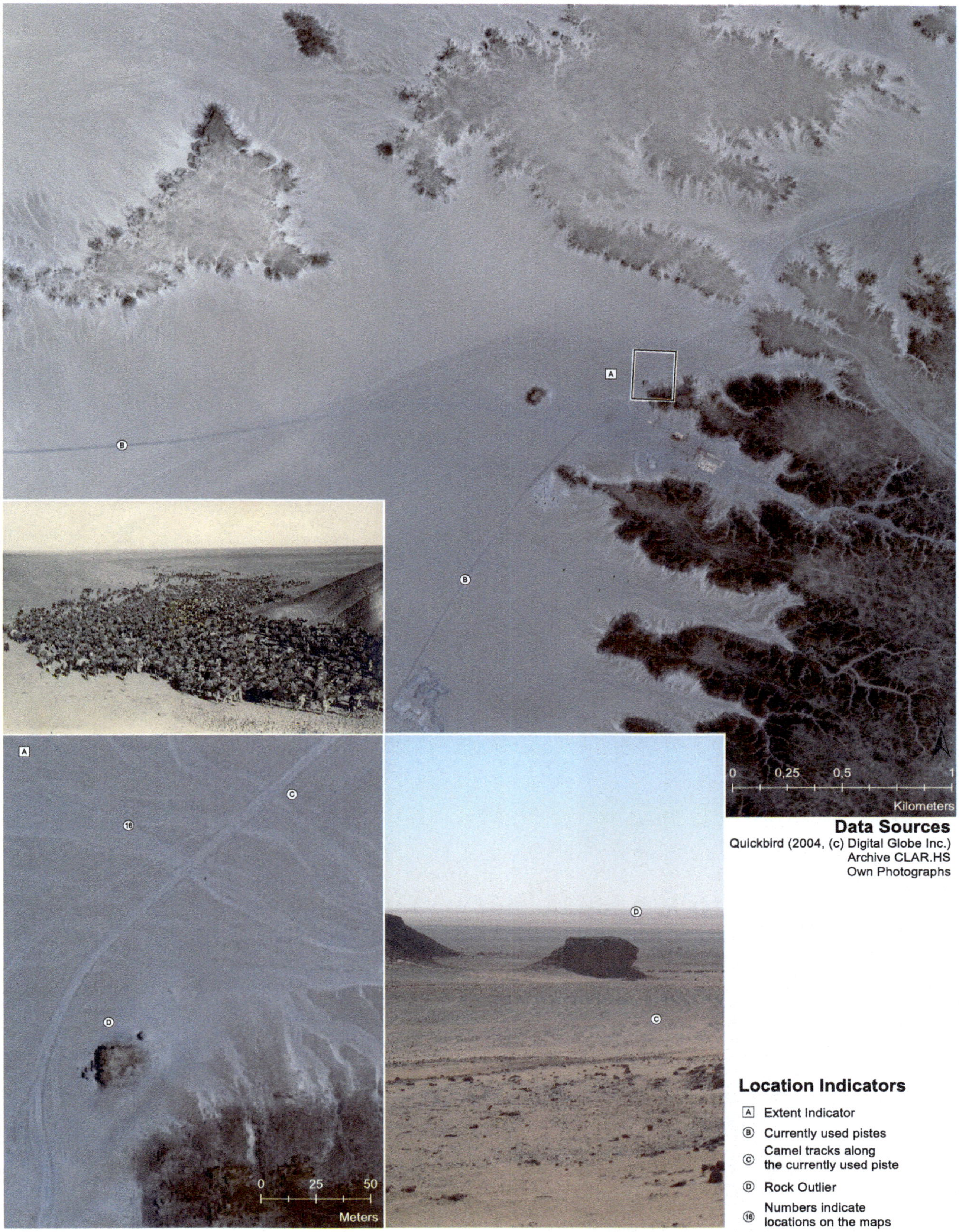

Fig. 5.73 Bornu Road: The Quickbird image from July 2004 covers a small part of the connection route between Libya and Bornu in the South. The so-called Bornu Road (see map ⑯) is known to be one of the most important and most famous caravan routes in the past. The section taken in the image is located between Majdul and Tajahri. It shows a variety of one-lane camel tracks running alongside one another where the photograph at the bottom depicts a subset of those traces in the sand. An early photograph of a caravan from an unknown author may help to imagine how a caravan might have looked like when crossing a similar terrain

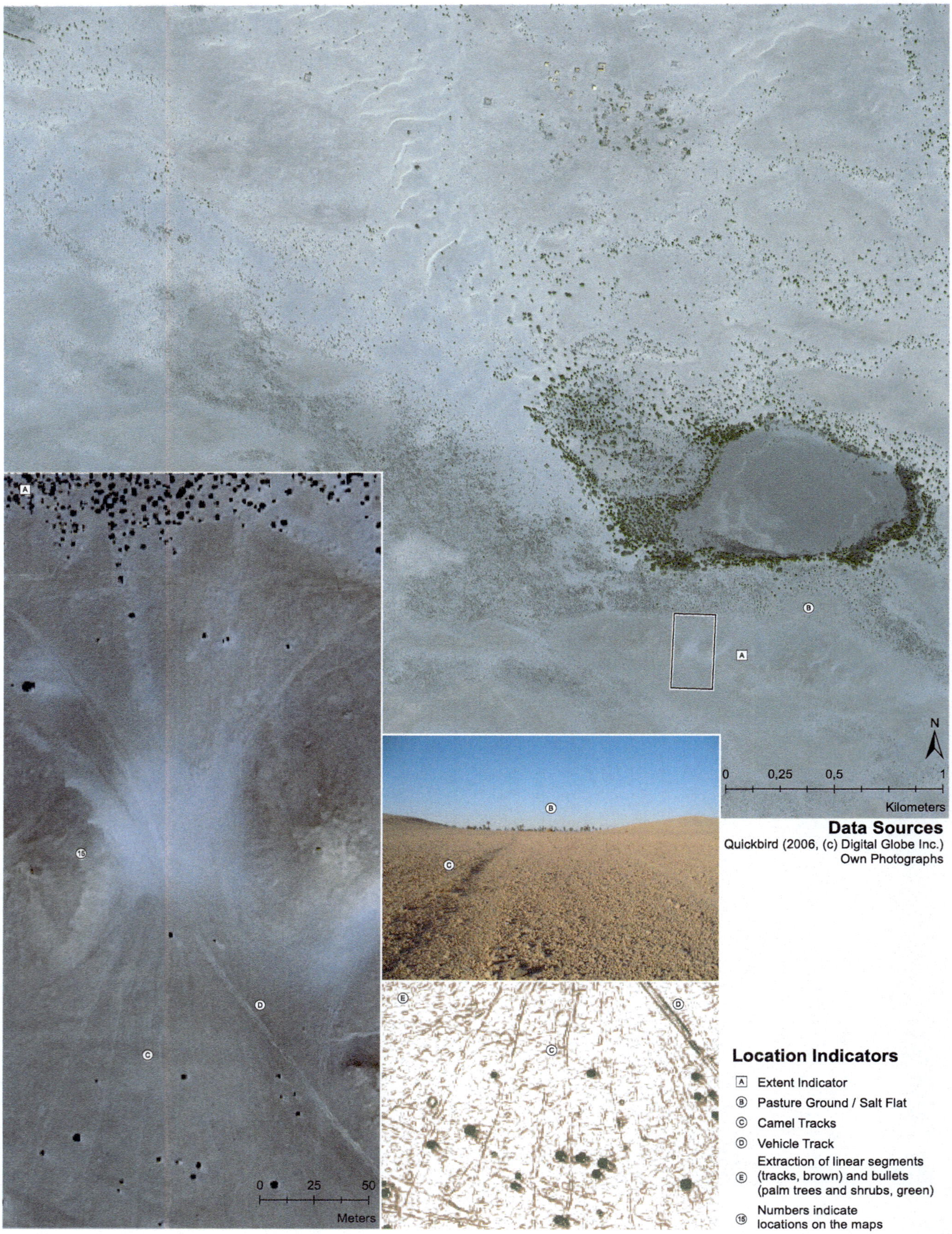

Fig. 5.74 South of Murzuq: The Quickbird satellite image from April 2006 shows a number of tracks Ⓒ in the south-east of Murzuq. On-site, those paths can be identified as caravan tracks (see map ⑮). Since the sixteenth-century Murzuq had become the most important trading centre within the Fezzan

Fig. 5.75 Caravans stopped nearby the oases of al Qatrun at a well

the greater Sahara desert region. However, this part would have been difficult to traverse by the caravans.

The littoral region of North Africa was linked with the different sub-Saharan territories, known as the Sudan, by a network of caravan routes. The shortest route ran from Tripoli to the Jabal Nefusah via the Fezzan oases into the Chad Basin. This ancient and highly frequented route was called the Bornu Road (Fig. 5.72). Ancient tradition leaves no doubt that this route had been in use since Roman times (and doubtless even earlier), it continued to be used probably by the Byzantines and Muslims.[66] Since Zawilah had lost its rank as a trading centre, the route started off in Murzuq and led first along an almost continuous strip of oases in an easterly direction as far as Umm al Aranib. About 10 kilometres to the east of the town, the path forked to the south. On the way to Majdul, the caravans had to overcome the eastern extension of the dune landscape Idhan Murzuq. To the south of Majdul traces of this famous road can be encountered. The satellite image shows a large number of one-lane paths running alongside one another. They appear to be very impressive in the area and can be interpreted as caravan trails (Fig. 5.73). The way further into the African interior continued via a chain of wells, desert castles (Figs. 5.78 and 5.79) and the oases al Qatrun, Bachi, Madrusah and Tajarhi.

The rise of several Western Sudanese kingdoms, like Songhay or Kanem in the period between the ninth and eleventh centuries, is connected with an increased scale in trade activities. A consequence of this development was a dense network of connections between the Mediterranean world and the Sudan by the end of the eleventh century. Although communications had been disrupted by the Muslim invasions of North Africa in the seventh century, it was not long before that the Berbers continued their traditional trading activities. Traffic along the Bornu Road continued then over the centuries. Until the end of the medieval period, trade was under the control of the Kanem Empire and then switched to the Bornu Empire, of which Kanem became a province. The new power of Western Sudan continued its trading contacts via the Kuwar–Fezzan route. It had been in touch with the different occupiers of Tripoli between 1475

Fig. 5.76 Traditional objects are exhibited in the medina of al Qatrun

and 1530. After the invasion of the Ottoman Turks in 1551, the contacts continued as they had before. According to several diplomatic documents, the Bornu trade continued in the seventeenth, eighteenth and nineteenth centuries under the rule of the Karamanlis (1711–1833) and during the second Ottoman period in Tripoli (1833–1911).[67] It seems that until the early 1820s, this route was the most active of all. The British explorers Dixon Denham and Hugh Clapperton found Kuka (Fig. 5.71), the new capital of Bornu, in 1823 in a state of prosperity. In various parts of Kuka, which was inhabited by probably about 60,000 people during its heyday, daily markets were held. The main market took place on Mondays outside the westernmost gate. Among the goods offered for sale were livestock, cloth and clothing of all kinds, foodstuffs, cooking and eating utensils, leather goods, basketware, carpentry and metalwork, weapons, even boats, firewood, fodder, charcoal, building materials, rope and other commodities. Slaves formed an important element in market transactions. Special slaves, such as eunuchs, dwarfs, deaf mutes and concubines, were generally sold privately, as were the best-quality horses, and not on the open market.[68] As already pointed out, since the eighth century, the most commonly traded goods were slaves. Lyon estimates for the transit trade through Murzuq in 1819 that 5000–5500 slaves passed through on the route from Bornu.[69] Denham and Clapperton give a strong insight into what it meant to travel along the Bornu Road: "*The horrid consequences of the slave trade were strongly brought to our mind; and, although its horrors are not equal to those of the European trade, still they are sufficient to call up every sympathy, and rouse up every spark of humanity. They are dragged over deserts, water often fails, and provisions scarcely provided for the long and dreary journey. The Moors* [The term "Moors" has been used in Europe in a broader sense to refer to anyone of Arab or African descent.] *ascribe the numbers to the cruelty of the Tibboo traders: there is, perhaps, too much truth in the accusation. Every few miles a skeleton was seen through the whole day; some were partially covered with sand, others with only a small mound, formed by the wind: one hand often lay under the head, and frequently both, as if in the act of compressing the head. The skin and membranous substance all shrivel up, and dry, from the state of the air: the thick muscular and internal parts only decay*".[70]

When wars raged among the empires of Bornu and Wadai, many merchants were caused to shift their operations to other

Fig. 5.77 Traditional objects are manufactured locally

routes across the Sahara. As a result, part of the traffic along the Bornu Road was diverted into the Wadai–Kufra–Benghazi route. Trade flowed also along the route between Kano and Ghadamis.[71] Continued raids during the 1850s rendered the Bornu–Fezzan–Tripoli route increasingly unsafe. Shortly beforehand, in 1849, 1600 slaves were transported along this route. Finally, the route was completely blocked between April 1851 and June 1852. Despite greater dangers from insecure political conditions between Bornu and Murzuq, the caravan traffic revived, but could not maintain the former level. Consequently, the volume of northbound exports from Bornu by the 1860s may have been only about one-third of that during the peak periods in the first half of the century. According to Rohlfs and Nachtigal, 5000 to 8000 slaves had formerly passed through Fezzan each year, but by 1869 the trade had been cut to one-third of this figure.[72]

The large trade caravans, which could comprise from 1000 up to 2000 camels or even more, did not stop directly at the oases. Al Qatrun (see *Al Qatrun-A Pearl in the Midst of Sand*), which was surrounded by sand hills, and mounds of earth covered with a small tree, called *athel* as reported by Denham and Clapperton, accommodated about 1500 inhabitants in the nineteenth century. It was once significant as a caravan stopover point on the famous Bornu Road. The caravans stopped at wells that were about 20 km away from the oasis. To the north and to the south of al Qatrun, still today two wells (Fig. 5.75) can be identified that are each about 20 km away from the oasis. The merchandise, which the merchants were convinced they could offload, was brought to the market inside the oasis. It was mostly textiles and paper that came from Tripoli by caravan. Leather and wooden bowls were transported from the south. Interior caravans came to al Qatrun also until the 1950s. Usually in the twentieth century, cereals from the north of the country were exchanged for dates. Money was used as a method of payment only by the wealthy people in the area.

As well as the caravan trade the property relationships at the oasis were organised too. As documents from the

Fig. 5.78 Desert castles (qsur) along the way to Tajarhi bear silent witness of times long past. One cannot but be fascinated both by their beauty and by the mysticism of times, lagging behind some hundreds of years

nineteenth century attest, the land of the palm groves, that was grouped around the oasis, belonged to a number of families. However, it could occur that a family other than the landowner owned the trees that were standing on it (see *Traders, Nomads and Slaves: Al Qatrun—Trading Post on the Bornu Route*). The fields that the inhabitants of al Qatrun, the so-called Qatruni, planted with barley, vegetables and forage plants were to be found in the wider surroundings of the oasis. For this reason, from February to the beginning of the summer, the men lived outside of town and only returned with the harvest.[73]

For a long time, al Qatrun resisted against a takeover by other powers. It was only in the nineteenth century that the Ottomans managed to penetrate into this region. The Ottoman governor lived in a fort located in the old town. At the beginning of the 1930s, the Italians invaded the oasis and changed the existing settlement patterns. The old town, which incorporated 180 houses, was partially cleared away and used as a quarry in order to build the Italian castle which today is still located on one of the town's hills. Finally, the medina was left behind in 1977 by the inhabitants in favour of convenient houses. However, they intended to maintain their oasis. With loving care they built a museum that presents written documents from the eighteenth and nineteenth centuries. The pieces of writing, that are displayed in the museum, provide local literature related to daily life. For example, some documents show guidelines for the decision of land or inheritance or the movements of constellations. Others provide information about traditional medicine. They describe curative medical knowledge passed down from generation to generation (Fig. 5.80) (see *Traders, Nomads and Slaves: Al Qatrun-Trading Post on the Bornu Route*) or suggest what one should eat on particular days.[74] The old town also accommodates an exhibition of objects for daily use that have been produced and used in al Qatrun. The objects are made primarily from palm branches and leather (Figs. 5.76 and 5.77). There are also pots and bowls. All the traditional products were manufactured by the women of al

Fig. 5.79 Mud brick remains of the castle-like complexes, with walls still standing up several metres high

Qatrun. Before the revolution, it was hoped that al Qatrun may participate in the upcoming tourism industry in Libya. Thus, the Italian castle had already been adapted for tourism purposes by a private initiative.

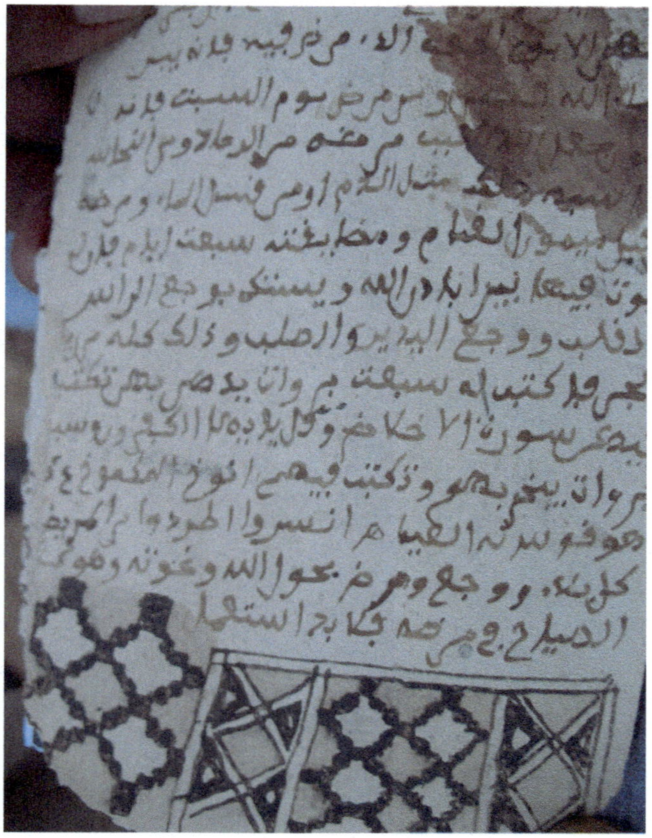

Fig. 5.80 'Folk medicine': The text says: A person came to me on Saturday to write him a prescription which consists of ashes and blood mixed with water to be anointed on his hands, face and anus and not to be removed for 7 days. He should use olibanum for a week and by God's will the pain will vanish

As in all of the Fezzan oases, it was for a long time sufficient to supply water from the flatly stored groundwater horizons from depths of up to 80 m with the help of various traditional techniques such as lever wells and drawing wells. As a result of population growth in more recent times, motor pumps have replaced the old techniques. Today, the groundwater on the water-carrying layers is not formed quickly enough anymore or is completely used up, which makes it necessary to dig a deep well of a depth of up to 400 m. The inhabitants of the desert complain that in the past 15 years it has become increasingly difficult to obtain water.[75]

Similar to the Wadi al Ajal (see *From Murzuq to Ghat*), the so-called *qsur* (desert castles; singular *qser*), fortresses built from clay bricks, line not only the way from al Qatrun to Tajarhi (Figs. 5.78 and 5.79), but can be also found along the Zawilah-Murzuq-Barjuj depression. The conception and the facilities are reminiscent of antique limes period forts. Facilities of a similar conception are known of in Egypt and the Near East (Fig. 5.80).[76]

Fig. 5.81 Murzuq: The Quickbird image from April 2006 shows the remains of an Islamic fortified villag Ⓖ, which is situated on the edge of a salina, and two desert castles Ⓔ, Ⓕ. The remains are located in the south-east of Murzuq (see map ⑭)

The earliest examples in the Fezzan date back to the Garamantian time, while others date from the Middle Ages or the Ottoman/Karamanli era. Systematic analysis conducted by the British archaeologists Mattingly and Sterry and their team has revealed the standing remains of hundreds of *qsur*, settlements, cemeteries and field systems along the Zawilah-Murzuq-Barjuj depression that date from the Garamantian to the early modern period. Imported Roman ceramics from the first few centuries AD were found on many of the sites.[77]

Along with the *qsur*, two nucleated Islamic fortified villages (*medina*) with citadels (*qasbah*) and mosques were identified south of Murzuq. One example can be seen in the Quickbird image (Fig. 5.81). The *medina*, which employs a mud brick construction (using salt-enriched mud lumps), is located adjacent to a dried-up lake (salina). According to the British archaeologists, it seems to date back to the sixteenth or seventeenth century and can be associated with the Banu Khurman and later on with the expansion of the Awlad Muhammad dynasty (see below), who tried to control Saharan trade in this area from the mid-sixteenth to the end of the eighteenth centuries. Two desert castles, which are situated north-west of the fortified village, can be detected in the Quickbird image, too (Fig. 5.81). The *qser* listed as Ⓔ was dated to the early fourteenth or fifteenth century. For the other desert castle Ⓕ there is no date as to how old this site is. The vast majority of the desert castles, situated in the Murzuq area, date to the late Garamantian period. It appears likely that the *qser* Ⓔ seems to be constructed in a similar period to that of the *medina* nearby and it appears to have been in use to at least the Ottoman period. In addition, a Garamantian origin cannot be ruled out due to the presence of imported Roman ceramic. Once established as an architectural form, these castles had a long lifespan. The British archaeologists argue that the location of these Islamic fortified villages on the edges of salinas and away from the field systems of the Garamantian *qsur* is evidence, that by this point many earlier settlements and agricultural infrastructure had fallen into disuse.[78]

The castles (*qsur*) had already been noted by Leo Africanus in the sixteenth century, who commented on this "*region, with great store of castles (…) inhabited by rich people*". He did not explain the function of these castles. For the moment, it appears that many of the desert castles seem to date back to Garamantian time. Later on, they were still in use. As suggested by Ayyub, the caravans paid passage tolls, for instance, at the numerous castles standing in the outskirts of Murzuq, in the district called Umm al-Hamam. The Libyan historian Ayyub also notes "*according to current tradition, these castles were the property of the Banu Khurman*".[79] When the domination of Fezzan by the Kaninimi kingdom ended, the sources mention a period of rule by the Banu Khurman. This was a dynasty who controlled parts of the Wadi al Ajal and the Murzuq region from around 1500 to 1550. Mattingly identified them as the descendants of the Garamantes, who had their principle centre at Jarmah. During a short period of time, they established their capital once again in Zawilah (see *From Zillah to Murzuq*). However, their attempt to consolidate their rule could have been marked by the construction of new fortified villages.[80] Ayyub also noticed that the desert castles were constantly at loggerheads with each other. "*The acumen of Muhammad al-Fasi* [founder of the Awlad Muhammad] *led him to use his goods and the armed guards of his caravan to help one of the Khurman amirs against another, until he made away with all of them in the district and ruled over it himself. He erected a new castle for himself at Murzuq (…). Then he extended his rule to Jarma, and united the entire Fezzan under his banner, setting himself up as Sultan of the Fezzan about 1550*". With the establishment of the new Awlad Muhammad dynasty at Murzuq, trade, the pilgrim traffic and the slave trade revived and expanded as never before.[81]

Along the route between al Qatrun and Tajarhi, which is part of the so-called Bornu Road, desert castles line the way to the south. Thus far, the lead archaeologist has recorded 27 desert castles in the immediate vicinity of al Qatrun alone.

What becomes evident after even a brief initial field survey is that the desert castles are situated at transnational caravan routes and water sources; they show spatial relationships. The castles were built within sight of each other, which suggest a strategic placement of the *qsur*. In an arid landscape, water means power and to control water is to control movement on the landscape. The caravans stopped at oases or meeting points to take on water and food. However, they also seem to have stopped at the numerous desert castles to take on water and food and to pay tolls.

There are many unanswered questions, but what emerges from the archaeological evidence is that on the one hand, the history of the Garamantes has to be rewritten, and this ancient people will probably have a more prominent place in the history of Libya.[82] On the other hand, it becomes evident that trans-Saharan trade and many of the oases settlements in Fezzan existed prior to Islam. Yet, there is much to suggest that intercontinental long-distance trade on these trans-Saharan trade routes went on from the fifth century BC[83] until the end of the nineteenth century—with interruptions at different times. The *qsur* were carefully situated at water sources and in the landscape to monitor important trade routes.

Investigations of the contemporary and cultural environment are necessary in order to make more definitive statements.

5.11 From Benghazi to al Kufra via Jalu

Meike Meerpohl

The map "Benghazi-Jalu" and "Kufra Oases" show parts of the Eastern connection of the caravan routes from Benghazi to al Kufra as well as their further course to the south. The cross-link roads to Jalu are also shown.

Trade links between Wadai and Benghazi started at the beginning of the nineteenth century. Wadai was founded in the seventeenth century as a sultanate by Abd al-Karim ibn Djame. The Wadai sultan controlled a vast area but was positioned between the two more powerful kingdoms of Darfur and Kanem-Bornu, located east and west of Wadai, respectively. Through flourishing economic ties with the exchange of slaves, ivory, ostrich feathers and gold against goods like sugar, clothes, tea and weapons, the Wadai neighbours had gained influence and wealth and were able to expand their power and dominate trade in the wider region. Hence, the Wadai sultan Abdel Karim Sabun ibn Saleh (sultan from 1803 to 1813) tried to establish trade links with the north in order to compete with the neighbouring kingdoms.[84] In 1809/1810, a trader called Schehaymah from the tribe of the Majabra who inhabited the oasis of Jalu, succeeded in finding a route to Warra, the capital of the Wadai, where he encountered the sultan, who was very interested in establishing new trade ties to the north. The contact between the sultan and the Majabra trader led to the formation of trade links between Wadai and Benghazi. Up to this time, trade between sub-Saharan regions and the Mediterranean Sea was carried out by *gallaba*, traders hailing from the Nile, who conducted trade mainly on the routes between Kuka, west of Lake Chad and Tripoli, between Kuka, Fezzan and Benghazi and between Benghazi, Murzuq and Warra. Wadai therefore only had indirect trade links to the north until this time.[85] After trade agreements were reached between the northern trader and the sultan of Wadai, members of the Bideyat were appointed as leaders of the first caravans, because the trade route mainly led through their homeland.[86] Later, many Majabra guides led various caravans through the desert.[87] Between 1820 and 1835, several ethnic conflicts caused numerous trade interruptions, since raids on caravans were profitable (Figs. 5.82 and 5.83).

In 1835, the Wadai sultan Mohammed al-Sharif ibn Sabun went on pilgrimage to Mecca. There he met Mohammed ibn al-Sanuss, who later founded the brotherhood of the Sanussiya. This contact was of great importance for the trade links between Wadai and the Mediterranean region during the following years. From 1836, trans-Saharan trade re-emerged, and every two to three years, a caravan with about 200–300 camel loads of ivory, skins and slaves reached the port of Benghazi.[88] Due to internal conflicts in the Wadai Sultanate, however, the revived trade links were again disrupted by raids on caravans. North of Wadai, the Zuwaya, who had occupied al Kufra since 1840, also raided caravans frequently[89] and the Awlad Sulayman, a nomadic tribe from the Fezzan, disrupted trade activities on the southern parts of this route from time to time as well.[90]

Fig. 5.82 A camel caravan on the way to al Kufrah. At the end of the day, hay is distributed among the camels. Where there is no pastureland, the camels need to be supplied by the herdsmen

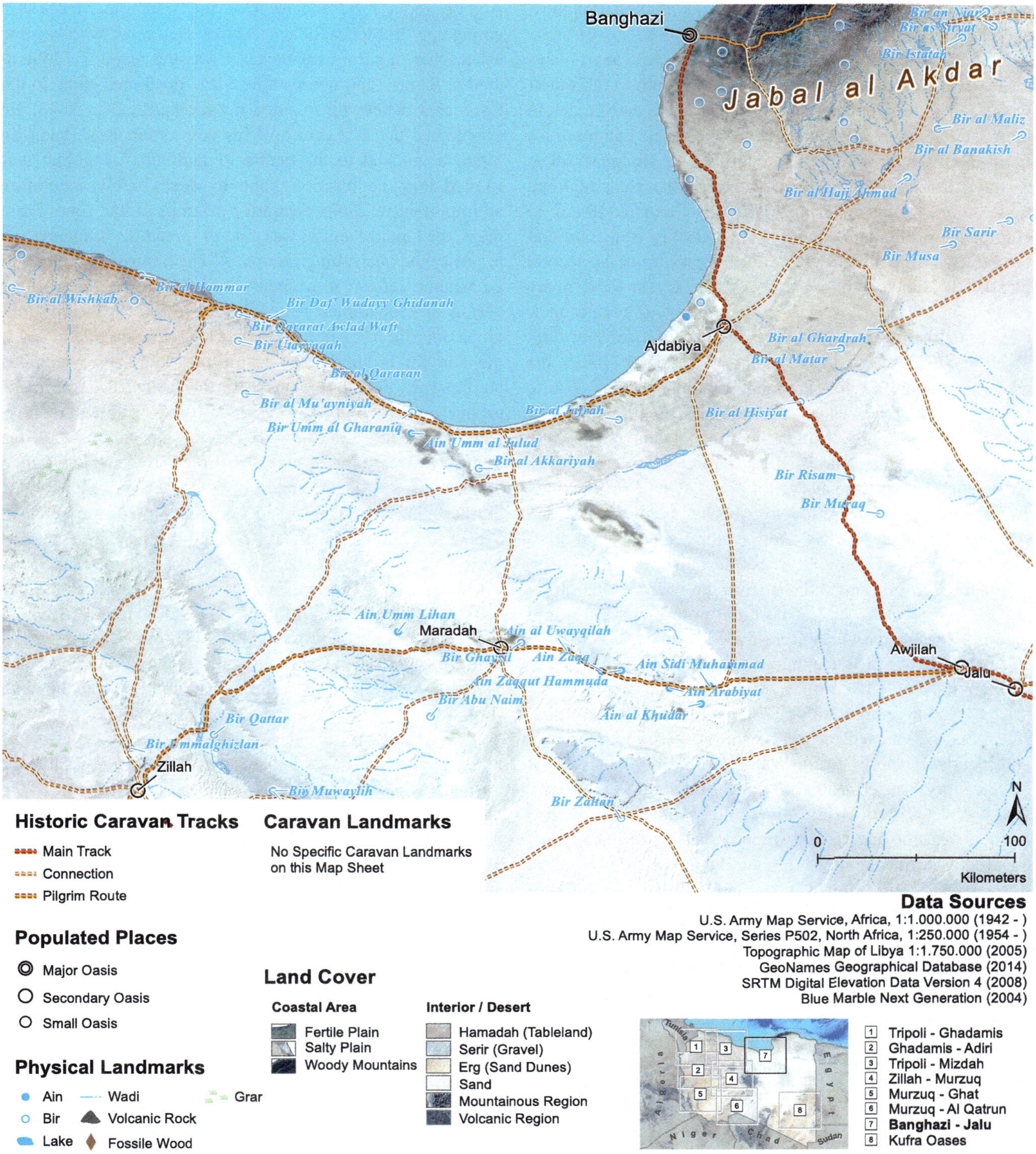

Fig. 5.83 Map: Benghazi–Jalu

In 1858, the Wadai sultan Mohammed al-Sharif died and his son Ali Mohammed was appointed to be the new sultan. During his rule, the Benghazi–Wadai route was revived as an important trade route, because the sultan invited several foreign traders to carry on trade in his country. His marriage to a *gallaba* woman strengthened his relations to influential *gallaba* traders.[91] The security on the routes improved after the raids by the Tubu and Awlad Sulayman decreased.

Between 1893 and 1894, seventeen caravans left Benghazi heading to Wadai.

On the way from Benghazi to Wadai, caravans always rested in al Kufra for some weeks to one month to refill their water and food supplies or to change camels and guides in the oasis. Some of the traders exchanged their commodities in al Kufra and returned straightway to Benghazi. The Zuwaya in al Kufra became agents of the trans-regional trade, their camels and their expertise enabled them to control the profitable economy. The Majabra from Jalu and the Zuwaya from al Kufra controlled the routes in the second half of the nineteenth century. As guides, they led many caravans to Wadai, worked as carriers and rented their camels to merchants.[92] From al Kufra, the caravans took about 40–70 days for the way to reach Warra or Nimro, the economic centre of the Wadaian capital north of Warra, which was later moved to Abeche. Caravans took a route from Benghazi to al Kufra via the oases of Jalu and Zighan. They took a route via Bishara, Sarra, Tekro, Ounianga Kebir, the Bedadi-Bideyat well, Funun, Wayta Serir, Wayta Kebir, Um Chaluba, Arada to Abeche.[93] After arrival in Abéché, the sultan received the travelling vendors first so as to offer them his commodities before they were allowed to negotiate with other merchants from the south. Goods from Wadai were usually sold at northern ports for the fourfold value.[94] The goods which were traded on the trans-Saharan trade route between Wadai and Benghazi were similar to the commodities of other routes. The imports primarily comprised weapons, ammunition, cotton, Muslim clothes, silk, sugar loaves, tea, coffee, drugs, spices, perfume, jewellery and pearls. Exports from Wadai primarily consisted of slaves, ivory, ostrich feathers, pepper and animal skin. Ivory was a coveted item in the nineteenth century, and ostrich feathers became increasingly popular after they were discovered as a prestige object among the female population in Europe after 1870. However, the largest trading volume of exports was the slave trade. From 1890 onwards, the route from Kanem-Bornu to Tripoli, west of the Wadai–Benghazi route, started to become less important. Also, to the east, in Sudan, al-Mahdi closed the *darb al-arbain* from Darfur to Asyut in 1885, and trade was diverted to the Wadai–Benghazi route,[95] which became the most important route of the region at the end of the nineteenth century. This region was far away from seaports, so that camels were still the cheapest means of transportation.[96]

The participation of the Zuwaya in the trade business and the increasing importance of the trade route at the end of the nineteenth century were closely linked to the brotherhood of the Sanussiya, who had settled in this region and who had reached their climax of power during that time.[97] The founder of the Sanussiya was Mohammed al-Sanussi al-Khattabi al-Hasani, later also called "Grand Sanussi" (al-Kebir). He was born in the region of Mostaganem in Algeria about 1787 and went on pilgrimage to Mecca at a young age. In Mecca, he met Ahmad Ben Idris, who became his teacher, a strong relationship was the result.[98] Mohammed al-Sanussi left Mecca after Idris' death and came to the Cyrenaica, where he founded his brotherhood in 1843. He established the first centre of his new order in al-Baida. In 1856, he moved to the centre to Jaghbub.[99] Jaghbub was located along the pilgrims' route to Mecca and was therefore an important religious centre for pilgrims at that time. From there, the Grand Sanussi was able to spread his influence and to pacify the quarrelling groups.[100] The Sanussi doctrine was an Islamic reform movement, modelled after the glorified period of Mohammed and designed to reform Islamic religious doctrine.[101] After the end of the Ottoman rule in 1911, the Sanussi brotherhood became the dominating religious confraternity in Libya.[102] Due to the fact that the nomadic groups, which controlled the trade routes in the Cyrenaica, were affiliated with the brotherhood, the leaders of the Sanussiya were able to enter into trade. Because the Grand Sanussi as a student met the Wadai Prince Mohammed al-Sherif in Mecca, who had been appointed the sultan of Wadai in 1838, the relation to Wadai was close. The contacts between them revived because they shared the same religious belief.[103] In order to strengthen their trade links, Sayyid Mohammed al-Mahdi moved his religious centre from Jaghbub to al Kufra in 1895. Al Kufra was strategically better located on the caravan route between Benghazi and Wadai, because trade was difficult to control from Jaghbub (Fig. 5.84).[104]

Before al Kufra became the centre of the Sanussiya, it was just a remote watering place that was visited by caravans on their way through the desert. With the institution of the Sanussiya, al Kufra became an important trading centre, where traders exchanged different goods and where several trade routes crossed.[105] In order to facilitate travelling from al Kufra to the south, the Sanussi built wells in Sarra and in Bishara in 1898 to shorten the waterless treks for caravans.[106] In 1902, the Sanussiya reached the climax of their influence, having established 147 *zawiya*.[107] *Zawiyas* were religious centres installed as focal points for the converted members to join the Muslim fraternity and to spread religious ideas throughout the region of Cyrenaica and beyond.[108] The different *zawiya* had been strategically established at crossroads and along the trade route to Wadai in order to serve as trade centres. For traders, it was advisable to affiliate with the Sanussiya to use their services,[109] because they were as a result released from paying tolls and gave tea, sugar, soap, candles or fabrics in return for their defence and protection and the hospitality in a *zawiya*.

The brotherhood of the Sanussiya made an important contribution to the continuance of the trans-Saharan trade routes and to the security on the routes. With the spreading of Sanussiya's doctrine, it was possible to solve problems

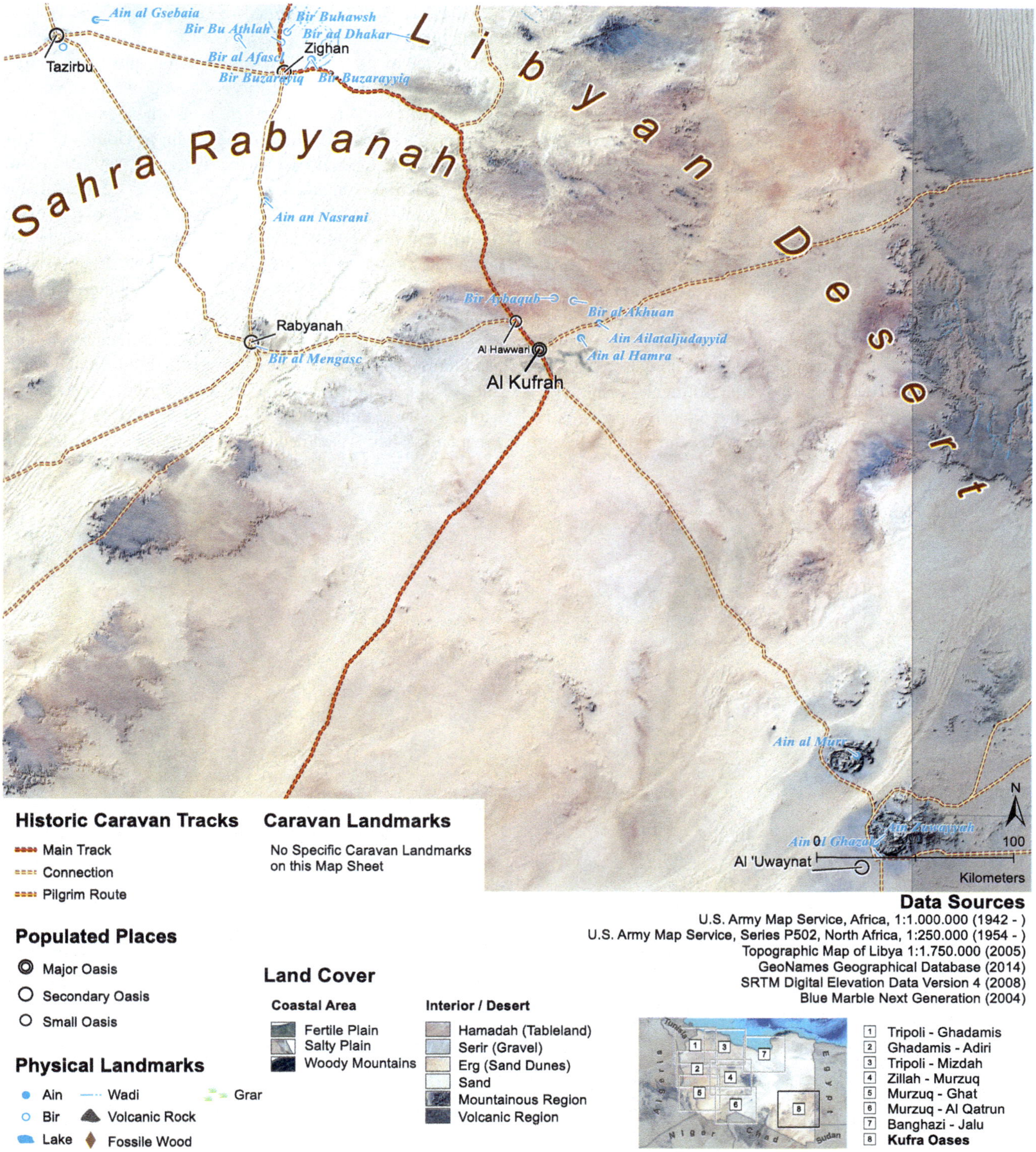

Fig. 5.84 Map: Kufra Oases

between different groups as well as to enable trade activities between them. The affiliation of the different groups into a brotherhood simplified trade activities.[110] The Sanussiya further transported commodities between north and south. They built trust between traders and agents. Groups of traders as well as retail dealers used the route between Wadai and the northern regions and made use of the brotherhood's current information regarding supply and demand. The

Sanussiya offered rules for loan systems, networks and business relations. The established *zawiyas* along the route were linked through postal systems in the late nineteenth century and enabled the flow of information and guaranteed the exchange of goods.[111] Many traders also filled trade orders from the Sanussi and brought ivory, ostrich feathers and slaves back from Wadai, which were exchanged for other goods in Benghazi or Cairo.[112] To sustain the relations with Wadai, the Sanussi sent the sultan different gifts, including fabrics, tea and sugar from the north.[113] From that time onwards, the various sultans of Wadai also sent caravans to the Sanussi leaders. Even when the Sanussiya did not control the region of Wadai, they were very influential there. From 1897, Mohammed al-Sunni was a representative of the Sanussi alongside the sultan in Abeche, who was installed to monitor trade and who was a Wadai political consultant.[114] With the help of the Wadai sultan, the Sanussi representative tried to unite the Libyan traders in Abeche against foreign intruders. He further appointed the sultan in Wadai as head of the brotherhood (Fig. 5.85).

In order to improve their organisation and their control over the different groups and to counteract the French, who had started to invade southern regions, the Sanussi moved their religious centre further to the south, to Gouro (or Qiru, Quru, Gouro), located east of the Tibesti in 1899.[115] However, al Kufra remained the main trade centre of the trans-Saharan route. At al Kufra's market, traders bargained on different commodities. In 1907, in the port of Benghazi, about 240,000 lb in import goods and about 304,000 lb in export goods were traded. Goods from southern regions covered about one-fifth to one-third of this trading volume.

Trans-Saharan trade between Wadai Sultanate and the Mediterranean coast had been very profitable for many years due to the extensive slave trade and the wide-ranging exchange of commodities. This changed in 1909 when the French took over Wadai and when the Italians extended their

Fig. 5.85 Once a remote watering place, al Kufra became the centre of the Sanussiya brotherhood. Today it is a modern settlement, where desert is made into fertile ground. Until 2011 a large demand for camel meat in Libya formed a stable foundation for new trade links

invasion of Libya between 1912 and 1914. After the French had started to invade Wadai, the Sanussi leader redirected his trade route to al Fasher and to Khartoum.[116] However, the era for caravans related to the Sanussiya and to the sultans of Wadai was over once the French occupied and finally controlled Wadai at the beginning of the twentieth century, and once the Italians invaded Libya and the Sanussiya reacted to Italy's encroachment.[117] This stopped all traffic on the trade route between Wadai and the Sanussi. In the following years, the colonial powers outlawed the slave trade, defeated and subjected the different ethnic groups, destroyed their traditional structures, drew borders and transformed the different autonomous sultanates into nation states. Years of instability followed that resulted in civil and tribal conflicts within the countries and between the states. Some of the tensions and hostilities exist to this day. The end of the trans-Saharan Wadai–Benghazi route was also signalled by a drought in Wadai between 1912 and 1914, which caused far-reaching migration from that region. That in return resulted in trade business becoming only locally focused.

With the establishment of Fort-Lamy (present-day N'Djamena) as the Chadian national capital by the French colonial power in 1900, the eastern economic centre lost its importance.[118] In the west, trade fell into the hands of the colonial power, and in the north, Libyan traders established trade links between the oases of al Kufra and Faya, while local groups were unable to participate in trade interactions. When Chad gained its independence in 1960, many of the Libyan traders left the northern regions of Chad[119] until the 1970s, when the commercial relations between the two countries broke down.[120] The Chadian civil war (1966–1978) and the Chadian–Libyan war (1978–1987) cut off all trade links, and conflicts, wars, crises and instability in the region made commerce between Chad and Libya impossible for many years.

Only in the 1990s, trade links between Libya and Chad re-emerged, mainly between al Kufra and the sub-Saharan zones of Chad.[121] As a result of adaptation to climate changes and ecological pressures, the semi-nomadic Zaghawa pastoralists managed to diversify their economic activities by entering the trade business and making use of family networks that resulted from broad migration across the region. The Zaghawa used remittances from migrated family members to restock their herds with camels. Consequently, a large camel population, a new focus on trade and a large demand for camel meat in Libya formed a stable foundation for new trade links. Additional technical innovations in the trans-Saharan endeavour, such as trucks functioning as mobile oases and the use of satellite communication, enabled a safer, faster and shorter desert crossing, leading to even more profitable trade in the last years,[122] at least until the series of popular revolts sparked in the Arab world in December 2010. The outbreak of the revolt in Libya in February 2011, that ended with the fall of Gaddafi's regime, reintroduced instability to the region and once more disrupted the trade links between the two countries.

Notes

1. Richardson (1848, 40).
2. Simon (2012)
3. McLachlan (2010), Roumani (2012), Simon (2012).
4. Richardson (1848, 77f).
5. Richardson (1848, 78f).
6. Rohlfs (1874).
7. Orhonlu (2010, 1955–2005), Rohlfs (1874).
8. Boahen (1961, 352).
9. Duveyrier (1864), Orhonlu (2012).
10. Richardson (1848, 395).
11. Richardson (1848, 404).
12. Richardson (1848, 386).
13. von Bary (1977), Duveyrier (1864), Krause (1882, 266–356), see mainly 328–335; Richardson (1848).
14. Duveyrier 1864.
15. Nachtigal (1879, 38–71), see mainly 47.
16. Barth (1857, 45).
17. Barth (1857, 45).
18. Barth (1857, 50f), Lyon (1821, 25f).
19. Passon (2009, 146)
20. Roumani (2012).
21. Barth (1857, 50).
22. Roumani (2012).
23. Lyon (1821, 25f)
24. Barth (1857, 134–140).
25. Barth (1857, 134).
26. Barth (1857, 135).
27. Barth (1857, 132f); see also Lyon (1821), Nachtigal (1879), Rohlfs (1871, 32), Rohlfs (1874, 21).
28. Barth (1857, 140).
29. Interviews with elders from al Fuqaha and al Qatrun conducted in November 2006.
30. Katz (2010).
31. Barth (1857, 148).
32. von Beurmann (1862/63) in Petermanns Mitt eilungen Vol. II, 68–78, Hornemann (1802).
33. Liverani (2000, 496–517).
34. Richardson (1848, 392).
35. al Idrisi in Hopkins and Levtzion (1981, 129).
36. von Beurmann (1862/63, 68–78).
37. Richardson (1848, 392).
38. von Beurmann (1862/63, 68–78).
39. Interviews with elders from al Fuqah, conducted in November 2006.
40. Richardson (1848, 396).

41. von Beurmann (1862/63, 68–78), Hornemann (1802, 66f).
42. Al Bakri (1913, 31).
43. von Beurmann (1862/63, 68–78), Hornemann (1802, 68–70).
44. Mattingly et al. (2015, 56–58).
45. Martin (1969, 15–27), Savage (1992, 351–368), Wright (1989, 35–54).
46. al Bakri in Hopkins and Levtzion (1981, 63f).
47. al Idrisi in Hopkins and Levtzion (1981, 130).
48. Despois (2010), Lewicki (1964, 296), Martin (1969, 15–27), Savage (1992, 351–368), Vikør (2010), Wright (1989, 35–54).
49. Hornemann (1802, 56), Mattingly et al. (2015, 36).
50. al Yaqubi in Hopkins & Levtzion 1981, 22.
51. al Istakhri in Hopkins & Levtzion 1981, 41.
52. Anonymous in Hopkins & Levtzion 1981, 138.
53. al Bakri in Hopkins & Levtzion 1981, 63f.
54. al Yaqubi in Hopkins and Levtzion (1981, 22).
55. Passon (2008, 145–152, 217f).
56. von Bary (1977), Krause (1882, 333), Nachtigal (1879), Westphal (2002).
57. Klenkler (2005, 110–127).
58. Mattingly (2000), Sterry et al. (2011).
59. Tacitus Histories 4.50.
60. Herodotus, Histories 1.183.
61. Pliny the Elder, Natural History 5.43–46.
62. Watts-Plumpkin (2005).
63. Watts-Plumpkin (2005), Mattingly (2011), Sterry et al. (2011).
64. Mattingly (2011).
65. Mattingly (2011).
66. Boahen (1962, 349–359), Liverani (2000, 496–517).
67. Martin (1969, 15–27).
68. Boahen (1962, 351), Denham et al. (1828, 249f), Fisher (2010).
69. Boahen (1962, 351–353), Lyon (1821, 188f), Lovejoy (1984, 85–116), Martin (1969, 15–27).
70. Denham et al. (1828, 124f).
71. Boahen (1962, 351–353), Lovejoy (1984, 85–116), see mainly 86–89; Martin (1969, 15–27).
72. Lovejoy 1984, 89-93; Nachtigal; Rohlfs.
73. Interviews with elders from al Qatrun conducted in November 2006 and documents from the museum of al Qatrun's old town.
74. Interviews with elders from al Qatrun conducted in November 2006; Passon (2008, 225f).
75. Interviews with elders from al Qatrun, Tajarhi, and with responsibilities from Murzuq conducted in November 2006.
76. Alhasanat et al. (2012, 343–359), Ruprechtsberger (1997, 51–69 and 77–81), Franz (1984).
77. Sterry et al. (2011, 103–116).
78. Sterry et al. (2012, 137–147).
79. Ayyub (1993), Willis (2005, 62).
80. Sterry et al. (2012, 137–147).
81. Ayyub (1993), Willis (2005, 62).
82. Mattingly (2011).
83. For trade and trade routes in ancient times; see: Mattingly (2011, 49–60), Liverani (2001, 2003); Sommer (2011, 61–63).
84. Ciammaichella (1987, 43).
85. Ciammaichella (1987, 36).
86. Cordell (1977, 22).
87. Cordell (1977, 27).
88. Cordell (1977, 23).
89. Rahma (1999, 49).
90. Cordell (1977, 24).
91. Cordell (1977, 24).
92. Ciammaichella (1987, 36).
93. Harding King (1913, 277).
94. Cordell (1977, 27).
95. Ciammaichella (1987, 31).
96. Cordell (1977).
97. Cordell (1977, 28).
98. Vikør and O'Fahey (1987, 72), Ziadeh (1958, 45).
99. Jennings (1993, 18).
100. Hassanein Bey (1925, 61).
101. Mattes (2007, 14).
102. Allaghi (1981, 100).
103. Ziadeh (1958, 50).
104. Evans-Pritchard (1949, 21).
105. Evans-Pritchard (1949, 21).
106. Hassanein Bey (1925, 64), Bagnold (1933, 1936, 299).
107. Evans-Pritchard (1945, 183).
108. Vandewalle (2006, 19).
109. Cordell (1977, 33).
110. Cordell (1977, 31).
111. Cordell (1977, 31).
112. Djian (1991, 126).
113. Djian (1991, 136).
114. Ciammaichella (1987, 45).
115. Rahma (1999, 47).
116. Cordell (1977, 32).
117. Cordell (1977, 33).
118. Arditi (1993, 184).
119. Arditi (2005, 24).
120. Arditi (1995, 851).
121. Bennafla (2004, 91).
122. Meerpohl (2007).

6 Places of Trade, Communication and Religion—Important Oases

Jacqueline Passon, Klaus Braun, Said Hamid, Salih al-Mahdi Khalifa, Najmiya as-Sadeq at-Tellisi, Mansour El Nayedh, and Joschua Metzger

6.1 Tripoli—Mediterranean Port and Trading Centre

Jacqueline Passon and Klaus Braun

Tripoli is located in the northwestern part of Libya on the edge of the desert, on a point of rocky land that projects into the Mediterranean and forms a bay. The city is known as Tarabulus al-Gharb (West) to distinguish it from its Phoenician sister city Tripoli in Lebanon which is known in Arabic as Tarabulus al-Sham.

6.1.1 On the Lee Side of a Cape

The African coast between Lesser Syrtis and Greater Syrtis is almost flat and sandy everywhere. In the eighth century BC—probably even earlier—Phoenician merchants began to establish commercial settlements on the North African coast which were used seasonally. The merchants established their commercial stations where their ships could safely anchor on the lee side of a cape, where trade routes from the south reached the coast and where there was also a fertile hinterland. Probably, they called one of these stations Macar Uiat. In the course of time, this trading centre developed into a city that should become known as Oea/Tripoli later on. The expansion of the settlement might have been connected with the rise of Carthage at the end of the sixth century BC when Carthage started to expand its influence in the Mediterranean and founded several colonies. Among these colonies are also the two sister cities of Leptis Magna and Sabratha. After the fall of Carthage, the Romans prepared themselves to subdue the Mediterranean countries in the second century BC. At first, the *emporia* (the Carthaginian trading towns were known as *emporia*)—which were completely integrated in the Mediterranean trade dominated by Rome—were left to the Numidians. Their subjugation was followed by the integration into the administrative structure of the Roman Empire. Henceforth, Oea belonged to the Roman province of Africa Proconsularis (Fig. 6.1).

In late antiquity, Oea, Sabratha and Leptis Magna experienced once again organised urban living under Byzantine rule. But only Oea, under its new name Tripoli (Arabic: Tarabulus), was to survive the different Arab attacks. Under the command of Amr Ibn al-As, Tripolitania was conquered in the years 642/643. Leptis Magna and Sabratha continued to exist until the eleventh century as Arab strongholds. However, subsequently they were buried by sand. Unlike in Egypt or Tunisia, the urban culture introduced by the Greeks, the Phoenicians and by the Romans found a gradual end in Tripolitania and Cyrenaica. Tripoli, as the only continuously used place of settlement, was rebuilt several times. The road network of the Medina, however, reveals the planning of the Roman *colonia*—the term denotes the highest status of a Roman city—in its main lines. As one can see in the map *Tripoli at the dawn of the twentieth century*, the decumanus that leads up from the harbour is still clearly visible in the modern road network. At the triumphal Arch of Marc Aurel and Lucius Verus ㉒ (Fig. 6.6), which remained as an architectural relict of the Roman era, the northern *decumanus* cuts a parallel axis of the *cardo*.[1] In Roman city planning, a *decumanus* was an east–west-oriented road and the *cardo* was a north–south-oriented street. There is every reason to suspect that this area constituted the principal public centre of the Roman settlement. It has been a constant feature in the urban fabric of Tripoli. According to archaeological findings and to the writings of Leo Africanus, a diplomat and author of the sixteenth century, as well as to the topography of the site, the Punic and later Roman settlement might be expected to lie in the area north and

J. Passon · K. Braun (✉) · S. Hamid · S. al-Mahdi Khalifa · N. as-Sadeq at-Tellisi · M. El Nayedh · J. Metzger
Dept. of Physical Geography, University of Freiburg, Freiburg, Baden-Württemberg, Germany
e-mail: klaus.braun@geographie.uni-freiburg.de

J. Passon
e-mail: jacqueline.passon@geographie.uni-freiburg.de

K. Braun and J. Passon (eds.), *Across the Sahara*,
https://doi.org/10.1007/978-3-030-00145-2_6

Fig. 6.1 Tripoli is called the "white town by the Mediterranean Sea." The photograph shows one of the first areal photographs taken from Tripoli in 1913 (Tittoni 1913, 95)

north-west of the Roman arch. Here, the promontory on which the city was built narrows considerably, and there are protected anchorages on both sides of it. Today, the highest point is marked by the water tower. On the eastern site of the headland, the harbour facilities of ancient Oea can be found (Fig. 6.2).[2]

6.1.2 The Arab-Islamic Medina Begins to Take Shape

After the successful attack of Amr Ibn al-As in the seventh century, Tripoli became an Arab-Islamic Medina and changed its appearance. The second decumanus gained significance and its intersection with the cardo became the public centre of the city. The intersection was highlighted and architecturally defined by the so-called Arbaa Arsat (four-column crossroads) [1] (Fig. 6.3), recalling the Tetrapylon by Marcus Aurelius. On the corners of the crossing, four remains of roman columns were used. Yet another change became visible: as in many Arab-Islamic cities, the layout of most of the streets followed an irregular pattern (Fig. 6.4).[3]

We learn from Ibn Khaldun that the time after the Arab invasion in the seventh century was characterised by numerous politico-religious revolts, in which mainly the Ibadi played a central role. This sect was able to win many followers among the Hawwara and Zanata Berbers, who formed the predominant element in the population. After the end of the Umayyad dynasty (661–750), the Islamic central state disintegrated into territories that gained independence and was ruled by dynasties. Tripolitania no longer was an independent power base, but was now dependent on the development of the more powerful neighbouring regions. Extremely detrimental to urban life was the invasion of the nomadic Arab tribes of Banu Hilal and Banu Sulaym in the eleventh century. During that time, Tripolitania experienced nomadisation again. Politically, Tripoli was controlled for a short intermezzo by the Normans (1146–1158) and then since the middle of the twelfth century by the Berber dynasty

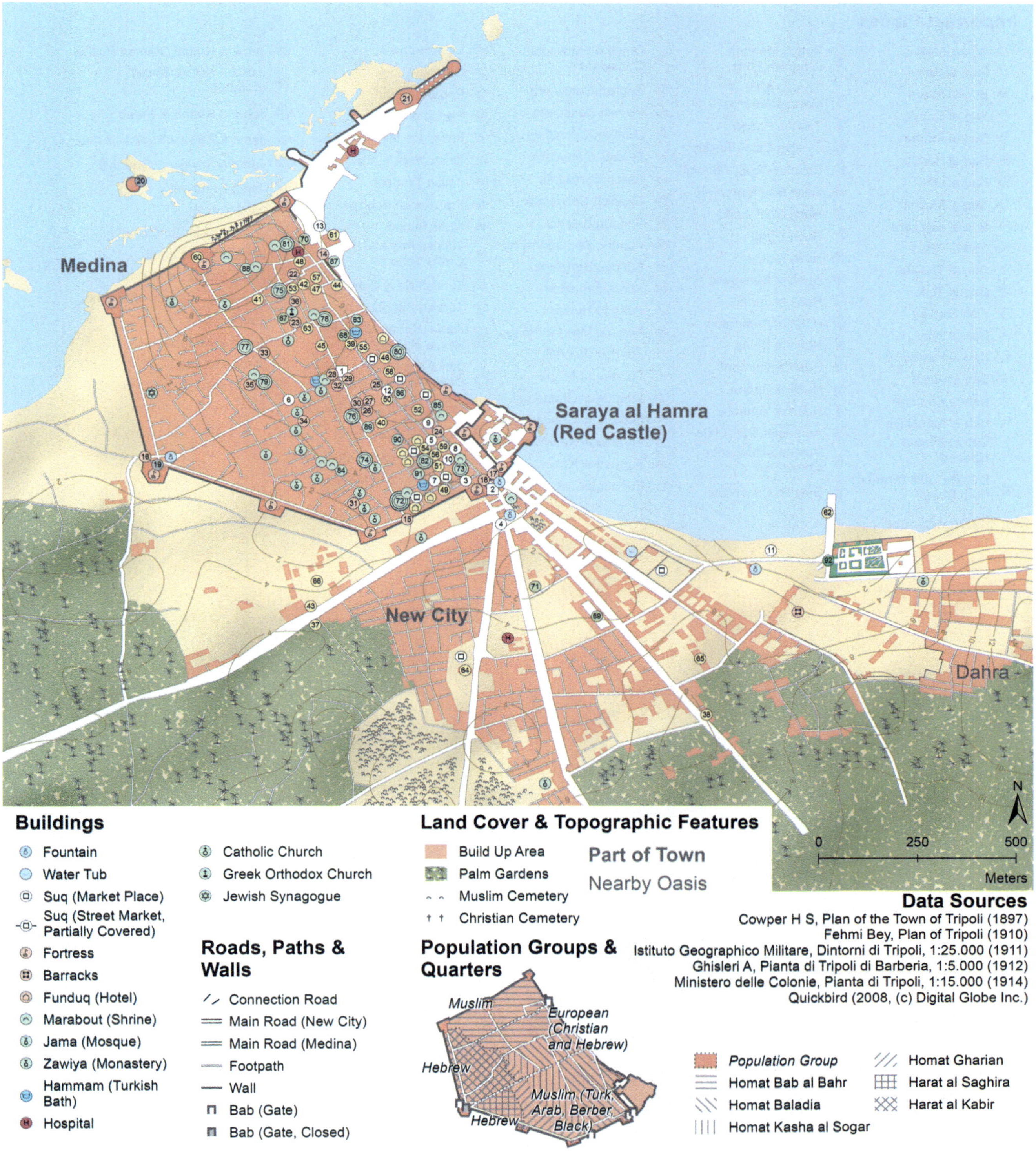

Fig. 6.2 Map of Tripoli: Tripoli at the dawn of the twentieth century

Important Places

- 1 Arbaa Arsat
- 2 Fum al Bab
- 3 Suq al Attara
- 4 Suq al Cobza
- 5 Suq al Framel
- 6 Suq al Harara
- 7 Suq al Liffa
- 8 Suq al Mushir
- 9 Suq al Qazdara
- 10 Suq al Rbaa
- 11 Suq al Thalata
- 12 Suq al Turk
- 13 Fish Market
- 14 Bab al Bahr
- 15 Bab al Hurriya
- 16 Bab al Jadid
- 17 Bab al Khandak
- 18 Bab al Manshiya
- 19 Bab al Zenata (Closed)
- 20 Burj Bu Leila (French Fort, Ruins)
- 21 Burj el Mendrik (Spanish Fort)
- 22 Roman Arch of Marcus Aurelius
- 23 Turkish Prison
- 24 Turkish Clock Tower
- 25 Baladia (Town Council)
- 26 Hush Ben Mahmud
- 27 Hush Ben Musa
- 28 Hush Gurgi
- 29 Hush Muhsin
- 30 Hush al Bashawat
- 31 Hush al Jamal
- 32 Hush al Karamanli
- 33 Hush al Khuja
- 34 Hush al Qarqani
- 35 Hush al Sabun
- 36 Hush al Yunani
- 37 American Consulate
- 38 Austrian-Hungarian Consulate
- 39 Belgish Consulate
- 40 Dutch & Norwegian Consulate
- 41 English Consulate
- 42 French Consulate
- 43 German Consulate
- 44 Greece Consulate
- 45 Italian Consulate
- 46 Spanish Consulate
- 47 Funduq Baishu
- 48 Funduq Bent al Sayyid
- 49 Funduq Benzekri
- 50 Funduq Hinshiri
- 51 Funduq Huwas
- 52 Funduq Mady al Hasan
- 53 Funduq Zummit
- 54 Funduq al Adluni
- 55 Funduq al Ghadamsi
- 56 Funduq al Ghadamsiya
- 57 Funduq al Karamanli
- 58 Funduq al Kebir
- 59 Funduq al Zahr
- 60 Light House
- 61 Custom Office
- 62 Esparto Wharf
- 63 Banco di Roma
- 64 Italian Ice Making Works
- 65 Italian Steam Mill
- 66 Italian Tannery
- 67 Franciscan School
- 68 Nuns School
- 69 Turkish Arts and Crafts School
- 70 Turkish Navy School
- 71 Turkish School
- 72 Jama, Kuttab & Zawiya Atiyya al Fallah
- 73 Jama & Kuttab Ahmad Pasha al Karamanli
- 74 Jama & Kuttab Ben Tabun
- 75 Jama & Kuttab Gurgi
- 76 Jama & Kuttab Huriyya
- 77 Jama & Kuttab Mahmud Khazandar
- 78 Jama & Kuttab Othman Pasha
- 79 Jama & Kuttab Shaikh al Hattab
- 80 Jama & Kuttab al Kateb
- 81 Jama & Zawiya Shaikh Ja'qub
- 82 Jama & Zawiya al Qadiriyah
- 83 Jama Darghout
- 84 Jama Ibn Muqil
- 85 Jama Shaikh Shanshan
- 86 Jama Shayeb al Ain
- 87 Jama Sidi Abdel Wahab
- 88 Jama Sidi Salem al Mashshat
- 89 Jama al Druj
- 90 Jama al Kharruba
- 91 Jama al Naqa
- 92 Pashas Public Gardens

Fig. 6.2 (continued)

Fig. 6.3 Four-column street, Arbaa Arsat 1 in the Medina of Tripoli

Fig. 6.4 Kasbah at the end of the Ottoman period (Laronde 1997)

of the Almohads (1158–1248). After their collapse, the Hafsids residing in Tunis managed to use their influence on Tripoli until the early sixteenth century (1248–1510).[4]

Al-Tijani, on his occasion of his travel to the holy lands in 1307–08, states in his travel accounts, *Rihlah*, that "*the streets in the city were clean, and larger than other places, extending from one end to the other of the city and intersected as a chessboard, and easy to go through.*"[5] His writings bear eloquent testimony to a flourishing place. He describes that Tripoli was administered by a Hafsid governor, living in the castle (*kasbah*), and a council of 10 notables (*shaikh*). He also observed a fine bath (*hammam*), a Great Mosque (*al-djami al-azam*), many shrines, and a madrasa (*al-madrasa al-mustansiriya*). He found strong walls in good repair, with a moat in some parts and he admired the Arch of Marcus Aurelius 22. In his mind, the city's intellectual life was flourishing at this time; cultivated people abounded.[6]

A unique building also hails from the Hafsid period (thirteenth–sixteenth century) which can be found in the Bab al Bahr area, opposite to the harbour in the immediate vicinity of the Marc Aurel arch. The mosque of Sidi Abdel Wahab al Qyasi ⑧⑦ (Fig. 6.7), which is distinguished by a unique architectural design compared to other mosques of the old town, originally served as a *ribat*. This is an Arabic expression for small fortifications that were built on the border of the Islamic sphere of influence. The construction of a fortress or the extension of an existing *ribat* was considered as pious work. The inhabitants of the fortresses were not only combat-ready soldiers, but also scholars who devoted themselves to the moral support of the troops. The military and religious character of such a building is also reflected in the architecture. Apart from accommodation and storage facilities, such buildings also had a mosque. The function of the *ribat* changed in the eleventh century. The fortifications then served to protect commercial routes, and as centres for isolated Muslim communities. The *ribat* tradition may be one of the early sources of the Sufi mystic brotherhoods, and a type of the later *zawiya*. Here, the homes of *marabouts*, who are religious teachers, are termed *ribat*. Concerning the building in Tripoli, which is indeed a small *ribat*, it is considered to be the marabout tomb of Sidi Abdel Wahab al Qyasi, who was buried at the tomb adjacent to the mosque, which is mentioned by Al-Tijani in his travel book.[7]

6.1.3 Spanish Intermezzo

In the fourteenth and fifteenth centuries, the balance of power in the Western Mediterranean shifted to the detriment of the Arabs. In 1492, the last Arab bastion on Spanish soil fell with Granada. In the course of the Reconquista, the Spanish rulers also spread to Northern Africa where they managed to gain several bases. One of these bases was Tripoli which was conquered on 25 July 1510 and managed by a Spanish governor. Pedro Navaro, a Spanish military engineer who led the Spanish forces during the conquest of Tripoli in 1510, describes the city as follows: "*This city is larger than what I thought, and although those who admired it spoke well about it, I can see that they were saying only half the truth; and among all the cities I have seen in the world, I do not find any that is comparable to it, both for its fortifications and its cleanliness. It seems rather an imperial city than a city that belongs to any king in particular.*"[8] Emperor Charles V gave the city to the Knights of St. John as fief; however, already in 1551, it was to be reconquered by the Ottomans. It is worth mentioning from this period that the Knights of St. John converted the *kasbah*. A castle of the Knights of St. John with Spanish and Turkish features was then built on the medieval fort. Its present structure dates back to the end of the fourteenth century, and it is called the *Saraya al hamra* (the red castle) because it used to be painted with an almost red colour. The castle is one of the major fortifications of the city and served as a defence centre against hostile naval attacks. Following the revolution in 1969, a museum was built in the tunnel of the castle to host Libyan antiquities and heritage throughout the ages.[9]

Leo Africanus who visited Northern Africa at the beginning of the sixteenth century (1518) identified Tripoli as an important and busy trading centre in the Mediterranean region: "*Tripoli was built by the Africans after the destruction of Old Tripoli. It has fine walls, but they are not very strong. It lies in a sandy plain where there are many stands of date-palms. The houses of this city are imposing compared with those of Tunis, and similarly the public spaces are orderly and well-provided with various trades, most notably weavers of cloth. ... Nevertheless, these people do much trade, because their city is close to Numidia and to Tunis, and because there is no other city but this one right up to Alexandria. It is also close to Sicily and Malta, and yearly visits to the port by the Venetian trading galleys are encouraged. The Venetians, as well as others who go there with the Venetian galleys, do much business with the Tripoline merchants.*"[10] But he is also reporting that Tripoli "*was ruined after its conquest by Christians. Nevertheless, they reinforced the castle with solid walls and important artillery.*" It is to presume that the town was rebuilt and the castle was transformed into a fortress with bastions.[11]

6.1.4 Tripoli Under Ottoman Rule

Since the beginning of the sixteenth century, Tripoli was the centre of power of the Turkish–Ottoman rulers, administrated by pashas and deys, followed by independent princes from the Karamanli family (1711–1835) and then again the Turkish and since 1911 by the Italian governors.

The first Ottoman period is characterised by a relative economic welfare due to important revenues coming from the trades and sea activities which were mainly connected to piracy. In particular, the al Saqizli rulers (1633–1672) were able to guarantee a period of relative peace and prosperity. It was at that time that the European states France and England signed commercial treaties with Tripoli to open up consulates inside the Medina. The French–Libyan relations were initiated during the 1630s under the reign of Mohamed al Saqizli. France sent a commercial representative to Tripoli, followed by the first French consul, who found his domicile in a building which still can be seen today. Until the outbreak of the Second World War, this house was home of the French consulate ④②, when it moved outside of the old town. The British consulate ④①, a house built in 1744 (Fig. 6.8), was situated nearby. The first British consulate was set up in Tripoli in 1658. The British consulate was used from the middle of the eighteenth century until 1940 and it used to be the starting point for geographical expeditions towards Africa. A hundred years later eight nations were represented in Tripoli including Denmark, Sweden, Holland ㊵, Tuscany, Spain ㊻ and the USA ㊲ (Fig. 6.5, 6.6, 6.7 and 6.8).[12]

6.1.5 Becoming a Multiracial and Multireligious Hub of Commerce

This first Ottoman period also evolves as an important period of construction, fortification and rehabilitation of the city. The map *Tripoli at the dawn of the twentieth century* shows the city as it was rebuilt and organised by the Ottomans at the dawn of the twentieth century. One can clearly recognise the *cardo* and the two principal *decumani* that show, along the strip parallel to the sea, a regular and wide

Fig. 6.5 View towards Babal Menshia. The menshia is the cultivated oasis which surrounded the walled city of Tripoli (Mathuisieulx 1912)

Fig. 6.6 Arch of Marcus Aurelius ㉒

Fig. 6.7 A unique building can be found in the Bab al-Bahr area opposite the harbour. The mosque of Sidi Abdelwahab al Qyasi (87), originally served as a ribat, which is an Arabic expression for small fortifications that were built on the border of the Islamic sphere of influence

Fig. 6.8 The first British consulate (41) was set up in Tripoli in 1658. The French had established diplomatic relations 45 years earlier. A hundred years later eight nations were represented, including Denmark, Sweden, the Netherlands, Spain and Tuscany

grid of streets, recalling the ancient Roman plan. These are the main arteries of the Medina. The area along these arteries became the city's political, economic and religious hub. West of these arteries, the streets are becoming curvilinear to follow progressively the polygonal geometry of the walls which represents the Arabic-Islamic Medina. In the neighbourhood of the harbour, a specific quarter, characterised by the presence of the European consulates with their warehouses, began to take form.

New markets (*aswaq*, sg. *suq*) and *fanadiq* (sg. *funduq*, hotel and warehouse) were realised.[13] With the construction of the Suq al Turk (13), the Suq al Qazdara (goldsmith's market) (9) and the Suq al Rbaa (10) the area became the densest commercial and artisan centre of the old town, integrating the already existing system of the *aswaq*. The Suq al Rbaa was built during the reign of Ahmed Pasaha al Saqizli (1649–1672). It represents the first market equipped with a roof in the city of Tripoli. The *suq*, which contains several columns of the Hafsid style, is still in existence today. Traditional costumes, hats, women's dresses and robes as well as men's traditional clothing are still sold here. Suq al Quaya and Suq al Kutub are part of al Rbaa old market. At Suq al Quaya, which is covered with a half-cylindrical vault, goods from Sudan such as ostrich feathers, leather and different kinds of dates were sold.[14]

Constantinople was far away, and therefore, a tight central control over Tripolitania was not possible. In Tripoli, a rule with strong local influence formed that was only nominally under the command of the Sultan and operated with the dreaded Corsair fleets in the Mediterranean. In 1711, the Turkish Pasha Ahmad Karamanli managed to become mainly independent, and he founded the dynasty of the Karamanli (1711–1835). The Karamanli family was of Turkish origin but related to the Arab-Berber people who formed the dominant group in Tripoli and its surroundings. Their advent with Ahmed Pasha al Karamanli (1711–1745) marked a long period of peace, which at least partially translated into a vivid urban life in Tripoli. In subsequent years, the multiracial and multireligious character of the Medina, which was inhabited by 14,000 people at that time, developed very fast. Due to the ever-present threat of piracy, new friendship agreements with European states were concluded. Regular consulates in some of the important houses of the Medina, located in the north-eastern urban area, were established.[15]

The dynasty has also left its trace in terms of urban development. The mosque of Ahmed Karamanli (73) that was built between 1737 and 1738 is considered the largest and most beautiful mosque of the old town which also has a *kuttab/madrasa* (school). The richly ornamented mosque, which was described by the traveller Abulabbas Ahmed Mohammed al Fasi during his journey in 1796, is facing the Saraya al Hamra. It is situated at Suq al Mushir (8), that is the work of Rayab Pasha, Governor of Tripoli between 1906 and 1908. Two of the main commercial and artisan centres of the old town branch out of this suq, the already mentioned Suq al Qazdara (goldsmiths' market) ⑨ and the Suq al Rbaa ⑩. Adjacent to Suq al Rbaa, Ahmed Pasha al Karamanli founded the al Rbaa new market, Suq al Liffa ⑦. It was similar to the al Rbaa old market also dedicated to the selling of quilts, covers, coatings, carpets and costumes for women and men.[16] Furthermore, the Karamanli house reflects the power and the living conditions of the former inhabitants. It was built in the second half of the eighteenth century. The Karamanli house ㉜ is situated at the four-column street (al Arbaa Arsat) [1], which was named as such due to the presence of four roman columns at the intersection of the four-columns street with the Suq al Harara (perfumers and spice dealers market, today the new goldsmiths' market) ⑥. During the reign of Yusuf al Karamanli, it was known as the "harem house". In the second Ottoman period of the nineteenth century, it was the domicile of the Tuscan consulate. The house, which is one of the most beautiful buildings of the old city, contains ornaments, columns, metal and woodwork and bears witness to the architectural development of Tripolitanian houses.[17]

The economic base of the Karamanli clan was mainly piracy which attracted the attention of the European powers. The French military campaign against Tripoli in 1728 ended with the almost complete destruction of the city. Nevertheless, only the French conquest of Algiers (1830) put an end to piracy also in Tripoli.[18] The new political situation led to a decision of the Sultan of Constantinople to dismiss the last Karamanli Ali Pasha and govern Tripoli directly again.[19]

The various sources leave no doubt that Tripoli was among the most important ports in the Mediterranean for many centuries. It was also a central transhipment point for goods coming from the North African coast, the Middle East and Europe, whereas slaves were the most important commodity. One also gets the impression that in early modern times, there was no change to its trade habits, as the somewhat different insight into the city and its hustle and bustle shows, that Miss Tully offered to her readers. She was the sister of the British Consul in Tripoli, Richard Tully, who served there in the 1770s and 1780s. In her letters, sent from Tripoli to an unknown correspondent in the years between 1783 and 1793, she recounts that in Tripoli's old town "*they do not excel here in shops, the best of these being little better than booths, though their contents are sometimes invaluable, consisting of pearls, gold gems and precious drugs. There were two covered 'bazaars', one of which is very large, and built in four aisles, meeting in a cross. These aisles are fitted up with shops, built on each side of them, containing every sort of merchandize, and having a way in the middle for purchasers to walk in. Several parts of this place are nearly dark, and the powerful smell of musk makes it very unpleasant to pass through it. The other bazaar is much smaller, and has no shops in it. Thither only black men and women are brought for sale!*" In her opinion, those slaves were brought and examined like cattle for sale.[20] She mentions that Tripoli was in a dilapidated and in a "*ruinous state*".

Slaves taken to Tripoli were either bought for local use or were re-exported mostly in French ships to other parts of the Ottoman Empire like Constantinople, Greece, Crete and the Aegean. In the eighteenth century, and probably also in earlier centuries, the Mediterranean trade in black slaves had been nearly monopolised by French ships, largely because they had the greatest immunity from attack by Barbary corsairs.[21] In Tripoli itself, slaves from Sudan were conscripted into the Pasha's black guard; others did the hardest work (particularly in the quarries) alongside with Christian slaves who were mostly taken during corsair raids in the Mediterranean. Women and girls were bought as domestics and concubines. Profits from the slave trade enabled the Pasha's administration to finance the Mediterranean corsairs.[22] Miss Tully tells how those "*unfortunate blacks*" were frequently brought to the city. "*They are carried to the bazaar, or marked house, where they are bought by the rich people of the place, who occasionally sell them immediately to merchants waiting to re-ship them to other parts.*"[23]

Nearly 20 years later, in February 1819, the English traveller Lyon is out and about in Tripoli and reports about the lively bustle in the streets of the Medina and the surrounding markets: "*There are two grand markets held weekly, one on the sands behind the town every Tuesday, and the other on Fridays, about four miles distance, amongst the gardens of the Meshea which form a stripe of about three or four miles in breadth, between the beach and the desert. In the town are Bazaars, which are open every day. These are streets, covered in overhead. The shops of merchants are ranged on each side, and are very small. Slaves and goods are carried about before the traders by auctioneers, who keep up a continual din, each calling the price last bidden. The Jews have a quarter of the town expressly to themselves, where they have their shops, and in which they are shut up every evening at sunset. (…) Several houses set apart for the reception of merchants and their goods are called Fondook* [funduq], *and answer to the description given of the Caravanserais of the East* (Figs. 6.9, 6.10, 6.11, 6.12, 6.13, 6.14 and 6.15)."[24]

Lyon already indicated that the Jews inhabited their own quarter in the city, equipped with a synagogue and school. Actually, the Medina of Tripoli was divided into individual districts, whereas the individual quarters were separated from one another according to religious or tribal affiliation (see map: *Population, Groups and Quarters*). In this way,

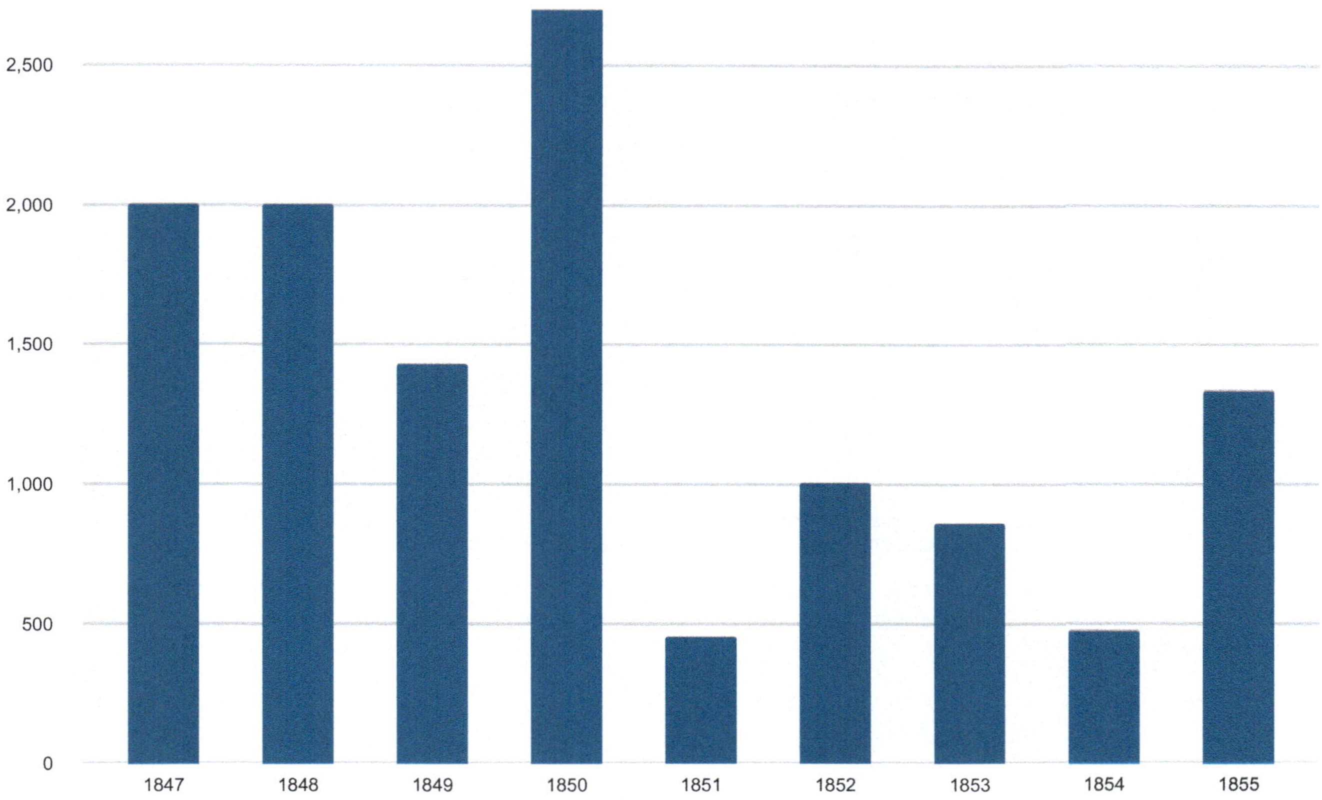

Fig. 6.9 Slave exports through Tripoli from 1847 to 1855 (see Lovejoy 1984)

Fig. 6.10 The Suq al Juma (photograph on the left Mathuisieulx 1912) was held every Friday. It is one of the oldest markets in the suburb of Tripoli. Grains, vegetables, fruits and all kind of animals were sold here (photograph on the right, Laronde 1997)

Muslims, Jews and Christians were physically separated. According to a nineteenth-century almanac, Tripoli had six quarters that were called Homat Gharyan, Homat al Baladia, Kasha al Sogar, Bab al Bahr, Harat al Kabir and Harat al Saghira. These were separated by void spaces. The *homas* inhabited by the Jewish community were called *harat* (Harat as Saghira the smaller one and Harat al Kabir the larger one). Homat Gharyan and Kasha al Sogar were mainly related to the caravan traffic. Homet al Baladia had a merchant character and was directly connected to the castle as the seat of the government. The particular character of a quarter was not only due to its different inhabitants, but also to its location. Bab al Bahr, where most of the Europeans lived, was located nearby the port and had purely marine features.[25]

Fig. 6.11 A caravan arrives at the oasis of Tripoli (Laronde 1997) to sell their products at the various markets. The Suq al Thalata was held every Tuesday ⑪. It was divided into several sections: one for vegetables and fruits, one for meat, another for cotton, wool and ropes, while another section was for the sale of cattle, camels, goats and donkeys

Fig. 6.12 A caravan arrived in the medina near the clock tower ㉔ (Laronde 1997). Today as then, the markets are busy with traders offering their fresh fare

For protection against the outside, a wall surrounded the Medina of Tripoli. In the end of the nineteenth century, there were four existing gates: Bab al Khandak ⑰ and Bab al Manshiya ⑱ were located on the southern side. According to Cowper, the Bab al Khandak was the chief one, being situated next to the castle. That adjoining it on the west was the Bab al Manshiya. He recalls that this one was double, with a small bazaar about sixty paces long between the inner and outer gates. This area was called Fum al Bab [2] ("the mouth of the gate"), and the chief article of sale was rope.[26] Bab al Jadid ⑯, which was erected in 1865 during the second Ottoman period to replace the Bab al Zenata ⑲, in the western part of the wall and Bab al Bahr ⑭ was situated at the northern side of the town. Of the latter, today, only small remains at the mosque near the Arch of Marc Aurelius can be found. Bab al Zenata was closed between 1832 and 1835 during the civil war between the Karamanli dynasty and the Ottomans and could not be opened later on. During

Fig. 6.13 In the streets of the medina in Tripoli: there were markets equipped with a roof to keep out the burning sun. It seems, that the markets were always crowded. In the background of the photograph on top right, one can notice an iconic monument in Tripoli, the Ottoman clock tower (24) (Mathuisieulx 1912). One of the most famous Tripoli`s byways is the so-called Arbaa Arsat (see photograph below on the left, Furlong 1909). The photograph below on the right shows the end of the great caravan route from the Sudan as it enters Tripoli (Furlong 1909, 198). The British tourist Edward Rae noted in 1877: "*The bazaars occupy the southern end, under the wing of the castle. To reach them from the harbour gate, one traverses the European quarter (…). We entered the long blank white alleys of this neighbourhood, where flying buttresses overhead cast broad shadows whenever the sun is not in the zenith. We came, after a few turns, to the Turkish bazaar, the chief and broadest thoroughfare of the city. White walls on either side carried a rude roof, under which vines trailed, and through which the sunlight streamed. (…) Many of the shopfronts were painted blue. Here were the barbers and grocers, the silk and cotton merchants. (…) Jews in dark blue turbans. Moors in white turbans, Turks in the fez, Arabs in brown rough barracans of undyed wool, with bare brown legs, wandered to and from. (…) The barbers' shops were especially neat, having gaily coloured racks for razors and combs, and clean matted divans. They had, too, old hand-mirrors, inlaid with ivory and mother-of-pearl, and jars of leeches. In the cafes sat Moors, with clean turbans of straw-coloured silk and white stockings, while the attendants moved quietly about with brass trays and the little cups of scalding coffee. (…) Next came the blacksmiths' bazaar, the entrance to the Djemma 'l Basha, and the apothecaries' bazaar. We watched the shops of dates and milk, one of the most common resorts of the poorer classes, who found there their breakfast, and too often their dinner. (…) Honey stood in vast jars much of it comes from Candia esparto baskets stood full of raisins, beans, red pepper, and ground corn for kouskousou. In an oil shop stood prodigious jars of olive oil (…). Much of the oil comes from Zleitun and Imsellatah, among the Gharyan hills. Sellers of oil, having asses laden with skins, passed us. Near the long colonnade of Djemma 'l Basha is the flower market. Close by was a cafe, and on seats placed along the white steps, a crowd of soldiers in white linen were enjoying themselves. Facing the mosque were shops of ironmongers, with sheep-shears, flat horse-shoes, tin powder-horns, primitive shot-pans, and strings of cowrie shells brought from Tomboukto. Men were selling coarse quilted linen skullcaps; boys carrying baskets of mulberry leaves and blossom were crying out for proprietors of silkworms. In an apothecary's shop hung ostrich eggs; a little farther was a leather-worker.*"[27]

§ I.

EINFUHR NACH TRIPOLIS I.J. 1898.

Silber in Barren	250000 fr.
Alkohol	100,000
Ambra und Agate	12000
Steinkohle	50,000
Cafe	200,000
Kerzen	80,000
Colonialdrogen-(Thee)	525,000
Stricke, Bindfaden	75,000
Modewaaren	50,000
Fuchsin	60,000
Häute, Felle	125,000
Eisen, Eisenwaaren	1,400,000
Galons von Goldund Silber	60,000
Streichhölzer	60,000
Wollengewebe	350,000
Trockene Gemüse	170,000
Holz, Holzkohle	150,000
Baumaterialien	200,000
Merceriewaaren	55,000
Möbel	50,000
Metall, Quincaillerien	200,000
Medikamente, Chmikalien	50,000
Parfümerien	20,000
Porzellan, Geschirr	25,000
Petroleum	50,000
Reis	200,000
Liqueure, Wein	50,000
Seife	100,000
Seidenwaaren	200,000
Tabak	820,000
Baumwollenwaaren	2,500,000
Zucker	500,000
Glaswaaren	75,000
Verschiedenes	215,000
Summa	9,327,000

A. 14305

§ II.

AUSFUHR AUS TRIPOLIS I.J. 1898.

Halfa-Gras	1,800,000
Bearbeitetes Silber	200,000
Vieh	450,000
Butter	80,000
Henna	50,000
Wolle	200,000
Matten	75,000
Orangen und Limonen	85,000
Gerste	550,000
Ziegen-Schaf-Kamelhäute	700,000
Thierhäute aus dem Sudan	800,000
Baumwollene Gewebe	75,000
Oel und Oliven	25,000
Schwämme	1,500,000
Straussenfedern	1,800,000
Eier	200,000
Pfefferfrüchte	25.000
Orangenwasser	25.000
Frische Früchte	40,000
Elfenbein	320,000
Wollgewebe	150,000
Früchte, Kerne, Datteln etc. Kartoffeln, Horn	268,000
Summa	9613750

§ 3.

EINFUHR DER WAAREN I.J. 1898 NACH LAENDERN VERTHEILT&

England	3,225,000
Oesterreich-Ungarn	1,000,000
Belgien	432,000
Frankreich	1,700,000
Italien	835,000
Uebertrag	7192,000
Türkei	1,000,000
Egypten	200,000
Tunesien	100,000
Holland	25500
Deutschland	527,000
Andere Staaten	283,000
Summa	9,327,500

§ IV.

WAAREN VON TRIPOLIS NACH DEM SUEDEN GESCHICKT I.J. 1898.

Englische Manufacturen	1,200,000
Seide, Cocons und Faden	123,000
Amulete	48,000
Zuckerhüte	300,000
Stoffe	6,000
Papier	16,000
Seidenwaaren	7,000
Ambra	5,000
Thee	150,000
Glaswaaren aus Venedig	7,000
dito Böhmen	14,000
Drogen, Quincaillerien	200,000
Feze	8,000
Summa	2,084,000

§ V.

WAAREN IN TRIPOLIS ANGEKOMMEN AUS DEM SUEDEN I.J. 1897.

Straussenfedern	1,014,000
Ziegenfelle gegerbt	1,050,000
Elephantenzähne	230,000
Summa	2,294,000

im Jahre 1898

Straussenfedern	1,800,000
Ziegenfelle gegerbt	800,000
Uebertrag	2,600,000
Elephantenzähne	320,000
Summa	2,920,000

§ VI.

AUS DEUTSCHLAND NACH TRIPOLIS IMPORTIRT I.J. 1898.

Gesponnene Seide	20,000
Ambra und Agat	12,000
Essenzen	5,000
Quincaillerien	80,000
Perlen	20,000
Tuche	50,000
Eisenwaaren	60,000
Gewebe, Sammt	62,000
Bier	4,000
Papier, weiss und bunt	7,000
Chemische Produkte u. Fuchsin	25,000
Stricke	8,000
Parfümerien	2000
Seidenwaaren	40,000
Liqueure, Spiritus, Medikamente	13,000
Galons aus Gold u. Silber	6,000
Spielzeug	2000
Verschiedenes	36,000
Summa	527,000

§ VII.

AUS TRIPOLIS NACH DEUTSCHLAND I.J. 1898.

Schaf-Lammfelle	50,000
Ziegenfelle	5,000
Schwämme	4000
Verschiedene Thiere	1,000
Summa	60,000

§ VIII.

REISEROUTE DER KARAWANEN NACH WADAI.

Tripolis nach Bengasi Schiffsverbindung Von

Fig. 6.14 Goods traded in Tripoli during the years 1897 and 1898 (German consular records)

the twentieth century, another three gates were built. On the 24 July 1908, the Bab al Hurriya ⑮ was inaugurated by the Turkish governor to remember the declaration of freedom. It was situated at the southern side of the wall. During the Italian time, Bab al Foulla, which means "hole in the wall", was built at the western side of the fortification. And in the 1990 s, Bab al Gaddafi was built opposite of the port in the Bab al Bahr area on the corner of the mosque of Darghut Pasha.[28]

The individual districts with their narrow streets are typical for an oriental city. To the eye of the European viewer, it might at first be a maze of alleys. The so-called cul-de-sac layout of the individual quarters is a typical feature and at the same time characteristic for the physical appearance of the old city of Tripoli. Only a few main roads run through the Medina, often branching into narrow culs-de-sac. Numerous short streets then lead from the culs-de-sac into the interior of the residential blocks. Here, the desire of the residents for protection of privacy becomes apparent. The numerous markets (sg. *suq*), which Lyon or Miss Tully call "*bazaar*", a term that derives from Persian, characterise the cityscape of the Medina of Tripoli. They form the traditional commercial, industrial and financial centre of the city. Now as then, you will find retail, wholesale and foreign trade in the narrowest of spaces. According to Micara, the Medina of Tripoli is not just another replica of the Arab-Islamic city. The urban system was developed on a pre-existing Roman town and was built by different populations such as Arab-Berber, European and Jewish. He suggests speaking of Tripoli as a Mediterranean Medina.[29]

Outside the Medina, there were two large markets which were held on different days. As Lyon reports, these markets were held each Tuesday and Friday. Both the Suq al Thalata ⑪ (Fig. 6.11) and the Suq al Juma (Fig. 6.10) are extremely persistent, not only in an eponymous function but also in

§ I.

Imports to Tripoli in 1898	Francs
Silver Bars	250,000
Alcohol	100,000
Amber and Agate	12,000
Coal	50,000
Coffee	200,000
Candles	80,000
Colonial Drugs-(Tea)	525,000
Ropes, Strings	75,000
Millinery	50,000
Fuchsin	60,000
Hides, Skins	125,000
Iron, Ironworks	1,400,000
Gold and Silver Gallons	60,000
Matches	60,000
Wool Fabrics	350,000
Dried Vegetables	170,000
Wood, Charcoal	150,000
Building Materials	200,000
Haberdashery Goods	55,000
Furniture	50,000
Iron, Metal Goods	200,000
Medicine, Chemicals	50,000
Perfume	20,000
Porcelain, Tableware	25,000
Petroleum	50,000
Rice	200,000
Liqueurs, Wine	50,000
Soap	100,000
Silk Products	200,000
Tobacco	820,000
Cotton Products	2,500,000
Sugar	500,000
Glassware	75,000
Miscellaneous	215,000
Total	9,327,000

§ II.

Exports from Tripoli in 1898	Francs
Alfa Gras	1,800,000
Finished Silver	200,000
Cattle	450,000
Butter	80,000
Henna	50,000
Wool	200,000
Mats	75,000
Oranges and Limes	85,000
Barley	550,000
Goat Sheep Camel Hides	700,000
Animal Hides from Sudan	800,000
Cotton Textiles	75,000
Oil and Olives	25,000
Sponges	1,500,000
Ostrich Feathers	1,800,000
Eggs	200,000
Peppper Fruits	25,000
Peppper Fruits	25,000
Fresh Fruits	40,000
Ivory	320,000
Woolen Fabrics	150,000
Fruits, Seeds, Dates etc. Potatoes, Horn	268,000
Total	9,613,750

§ 3.

Import of Goods in 1898 Ordered by Country	Francs
England	3,255,000
Austria-Hungary	1,000,000
Belgium	432,000
France	1,700,000
Italy	835,000
Amount carried forward	7,192,000
Amount carried forward	7,192,000
Turkey	1,000,000
Egypt	200,000
Tunisia	100,000
Holland	25,500
Germany	527,000
Other Countries	283,000
Total	9,327,500

§ IV.

Goods Sent from Tripoli send South in 1898	Francs
English Manufactures	1,200,000
Silk, Cocoons and Strings	123,000
Amulets	48,000
Sugar Loaves	300,000
Fabrics	6,000
Paper	16,000
Silk Products	7,000
Amber	5,000
Tea	150,000
Glassware from Venice	7,000
ditto Bohemia	14,000
Drugs, Ironware	200,000
Fezzes	8,000
Total	2,084,000

§ V.

Goods Arriving in Tripoli from South in 1897	Francs
Ostrich Feathers	1,014,000
Tanned Goat Skins	1,050,000
Elephant Tooths	230,000
Total	2,294,000
In 1898	
Ostrich Feathers	1,800,000
Tanned Goat Skins	800,000
Amount carried forward	2,600,000
Amount carried forward	2,600,000
Elephant Tooths	320,000
Total	2,920,000

Fig. 6.15 Translation of goods traded in Tripoli during the years 1897 and 1898 (German consular records)

their use. The Suq al Juma is one of the oldest markets in the suburb of Tripoli, about ten kilometres from the old town. It was named as such after the *suq* that was held every Friday in the oasis nearby. Grains, vegetables and fruits were sold here. The market has expanded to include the selling of all kinds of animals.[30] While today there still is a popular market in the quarter of Suq al Juma, the Suq al Thalata ⑪, had to make way for an ambitious "*Green Belt Project*" in the year 2009 planned by the late al Kadhafi administration. In the past, the Suq al Thalata ⑪ was held every Tuesday. Tin sheet huts were built at night in the open air to become a busy market in next morning. The *suq* was divided into several sections: one for vegetables and fruits, one for meat, another for cotton, wool and ropes, while another section was for the sale of cattle, camels, goats and donkeys. The *suq* was moved to several places at various times throughout history. In the nineteenth century, it was held at the open space between the Saraya al hamra (Fig. 6.4) and the municipal gardens, then it was moved to the area between Bab al Jadid and al Maarri Street, and later, it was moved to the site Bab al Azizia.[31] This market stretched over an area of about four hectares until the end of 2009. It primarily housed retail shops of various industries such as electronics, toys or accessories. In the course of the implementation of the green belt, the local dealers were resettled, a fact which has led to fierce debates in public. Starting point for these measures was a decision of the government to establish an urban green belt. The dealers affected by the resettlement were compensated with the provision of areas on the outskirts, which, however, could not satisfy a large number of the affected traders. The resistance was sparked mainly at the new location which is far outside of the city centre.

Fig. 6.16 Al Zahr funduq (59) is situated at Suq al Mushir

Fig. 6.17 A funduq is a characteristic phenomenon for the old town of Tripoli. Miss Tully, the sister of the British Consul in Tripoli, Richard Tully, gives a vivid insight into everyday life in a funduq in her letter from 9 September 1783: "This building is very huge, with a square area in which is a well and a gebia, or marble reservoir for water, for the convenience of the Moors to wash in before prayers and meals. Round the area is a number of small rooms, each for the goods or merchandize of the person or persona who may sleep in the apartments over it. The camels, horses, and mules of the travellers are ranged round the yard. When a stranger arrives, a Moor dusts the floor of an empty room, and spreading a mat, which is all the furniture allowed, leaves the guest in quiet possession of it. Those who can afford it, are expected, on quitting it, to leave a small gratuity to the porter, and none can get out or into the fenduke till the adan, or dawn of day, when a Moor unlocks the gates." (Tully 2009, 27–28)

Lyon, however, also describes another important "institution" of Tripoli, the *funduq*. The *funduq,* the equivalent of which is the caravansary in the Middle East, not only is in structural terms one of the most striking building types in the Medina in Tripoli, but can also be found in a great number there. According to records, there were approximately 35 *fanadiq* in the Medina. Twenty-one of them are known by name. They were located along the big markets, west of the cardo and near the seaside. al Zahr funduq (59) (Fig. 6.16) situated at Suq al Mushir (8) is considered to be one of the best examples. It is most likely that it was built during the reign of Yusuf al Karamanli by Mustafa Gurgi, the brother-in-law of the pasha. The place was named al Zahr, which in Arabic means orange flower, because this kind of flower and other flowers and essences were stored and sold in the open courtyard. Fragrant products such as orange flowers, henna and volatile oils were also distilled and then shipped to Turkey. There is a large wooden door, known as the Bukhukha door, leading towards the entrance which today accommodates a café. The entrance also has a circular vault with a prominent architectural ornament at the top. As usual, the *funduq* consists of two floors. To the left is a staircase leading to the upper *riwaq*, which is an arcade or portico open on at least one side. It has four galleries with 28 rooms opening onto them. In the past, the rooms on the upper floor were used as accommodation facilities for merchants or other guests. The rooms in the lower *riwaq* served as storage rooms of the caravans. Today, there are numerous shops. Many of these facilities have a persistent structure by retaining their type of use, and until the end of the regime of Gaddafi, they served as hotels to accommodate mainly tourists and business travellers (Fig. 6.17).[32]

In 1835, the Turkish sultan felt disposed to interfere because of inner disruptions in Tripoli and put an end to the rule of the Karamanli family. As a result, Tripoli was incorporated into the Ottoman Empire as *vilayet*. However, the splendour of the bygone days seemed lost. According to Robert Greenhows' "*Sketches of the history and present condition of Tripoli*", the town was run down at that time. "*Tripoli, the other and least important of the States of Barbary* (he compares it to Algiers and Tunis), *had until lately pursued a course similar to that of Tunis, and its condition was highly prosperous; (…).*" But now, "*there is but little appearance of wealth in Tripoli; the Moorish population amounting to about fourteen thousand are in general very poor, the trade being almost exclusively in the hand of the Jews, whose number is about two thousand. The palace contains some apartments possessing a certain degree of grandeur and furnished in a costly manner principally with French articles; in the town there are a few good stone buildings, with courts and arcades in the Italian style; these are however chiefly occupied by the foreign Consuls and merchants, the greater part of the inhabitants dwelling in mere hovels of mud but one story high. The roofs of the houses are all flat, and great care is taken to have the rain conveyed from them into cisterns, as there is not a well or spring of fresh water in the place. (…) the immediate environs of Tripoli are desert; about two or three miles to the eastward is a rich and highly cultivated plain called the*

Messeah where the Foreign Consuls and the wealthy inhabitants of the town have their villas".[33]

The ruling powers in Tripoli were the first to desist from piratical cruises during the Barbary wars. Dissensions in the family of the sovereign had also led to numerous revolts in different regions, repressed more or less successfully. In view of the French pressure that was being exerted on the Karamanlis and on account of the presence of the French in Algeria, the Ottomans decided to display the suzerainty of the Sultan over Tripolitania.[34]

In the course of the nineteenth century, the town has apparently recovered. According to Rohlfs, "*the streets of Tripoli are wide and clean (of course always in comparison with other Muslim cities) and some of them have been paved lately and provided with lanterns*".[35] He further states that Tripoli continued to expand its trade every year. During the years of 1868, 1869 and 1870, an annual average of 427 ships entered the port and 414 ships left it. In 1875, the number of the outgoing and incoming ships had more than doubled, according to Rohlfs.[36] According to an almanac from 1882 to 1883, 25,000 inhabitants, among which were 4000 Jewish and 3000 foreigners, lived in the capital of the vilayet of Tripoli. This is an indication that trade in the late nineteenth century was still flourishing; however, it entered a state of stagnation with the progression of the Ottoman rule. Economic life stagnated towards the end of the Ottoman era which affected the frequenting of the port. At the beginning of the twentieth century, the German geographer Banse noticed an "*empty port*": "*Until 1911 three steamers at once were a great event.*"[37] He describes Tripoli at the beginning of the twentieth century as an economically depressed city.

There are many reasons for this stagnation: the fall of the Ottoman Empire had already advanced, and the caravan trade stagnated, as one of the most important acquisitions from the south, slaves, could no longer be traded. Until well into the nineteenth century, slaves were an important commodity despite the signing of an anti-slavery agreement on the part of the Ottoman Empire. Until the 1850s, the Ottoman authorities regarded the slave trade as an open, acceptable and legal business, and appropriate taxes were levied and paid on it. Furthermore, political events, both outside of Africa and inside the caravan trade, had a strong influence and triggered the rise and fall of the terminal port cities. Many researchers make it clear that Tripoli was almost entirely dependent on the trade with the sub-Saharan hinterland in the interior of Africa. Around the turn of the century, a lively caravan trade was part of the foundations of life of many North African port cities. The situation of Tripoli rapidly deteriorated towards the end of the nineteenth century as the French annexed more and more territories between Niger and the Tibesti mountain range to their colonial empire. Since then, a border separates Tripoli from its "hinterland", a state that was not to change by the Italian seizure of power.[38]

Unlike European cities in the nineteenth century, cities in the Arabic-Islamic area were not exposed to the radical growth and conversion process. As in Tripoli, in many cases, the medieval city fortifications had maintained their character as physical city limits. Only with the annexation of Libya by the Italians, the city experienced a period of rapid urban growth. Since about 1880, the city started to expand slowly across the surrounding walls to the south-east. At the end of the Ottoman rule, the expansion of the city concentrated along the main arterial roads that extend in a radial direction to the south and south-east starting from the market place in front of the old town in the south-east. Previously, only a few hotels and stalls were found outside the old city walls. At first, barracks were built. As a result, many officers left the old city. Rather quickly the new city of Tripoli established itself around the turn of the century (see map: *Core area of Tripoli Development of settlement area between 1909 and 2002*).

6.1.6 Tripoli on the Eve of the Italian Invasion

Gottlob Adolf Krause, a German Africanist who travelled to Tripoli for the first time back in 1868, took up residence in the city again from 1907 to 1912 and thereby bore witness to the Italian preparations for colonial expansion. He chronicled developments including the numerous business connections established at this time between Tripoli and Italy as well as other European nations. Here, particular mention must be made of a steam-driven mill owned by the Banco di Roma, which was supplied with grain from Italy and generated profits through the sale of flour, an esparto grass press which also belonged to the Banco di Roma, and a factory for the production of ice on which the Italian operating company had an apparent monopoly.[39]

As well as describing these economic activities and the associated geopolitical interests of the European powers, Krause also devoted himself to documenting the day-to-day characteristics of life in Tripoli. He gave detailed descriptions of the happenings and the goods available at the various markets, in particular the Suq al Thalata, which was held every Tuesday at the entrance to the city. According to Krause, the market had "*gained a certain renown thanks to its vibrant, oriental image*"[40] and the authentic Middle Eastern way of life shows there regularly drew French visitors arriving on the steamer from Tunisia.

A crucial and central concern for Tripoli's residents in addition to trade and providing the growing population with all manner of goods was the supply of water. As was the case in the entire Mediterranean region, it was necessary to establish water supply systems in order to survive the dry summer months. These systems traditionally use cisterns to store precipitation collected during the winter. An alternative

Fig. 6.18 Pashas fountain fed by the Bir Bu Meliana at the south-eastern entrance to the Medina at Bab al Khandak ⑰

Fig. 6.19 Suq al Cobza (bread market) ④ with a fountain in 1916

Fig. 6.20 Fountain at Bab al Jadid ⑯ is still visible today

approach is the construction of wells to tap groundwater from deeper underground. However, this proved difficult in the immediate vicinity of Tripoli as salt water from the Mediterranean entered the water-bearing strata, rendering the water undrinkable. To guarantee the water supply in spite of this, according to Krause, the Turks installed wells at three locations in the city which drew water from a source via a small water conduit. The source is located around two and a half kilometres south–southwest of the city and is known as Bir Bu Miliana (Fig. 6.18), which literally means "*father of abundance well,*" a reference to its inexhaustible water reserves.[41] The well was reached by following the street which ran due south from the bread market ④ (Fig. 6.19). After crossing the "new town" built during the Ottoman period, the path then arrived at the gardens full of date palms and olive trees. The well itself was situated at the southern edge of these gardens, which marked the relatively abrupt transition to the adjacent barren semi-desert.

According to Krause, the water channelled from there into the city could be collected at three "*schisms*" or distribution points by anyone who wanted it, though he adds that in reality the water was "*transported to customers in the city and suburbs by professional or casual water vendors*".[42] It remains unclear whether people had water brought to them out of convenience or whether a payment was required which was collected by the water vendors.

These public "water filling stations" were all located on the edge of the medina. Two of them were built in close proximity to each other. The fountain known as "Pascha's Fountain" was situated directly at the entrance to the medina at the Bab al Khandak ⑰, and the other in the middle of the bread market. The third fountain was located inside the medina, directly at the city wall at the Bab al Jadid ⑯ (Fig. 6.20). The Bir Bu Miliana well supplied a fourth facility built by the Turks in the middle of the Suq al Thalata ⑪, which ran along the coast to the east and south-east of the medina and joined the Pasha's Public Gardens at its eastern end ㊲. Krause reported on this as "*a new creation by the Turks*" (Fig. 6.21).[43]

In 1911, Tripolitania was annexed by Italy. The seizure was carried out with military means as the Italians encountered heavy resistance not only by Turkish troops but also by the Libyan tribes. Only in 1934, the Italian colony Libya was set up from the conquered territory, and in the years that

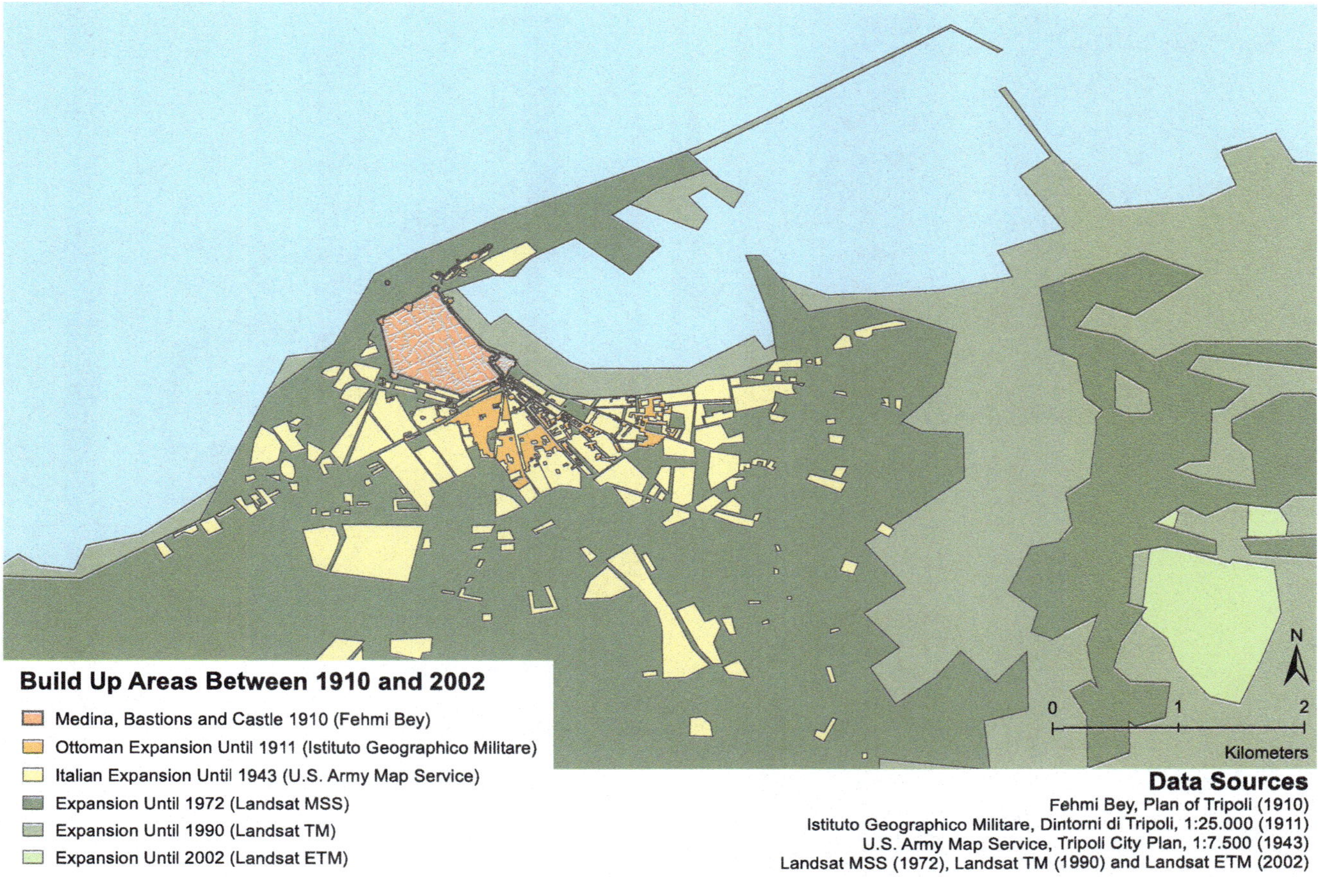

Fig. 6.21 Map: Core area of Tripoli—Development of settlement area between 1909 and 2002: like most cities of the Arab region, Tripoli did not undergo that much growth and transformation during the nineteenth century as the European cities did. However, it provides a good example of modern urban development taking place immediately adjacent to the historic nucleus. The origin of the new town dates back to 1880 when the Ottoman administration started developing a number of new technical and residential districts extra muros across the south-east. The expansion of the city concentrated along the main arterial roads that started in the centre of the marketplace in the south-east, opposite to the medina and extending to the south and south-east. With the annexation of Libya by the Italians in 1911, the period of rapid growth for the city began, which clearly reflected contemporary European town-planning influences

followed, 100,000 colonists were settled there. Apart from the existing settlement, a planned European new city started to grow.

6.1.7 Roman Legacy Versus Indigenous Traditions

With the annexation of Libya by the Italians in 1911, the period of rapid growth for the city began. This expansion clearly reflected contemporary European town-planning influences—although the colonial planners tried to use an architectural language that offered a Western conception of Eastern architecture.[44] In doing so, urban architecture was shifted from the modernist appropriation of the early 1930s, in which the indigenous culture is fused with modern aesthetic sensibility, to a historicist one in the latter part of the decade, when the indigenous is re-enacted as a racially encoded project of historic preservation. The new town limited the old town of Tripoli soon to the role of a historical core. Urban growth was concentrated along the development axes. Being influenced by the Ottoman period, it followed the main arterial roads to the south-east. Further development took place subsequently in concentric rings around the current centre. Moreover, the Italian authority began remodelling some parts of the old town with the demolition of parts of the walls in order to construct a modern port with the material.[45] But it was not long before the value of preserving the fabric of the Medina was recognised when tourism became a major key aspect in the development of the colony. Among the first projects to be undertaken was the restoration of the wall of the old city in 1923. But the decision of what would be preserved and what not was solely in the hands of the Italian authority, viz. the building

Fig. 6.22 At the beginning of the Italian rule there was a cinema in the arch ㉗ (Laronde 1997)

Fig. 6.23 Arch of Marc Aurel in 1937 ㉗ (Laronde 1997)

Fig. 6.24 Arch of Marc Aurel today ㉗

commission, which on the one hand used modern principles of hygiene and urban order and, on the other hand, chose to value one history—of course, the Roman one—over the other.[46] The conflict between Libya's Roman legacy and its indigenous traditions was resolved in a manner that was entirely consistent with the values that informed Fascist planning in Rome. Monuments were "liberated" from centuries of anonymous built fabric.[47] The restructuring of the area around the Arch of Marcus Aurelius ㉗ (Figs. 6.22, 6.23 and 6.24) actually created considerable discontinuity in the adjacent fabric, requiring the demolition and restructuring of substantial portions of three of the city's most significant eighteenth-century *fanadiq* (hotels).[48] From a historic-cultural point of view, the dealing with the triumphal arch clearly illustrates the interpretation sovereignty and patterns of this most obvious relic of ancient Oea. During his first visit to the city in 1846, Heinrich Barth, a German historian, noticed that "*this arc is not without interest, however its four arched openings, [...], are now bricked up and unworthy of the venerable Emperor, a Malti (Maltese) established a wine bar here. Even from the outside the building is partly obstructed by adjacent houses [...].*"[49] The German researcher presents himself as an European beholder of the scene who places ancient tradition before all other urban developments. Gerhard Rohlfs, who travelled to Tripoli in 1868, also stressed the superiority of antiquity: "*What strange fate had befallen this city and what does the future hold for it if it should be placed under the rule of an enlightened government just like Algeria. Was not the old Tripoli, that unity of three cities Leptis Magna, Oea and Sabratha, once one of the most flourishing and richest colonies on the northern shores of Africa?*"[50] The triumphal arch symbolises to him the glorious past of ancient Tripoli that should be revived again. Furthermore, he pointed out the conversion of the arch. In the meantime, a wine bar operated by a Maltese was removed from the arch, while Rohlfs makes it clear that this "*[did not happen] out of reverence for a work of art from antiquity, but as there is a Turkish law according to which liquor bars can be established only at a certain distance from a mosque*".[51] The arc is located in direct vicinity of the Gurgi Mosque (75) that was built by Mustafa Gurgi in 1833/1834, who was the commander of the Tripolitanian navy, and who later became brother-in-law of Yusuf al Karamanli, the governor of Tripoli between 1795

Fig. 6.25 Sea front of Tripoli at the end of the Ottoman era (Laronde 1997)

Fig. 6.26 Pilgrims in the port of Tripoli on their journey to Mecca (Mathuisieulx 1912)

Fig. 6.27 Medina of Tripoli and part of the port as seen today

and 1832. The mosque is similar in structure to the Ahmed Pasha mosque ⑦③.[52] Both researchers overlook that over centuries, the arch had been granted a role in the context of the Medina that, in the end, prevented it from disappearing unlike all other monuments of the Roman city.[53] Even the Arab pilgrim traveller Mohammed al Abdari expressed his enthusiasm in 1289 about the beauty of the arch and the used marble, building material which awakened desires as al

Tigiani reported in 1307. A wealthy man of the city intended to use the arch as a quarry, whereupon the citizens of Tripoli converted the building to a mosque without further ado. At the end of the seventeenth century, the arch was threatened again with destruction. Mohammed Pasha Saqizli, governor of Tripoli between 1631 and 1649, tried in 1678 to profit from the building. Once again, the citizens of the city fought back referring to the fact that such an act would be a crime and that the building had been respected and honoured by the people of Tripoli for centuries. Thereafter, the arch seems to have been used for different purposes. It is recorded that between 1697 and 1818, it accommodated various shops. According to Rohlfs, even in the time afterwards, the arch had been continuously used as sources confirm, whereas the type of use had been subject to constant change. In 1887, the documented use of the arch was as a bar. In 1896, a Greek was said to have used the arch as a cheap restaurant. At the turn of the century, it was said to have accommodated again a bar, then a night club and finally a fruit shop.[54] At the beginning of the Italian rule, there was a cinema in the arch. Entirely in keeping with Barths or Rohlfs, the Italian colonists finally removed all subsequent installations from the arch in 1912, wiping out all traces of its use after antiquity. During the colonisation of Libya, the Italians pursued a policy of an optically visible Italianisation of the country which was particularly visible in the promotion of archaeological excavations.[55] The primary objective was the staging of two historical poles, the period of North Africa under Roman rule and the country occupied by the Italians. These measures went far beyond a mere archaeological conservation or modern restoration of the ancient monuments. After having removed, in the eyes of the Italians, evidence of the less glorious historical phases, antiquity was represented in line with fascist ideas. The complete excavation of the arch and its reconstruction led to its release from the sphere of regular historic preservation. The archaeological conservation took second place to the political staging. In this sense, the building density of the Arab-Ottoman Medina surrounding the arc was gradually pushed back to create a free zone around this most important relic of Roman Oea.[56]

Today, Tripoli and its surroundings form the largest agglomeration in Libya which consists of the *shabiyat* of Tripoli, Tajura and al Jfara and currently extends over an area of almost 3800 km^2. According to the census of 2006, Tripoli is occupied by two million inhabitants. Today, Tripoli has become two cities in one, with fundamental differences in its structural order similar to what can be found in many other cities of the Arab region (Figs. 6.25, 6.26 and 6.27).[57]

6.2 Nalut—a Fortified Granary in the Nafusa Mountains

Said Hamid

The city of Nalut is located to the south-west of Tripoli. It lies approximately halfway between Tripoli and Ghadamis, at the western end of the Nafusa Mountains coastal range, and is situated at latitude 31.45° north and longitude 11.05° east.

6.2.1 Desert Transit Centre

Nalut owes its importance to its distinct geographic location, poised as it is on the caravan routes linking Tripoli, Ghadamis and several Tunisian cities. It was one of the desert transit centres, where caravan traders used to halt for periods of rest. Water and pasture were not only available in the neighbouring valleys but also within the city limits itself. There, one would find numerous springs of fresh water, such as the Tala and the Tighlis springs. Moreover, the surrounding region produced figs, olive oil and cereals. The region also became famous for some traditional products, most important of which is the item of Libyan traditional clothing known as the Naluti *huli* or *jird*. This was made of

Fig. 6.28 Nalut is home to a qsar, which is a fortified granary with a small market, a special place for meetings and storage rooms

Fig. 6.29 Nalut once linked the routes from Tripoli, Ghadamis and Tunisian cities. The qsar Nalut was a communal building, which housed more than 400 rooms, where local families could store their grain, olive oil and dried fruits. Next to the qsar, there are still houses in local style and several mosques

pure white wool and famous for its high-quality manufacture. In addition, the people of Nalut were engaged in some traditional trades, such as the production of clay utensils and leather products. Among the most important landmarks of Nalut, one could mention its huge palace, built in the eleventh century. This palace housed more than 400 rooms, which were used for the storage of olive oil and cereals. In the settlement itself, there were many traditional oil presses and old mosques as well as underground chambers in which people used to live (Figs. 6.28 and 6.29).

Nalut was linked to the caravan route from Jadu, or from the cities of the Jafara plain. Caravans took the route from Tripoli, al Aziziya, Bir al-Ghanam, al-Jush and Tiji to arrive at Nalut. From there, the route proceeded towards Sinawin, Dirj and Matris until it reached Ghadamis. From Ghadamis, the route proceeded to Ghat and from there to the central and western lands of the Sudan. This route witnessed considerable caravan commercial activity, and merchandise was exchanged between the Mediterranean and the African ports during periods of security and stability which characterised the eighteenth and nineteenth centuries. Towards the end of the nineteenth and the beginning of the twentieth centuries, the route was influenced by the commercial recession that was a result of the drop in demand for sub-Saharan products. This led to a demise of the economy of the town. The adoption of alternative maritime, riverine and land routes leading to the interior of the African continent, precipitated the end of caravan trading.

Nalut was linked to another route which witnessed considerable commercial activity in the beginning of the nineteenth century. It passed through the city of Wazin in the direction of the city of Dheybat. From there, the route proceeded towards the other Tunisian cities in the north. It was one of the important commercial bridges linking the traders of the Nafusa Mountains and Ghadamis with Tunisia and the countries of the Arab Maghreb. This route was impacted by the French occupation of Tunisia in 1881 and consequently lost its commercial importance.[58]

6.3 Ghadamis—"Pearl of the Sahara"

Jacqueline Passon

Ghadamis, known as "the pearl of the desert", is located at the edge of the Sahara Desert, close to the Libyan–Algerian border. It is an outstanding example of a traditional settlement.

The oasis town of Ghadamis lies in the northern part of the Sahara, surrounded by a ring of date palms. In the past, the town performed an important function as a transportation hub for trans-Saharan trade. Its position on the northern edge of the Sahara lends Ghadamis its strategic significance: five caravan routes have their crossroads here. Several springs formed the foundation of the place, of which the best known is the Ain al Faras, which has almost run dry today. Intensive market gardening was made possible by this water. Until well into the nineteenth century, Ghadamis was an important staging post for trans-Saharan trade.

6.3.1 Ain al Faras—Source of Life in the "Most Barren Desolation"

The first reports of Ghadamis (Fig. 6.30) are from ancient times. At that time, Roman troops were stationed temporarily in Cydamus, as it was known in Latin.[59] According to the English historian Goodchild, the Romans constructed a fortress in Ghadamis between 201 and 235.[60] In Byzantine times, in the sixth century, the oasis had a church and was the seat of a bishop following the conversion of the population to Christianity. The "idols" which stand nearby the old town are of ancient character, and it is supposed that they are Byzantine mausolea. In the seventh century, Ghadamis came under the domination of the Muslim Arabs and the population quickly accepted Islam as their religion. The Arab conqueror, Uqba ibn Nafi, had occupied the oasis with a detachment of cavalry between his conquest of Fezzan and his march on Gafsa in 667. Between the eighth and tenth centuries, the oasis became Ibadi and developed into a trans-Saharan port.[61] Consequently, it is not surprising that fourteenth-century historian Ibn Khaldun reported the significance and prosperity of Ghadamis: "*...today Ghadamis is a greatly populated city. It is a stopping place for pilgrims from the Sudan and for traders on their way from the Maghreb to Alexandria and Cairo.*"[62]

Leo Africanus, the papal historian of the Orient, located sixteenth-century Ghadamis in a desert that was "*very dangerous for travelling salespersons*".[63] Travellers from Europe first arrived in the town in the nineteenth century. Foreigners, like the Christians from Europe, were eyed with suspicion and curiosity. Ever since the 1860s, it had not been a matter of course to let Christians into the desert city. Since the French conquests in Algeria in that century, sufficient bad experiences had been collected of the colonial ambitions of Europeans.[64]

To James Richardson, an English explorer in Africa, the view of the city seemed like deliverance from the desert. He noted on his 23 days lasting journey on the route from coastal Tripoli to Ghadamis in 1845: "*Just when day had broken over half the havens, I saw Ghadamis! Witch appeared like a thick streak of black on the pale circle of the horizon. This was its date-woods. I now fancied I had discovered a new world*".[65] The landscape around Ghadamis seemed inhospitable, not only to the English explorer. The Austrian geographer and world traveller Josef Chavanne spoke of the "*most barren desolation*".[66] The reason that humans have settled here over such a long period of time can be seen in the variety of trade routes between the

Fig. 6.30 View on one of the mosques of the oasis. During the sixth century, the population had been converted to Christianity by Byzantine missionaries. But in the seventh century, Ghadamis was conquered by Uqba ibn Nafi and the population quickly converted to Islam. Then, Ghadamis played an important role as base for the trans-Saharan trade until the nineteenth century

Mediterranean coast and the African interior. This may have been a very important (if not the most important) circumstance in the prosperity and growth of the area.

The survival of humans in such a hostile place was and still is first and foremost dependent on access to the resource of water. The houses stand closely together with their flat roofs and accessible terraces. Dark, shady alleys all lead to the large square in the centre of the city, where the source of life in Ghadamis is found. This spring, according to Rohlfs, is "*the creator of the oasis and therefore the founder of the city (…)*".[67] The town is laid out around the artesian Ain al Faras ① (Figs. 6.31 and 6.32). This limited water supply necessitated the functional arrangement and formal construction of the oasis. The old core of the oasis, which is completely surrounded by a wall, measures around two kilometres in diameter.[68]

The irrigated area would not have been greater than 75 ha and would have been divided into countless small parcels. The spring "*is collected in a elongated square basin, 25 m long [and] 15 m wide, from the floor of which one can see the water springing up in several places*"[69], reported Chavanne in 1878. After the water had been collected in this way, it was channelled to the gardens in three ditches towards the east, north and south. In order to guarantee exact sharing of the water among the individual parcels of land, special clepsydras were used, known as al Quadus ②. From Rohlfs, we know how they worked: "*An iron pan with a round opening in the bottom, through which the water, when filled to the top, will flow out in about three minutes. Every time a Gaddus is passed through, a lad employed for this purpose, who after a certain time will be relieved by another, makes a knot in a palm leaf. Seven Gaddus is called a Dermissa and this gives around twenty minutes irrigation, enough for a garden with sixty palms.*"[70]

The permanent population of the oasis was around 7000 according to Chavanne's data. After the arrival of a caravan, this could however grow by a further 1000.[71] Other sources even suggest a population figure of 12,000 in the year 1880, which would have made Ghadamis one of the largest oases in the Sahara.[72] Each of the seven clans living in Ghadamis had its own urban district with its own social institutions like celebration and meeting areas (see map: *Population groups*). Inhabitants of the oasis were mostly Berber of the Bani Wazit and Bani Ulid. They resided in six of the seven district areas. The Awlad Bellil, who considered themselves of Arab origin, lived in one district. They used to live as "enemies", shut up in seven districts isolated from one another by walls whose gates were shut at night. But despite these divisions, the little urban centre of Ghadamis has been able to maintain trade successfully across the centuries and, for a long time, its independence.[73] Besides the caravan trade, the main source of income for Ghadamis residents was agriculture. The town lay in the middle of the desert, not in a highly favoured place—like Tripoli on the Mediterranean coast—but it still managed to offer its inhabitants a prosperous life.[74] Agriculture comprised for the most part fruit, vegetable and grain cropping.

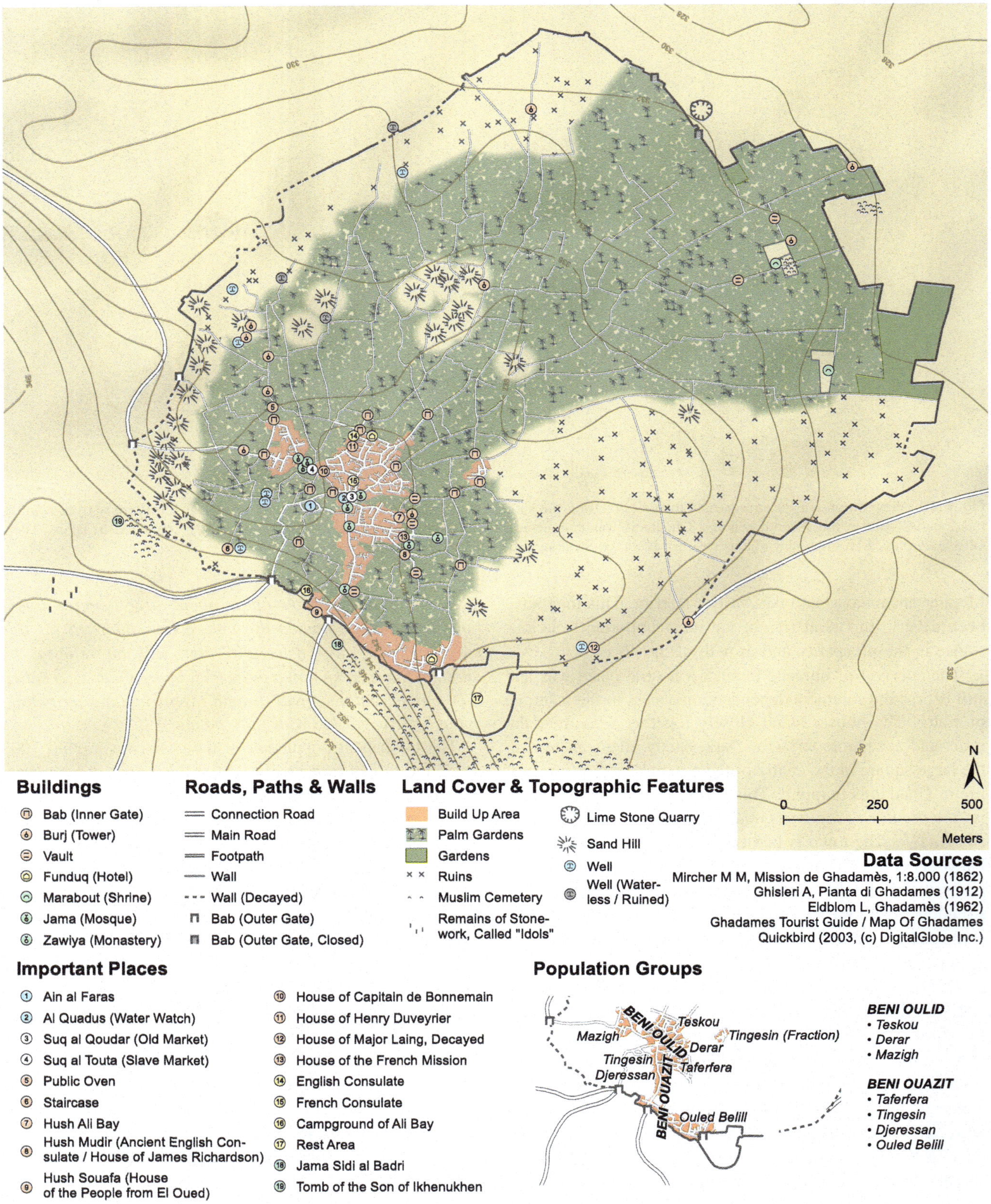

Fig. 6.31 Map of Ghadamis in 1862

Fig. 6.32 Ain al Faras as seen by the Austrian traveller Josef Chavanne (1879, 223) in the 19th century and from a satellite perspective (Quickbird)

The gardens can be imagined as a house with three floors: on the lowest level field crops such as wheat, barley and/or millet were grown; above them, fruit trees and/or olive trees were cultivated; finally, the palms were on the highest level, which gave shade and offer protection.[75]

6.3.2 Desert Architecture

The climate in this region is characterised by extreme temperature contrasts. Due to the climatic conditions and the lack of building materials, the residents developed an idiosyncratic desert architecture that, according to reports of travellers, had a rather unfriendly effect. In 1867, Rohlfs described it as a fortress: "*The town of Rhadames compares externally (…) with its close-built, many-storey houses, whose naked walls are only here and there at the highest level punctured with a tiny window opening, with a compact, irregular, walled-up fortress*".[76]

James Richardson also remarked that the walls looked as if they already bore the weight of thousands of years, and he mentioned the narrow, oppressive alleys.[77] The houses themselves were constructed from bricks formed from clay, palm leaves and lime. Due to its excellent adaptation to the climatic conditions, this construction method is still used today, for example, in the Egyptian oasis of Farafra. Nowhere in Libya is a better adaptation of housing to local climatic conditions to be found than in Ghadamis.[78] The houses are so arranged that a house often shares all four walls with its neighbours, lending the town an interlaced, labyrinthine appearance. Even the streets are covered, and the covers are furnished with shafts to provide light and ventilation. They not only performed the function of giving shade, but they also served purposes of defence and gender segregation. Rohlfs reported that the women of the town did not choose the streets as their paths; like a "higher town", the flat roofs were reserved for them. These served so to speak as "women's roads", at which men could hardly even glance.[79] Eldblom came to the conclusion that, due to the inhospitable surroundings, a very specific social and economic structure had developed in the isolated oasis.[80] Ultimately, he took a natural deterministic view here, postulating the specific qualities of the natural environment as the origin of development for different forms of organisation in a community. But the image of a closed society was also conveyed through nineteenth-century sources, a society which barely accepted foreign influences, the power of the central government or change (Figs. 6.33, 6.34, 6.35, 6.36, 6.37, 6.38 and 6.39).

Ghadamis might have consisted of several *qsur* in the sixteenth century. But then, it seems to have become concentrated into a single village which has preserved its appearance until today (Fig. 6.41). Its inhabitants were able to stay independent. This independence was only limited by the obligatory association with the Tuareg Adjer, at the extreme limit of whose territory Ghadamis was situated. Moreover, the Ghadamis people always maintained close ties to Tunis and Tripoli. The oasis suffered several attacks by Hafsid and Turkish troops but always managed to free itself rapidly from the taxes imposed by Tunis. It was nonetheless obliged to recognise the authority of the Turks of Tripoli in 1860. After 1874, it was given a little garrison; however, it continued to administer itself with a *shaykh* formed from the heads of noble families.[81]

Fig. 6.33 Ain al Faras as seen today

Fig. 6.34 From Ain al Faras, the water runs via water lines to the houses and gardens

Fig. 6.35 View from the top of one of the houses. The open-air terraces were reserved strictly for women

Fig. 6.36 Ghadamis in the view of Chavanne (1879, 225) and Richardson (1849, 225, 268). Ghadamis is one of the oldest Saharan settlements. The oasis settlement is built entirely out of mud and it is almost completely covered. It is not surprising that this ancient design has been invented in the Sahara: the mud houses maintain a cool town in summer and a warm habitat during cold winter nights. In addition to protection from the burning desert sun, the roof of the city acts like a separate city with complete open streets and lanes, exclusively used by women. This gave them the chance to move undisturbed from one house to another. Ghadamis has been declared a World Heritage Site by the UNESCO in 1986

Fig. 6.37 The gardens were organised vertically, palm tree (date palms), then fruits, creating the necessary shade for crops

Fig. 6.38 Interior wall designs and house decorations from Ghadamis: the interiors of the houses are painted with unique Berber designs, using red paint on white walls. The inhabitants also decorated their rooms with hanging ornaments. This gives the rooms a fascinating atmosphere, which makes Ghadamis unique. The traditional Berber designs used to include the triangle, the diamond, the sun, the moon, the palm, the eye, the hand and the Tuareg cross. The objects used to decorate the rooms include mirrors, ornaments made of palm straw, brass and copper, as well as Berber carpets, rugs and cushion

6.3.3 "Trade Negotiator" Between the Mediterranean and the African Interior

Trade was a primary source of income for the city. Traders from North and Central Africa met in Ghadamis. Caravan routes led to the north towards Tunis and Tripoli. In a southerly direction, the routes led via the deserts towns of Ghat and Murzuq on the one hand to the legendary Timbuktu and on the other hand to the Bornu Empire. Ghadamis was more than just a transportation hub in this trade network. The inhabitants themselves travelled as sales people throughout North Africa. The extreme points of their journeys were Tunis and Tripoli in the north, Agades, Kano and Timbuktu in the south. Of a total of 7000 people living in the town in 1867, around 1000 were constantly on the road with various caravans. Chavanne described the role of the traders from Ghadamis as the "*negotiator of trade between the Mediterranean and Central Africa*".[82] Textiles, glass beads, essences and other goods from the north were exchanged for slaves, ivory, ostrich feathers, leather, hides and gold from the south. Gold and slaves were among the most significant trading goods.[83] However, since the time of the African explorers of the nineteenth century, the gold trade from the African interior had developed to a greater extent via the European trade settlements on the African west coast. At the start of the nineteenth century, Ghadamis was primarily an important post for the trans-Saharan slave trade (Fig. 6.40). This was also attributable to the good relationship of the town with the surrounding Adjer-Tuareg, who provided camels and guaranteed protection during transport, as well as being hired as travel guides for caravans. With the French conquests in North Algeria (1830–1840) and the banning of the slave trade in Tunisia (1840s), Ghadamis saw itself

Fig. 6.39 One of the ancient streets of Ghadamis. Shade is bestowed by the palm trees

Fig. 6.40 Former slave market of Ghadamis ④

robbed of its most important export markets. Added to this came high Ottoman taxes and an increasing insecurity of the trade routes, especially the route to Tripoli. In the last third of the nineteenth century, the decline of trans-Saharan trade in Ghadamis became apparent.

In the mid-twentieth century, there was still trade in Ghadamis, although it only took place on a local level. The negative developments also impacted craft activities as well as agriculture. The artisans, once very affluent, and since the eleventh century famous for their hides, have almost ceased to exist, they lacked raw materials.[84] Besides this, agriculture was in decline at that time. The extremely unequal property situation had a particularly limiting effect on cultivation. The sale of property was not permitted. Furthermore, income had to be divided within each family, which considerably minimised personal income. Finally, families consisted of up to 100 people.

The mechanisation of cultivation was hindered by the walls that surrounded each little garden. In fact, the traditional cultivation areas within the old borders of the oasis were more and more given up and new fields were cultivated. The intensification of cultivation in these new areas did not, however, lead to the self-sufficiency of the oasis inhabitants.[85] The developments influenced the structure of the oasis, which had always been characterised by clear

Fig. 6.41 The oasis town of Ghadamis is surrounded by a ring of date palms. The areal photograph of Ghadamis was made during the Italian colonial period

social differentiation. The social structure of many oases underwent a change through the abolition of slavery. New ways of earning a living became apparent, which prompted former slaves to leave the oases. Now farmhands and tenants gained access to education for their children. For themselves, they found rather better paid work, for example in the oil industry. Whilst these groups profited from wide-reaching social and economic changes, nomads found it hard to adapt to the upheavals because of their mentality. The result was the social decline of this group.[86]

6.3.4 Ghadamis—A Border Town in a Border Triangle

Today, about 7000 inhabitants of the oasis—primarily Berbers and Tuaregs—live in the new town with modern new-build areas, streets, schools and social facilities, which were built outside the old town by the government in the late 1970s. Many inhabitants return to the old town in the summer as the architecture here offers better protection from the heat. Furthermore, many Ghadamis people try to preserve the apartments and gardens of their forefathers.

Today Ghadamis is a border town in the triangle of Libya, Tunisia and Algeria. The inhabitants of the town make their living primarily for working as state officials and rather less from agriculture. Between the years 2006 and 2010, in the course of al Gaddafi's policy of openness, the Ghadamis people had attempted to build up a new economic basis for their town—tourism. Countless travel agencies were established. A circumstance that was very helpful for its marketing was that in 1986 the walled centre of the old town was given the title of UNESCO World Heritage Site. This decision was based among other things on the long tradition of settlement in this place. Ghadamis, whose heyday of trans-Saharan trade had come to an end and could not compensate for the shortfall in its economic base, could only profit from tourism. Tourism brought travellers to the town again and with them recovery and new job opportunities.[87] This promising development, however, was halted by the end of the al Gaddafi regime.

6.4 Mizdah—Important Crossroads in Western Libya

Salih al-Mahdi Khalifa

Mizdah is situated 85 km to the south of Gharyan. It is considered to be the first desert centre to the south of the western mountain, connecting al-Qibla, al-Hammada and the Fezzan with the civilizational areas on the Mediterranean coast. The city is located on one of the bifurcations of the Suf al-Jin valley, at the intersection of latitude 31.25° north with longitude 13.05° east, and at a height of about 400 m above sea level.

6.4.1 Divided City

Throughout history, the city is divided into two populated conglomerations, namely Upper Mizdah, the biggest of the two, and Lower Mizdah which is situated to the east of Upper Mizdah. Since ancient times, the city, with its two conglomerations, constituted an intersection point for two Roman caravan routes. To the south of Mizdah, and along the main road towards the Fezzan, there is a chain of small fortresses and castles that had been erected with the aim of defending the caravan routes. This indicates that Mizdah, since ancient times, was one of the important dynamic areas which linked the coastal cities to the southern regions. Moreover, around the city, one comes across the remains of an old bastion with openings in its higher sections, along with three semi-conical towers called *"qasaba"*. The so-called qasaba of the Gazelle is 14 m high and is situated to the far north; the qasaba of Umar Belhaj is seven metres high, whereas the one called Qasabat Hnesh is nine metres high and is situated in the far south. These towers serve to control and defend the city.[88]

6.4.2 Crossroads of the Main Trade Caravans

Mizdah owes its importance to the fact that it is located at the intersection of a great number of caravan routes directed south, towards central Africa, across the Fezzan, Murzuq and Ghat regions. One of the most important routes starts in Tripoli and proceeds towards al-Kidwa (al Aziziyah), Bughilan, Awsaden, Ain Ghan, Gharyan, Birtsha, Mizdah, Wadi Tajeflet, abounding in the plant genista raetam, which is a desert shrub having small white flowers. The route proceeds via Umm as-Sabt, the Taskija valley and the Zamzam valley, which is rich in pasture, to Mudallal, renowned for its amaranths, and Busif, famous for its pasture, to Bir at-Tabuniya, an important meeting point on the caravan routes. From here, numerous paths branch out in the direction of the Fezzan oases, or the Algerian desert cities, and from there towards West Africa. From Bir at-Tabuniya, the route proceeds towards Bir at-Tayih, and from there towards Bir al-Hasi, whose depth does not exceed five feet. The route continues until it reaches Wanzrik, in the Wadi ash-Shati (Fig. 6.42).

There are also two routes from Tripoli to Murzuq, the shortest of which proceeds south of Gharyan in the direction of Mizdah, and then, it branches out into two routes. The first route stretches out towards the west, whereas the second one, the longest, goes towards the east. This was considered the usual route taken by caravans and by the postal service, in view of the availability of water resources in well-organised stations. This route also offered adequate means of rest, as well as numerous secure population centres, such as Bani Walid, Abu Nujaym, Soknahh, al Fugaha, and the oases and cities of the Fezzan. The trip usually lasted thirty days, whereas for the western route one needed a little more than twenty days. However, the caravans did opt for the latter route due to the lack of cities and villages for long distances along the route, as well as because of the lack of water resources. In view of the availability of many pastures in the nearby valleys, such as Suf al-Jin, Wamis, Mirsit, Imran valley, al-Jalila, al-Baqla and Dreydar valley, especially during spring and autumn, the caravans tended to direct themselves towards such places which abound in grass, tamarisk, acacia and terebinth trees, as well as sesame, which constituted the nutrition of the caravans' camels.[89]

A number of minor routes branch out from these two major ones, the most important of which being:

- The Zintan, Aqla Jifrat, Mizdah route.
- The Qsar Yefren, Mizdah route: The route passes through Yefren, ar-Rumiyya, Mawqa al-Hosh, Qsar Wamis, as-Shaqiqa. Here, one finds seven wells abounding in sweet water and whose depth is only one metre.
- The Mizdah, Fasanu waterwheels' route passing by Dirj and Ghadamis: The water resources on the route consist of wells, reservoirs and natural marshes. These are sometimes situated wide apart from each other and are known only to the most experienced. Along the caravan route, one comes across the Wadi al-Khayl, the an-Nasira well, a series of swamps marshes with salty waters, the most important of which being the Atwa marsh and the Atwa depression. In their vicinity, there is a well of sweet water three metres deep.
- The Nalut route which passes by ar-Rajil waterwheel, al-Kilab well, Fasanu waterwheels and Mizdah.
- The Mizdah route, via Nasma well, passing by the al-Mizraq spring and Sqafnana well ceeds via Umm as-Sabt, the Taskija valley and the Zamzam valley, which is rich in pasture, to Mudallal, renowned for its amaranths, and Busif, famous for its pasture, to Wadi Tabunia, an important meeting point on the caravan routes (see f*rom Tripoli to Mizdah*). From here, numerous paths branch out in the direction of the Fezzan oases, or the Algerian desert cities, and from there towards West Africa. From Wadi Tabunia, the route proceeds towards Bir at-Tayih, and from there towards Bir al-Hasi, whose depth does not exceed five feet. The route continues until it reaches Wanzrik, in the Wadi ash-Shati (Fig. 6.43).

There are also two routes from Tripoli to Murzuq, the shortest of which proceeds south of Gharyan in the direction of Mizdah, and then, it branches out into two routes. The first route stretches out towards the west, whereas the second one, the longest, goes towards the east. This was considered the

Fig. 6.42 Mizdah was probably established in the mid-second century BC and was called Musti Come by the Romans. The region around Mizdah is penetrated by a large number of valleys. Inscriptions on rock walls found in Wadi al-Khail, south-east of the oasis, show that the region had fertile grazing land for thousands of years

Fig. 6.43 Mizdah has some old houses that mark the old town (photograph on left). A well situated right on the edge of the old city

Fig. 6.44 German traveller Heinrich Barth stayed in Mizdah for some time on his way to the south (Barth 1977, 198)

usual route taken by the caravans and by the postal service, given the availability of water resources in well-organised stations. This route also offered adequate means of rest, as well as numerous settlements, such as Bani Walid, Abu Nujaym, Soknah, al Fugaha and the oases of the Fezzan. The trip usually lasted thirty days, whereas for the western route one needed a little more than twenty days. However, the caravans did not opt for the latter route due to the lack of cities and villages, as well as because of the lack of water resources. In view of the availability of many pastures in the nearby valleys, such as Suf al-Jin, Wamis, Mirsit, Imran valley, al-Jalila, al-Baqla and Dreydar valley, especially during spring and autumn, the caravans tended to direct themselves towards such places which were abound in grass, tamarisk, acacia and terebinth trees, as well as sesame, which constituted the nutrition of the caravans' camels (Fig. 6.44).[90]

6.5 Al Qaryah—Southern Outpost of the Roman Empire

Said Hamid

Al Qaryah al-Gharbiya, or the Western Village, is situated in the red rocky desert, about 310 km southeast of Tripoli. It is located at latitude 30.20° north and longitude 13.45° east.

6.5.1 Southern Outpost

The importance of al Qaryah al-Gharbiya can be dated back to Roman times. During the reign of Septimus Severus, in the year 201, a fortress was built there to protect the Tripolitanian borderline. Moreover, the fortress enabled the Romans to extend their rule over the southern regions, thus penetrating the desert shield formed by the Libyan tribes. The Romans were able to control the commercial route that linked Tripoli to the city of Garama (Jarmah), whose people controlled the desert caravan trade. One may state that the western village was like a bridge that linked the economy of the Mediterranean region, on the one hand, and the economies of the oases, on the other hand.

The western village witnessed the excavation of some old archaeological sites located between the eastern depressions, the area of the fortress and the present-day oasis. These sites spread over an area of about six to ten hectares, and they pointed towards considerable human habitation. Such habitation was facilitated by the availability of water and date palm groves. The western village owed its importance to its location on the caravan trade routes linking Tripoli to the sub-Saharan countries. Trade caravans used to carry merchandise from the Mediterranean regions towards the south, proceeding from Tripoli towards al Aziziyah, Gharyan, Mizdah, until they reached al Qaryah. From this

Fig. 6.45 Remains of a building and a mosque in al Qaryah al-Gharbiya

Fig. 6.46 During Antiquity and the Middle Ages, al Qaryah al-Gharbiya had been a transit centre

village, the caravans would leave for Tamzawa and, at this point, two sub-routes branched out, one towards the west, to Idri, and then southwards towards Jarmah, Awbari, Ghat, and from there to Timbuktu or to Kano. The second sub-route, in turn, branched out in two other sub-routes. The first went towards Sabha and then south-eastwards in the direction of Zawilah, Tmissah, al Kufrah, al Awaynat and al Fashir, whereas the second route left Sabha towards the south-west, to Jarmah and Tasawah. From Tasawah, two routes branched out, one proceeding to Ghat, the other to Murzuq and Taraghin. At this point, the route headed southwards to al Qatrun, Madrusah, Tajarhi, Kuwar and Kuka. In Roman times, there was a route that served military purposes. It linked Abu Njim, al Qaryah al-Gharbiya and Ghadamis. During the third century, three Roman fortresses were built in Ghadamis. These constituted the southern border of the area subjected to Roman control. The road between the western village and Ghadamis was about 412 km long.

During Antiquity and the Middle Ages, al Qaryah had been a transit centre where caravan traders would stop for their rest and for the replenishment of their water supplies from the available wells and water sources. Moreover, the traders also procured dates and cereals, particularly during seasons of abundant rainfall. The village gradually declined as well as its population, and during the nineteenth century, only a handful of people lived here. According to Barth, the place then was avoided by the caravans (Figs. 6.45 and 6.46).[91]

6.6 Adiri—a Transit Station and Meeting Point of Commercial and Pilgrim Caravans

Najmiya as-Sadeq at-Tellisi and Mansour El Nayedh

The oasis of Adiri is situated in the remote west of Wadi ash Shati, one of the great dry valleys which cross through the Fezzan region. It is about 180 km away from the city of Sabha. Adiri is located at longitude 27.30° north and latitude 13.05° east.

6.6.1 Transit Station in a Semi-Arid Area

Geographically, Adiri is situated in a semi-isolated area surrounded by sand dunes on all sides. Idhan Awbari borders Adiri on the west and the south-west, separating it from Wadi al Hayat in the south. In the north, it is bordered by the hills and valleys of the Hasawinah Mountain, whereas in the east it is bordered by Wanzrik and Tmisan. By virtue of its geographical location, Adiri constituted an important strategic centre on the caravan routes directed northwards and southwards. It was also important from the historical and cultural points of view, as confirmed by the remains of castles, mosques and tombs, the construction of some of which goes back to historical periods following the Islamic conquest of the Fezzan region (Figs. 6.47 and 6.48).

Adiri's desert climate is characterised by soaring temperatures in summer and the scarcity in rainfall over winter. During spring, it is exposed to north-westerly and southerly winds laden with dust. These are the famous *qibli* winds. Adiri owed its commercial and economic importance to the availability of underground waters as well as to the abundance of orchards and date palm trees. Moreover, a great number of wells are scattered around the city and along the roads leading to it. These are shallow wells. Their depths range very often between eight and ten feet. In 1998, it was calculated that there were about twenty wells and fifty springs whose waters are fit both for drinking and for agricultural purposes.

Adiri's main economic activity consists of agriculture. Fresh and dried dates are among the city's most important products, apart from barley, grain, canes, and limited quantities of corn which suffice for the local consumption. Moreover, vegetables are also cultivated in winter and in summer. Today, these include, for example, tomatoes, melons, peppers, onions and garlic. The inhabitants of the area are also engaging themselves in the herding and rearing of camels, goats and sheep.

6.6.2 Commercial Activity

Adiri witnessed a thriving commercial and economic activity during the eighteenth and the nineteenth centuries. This was due to the favourable security conditions and stability in the region as well as to the presence of water resources, including running springs and wells situated along the routes leading to Adiri. During this period of time, Adiri witnessed considerable commercial caravan activity; it became an important centre and station on the caravan trail. For this reason, Adiri turned into an important station that supplied caravans with water, fresh and dried dates and other necessary food. By the end of the nineteenth century, the caravan activity declined because of the discovery and adoption of alternative routes. The decline was also due to the expansion of the colonialist movement across the sub-Saharan countries. This development led to most commercial routes being re-routed, favouring rivers and other routes heading towards Western Africa.[92]

Adiri represented a transit station and a meeting point for a number of important commercial caravans as well as pilgrims' caravans hailing from the Arab Maghreb which headed northwards, southwards or eastwards towards the

Fig. 6.47 Modern Adiri is a town in the Wadi al Shati district, today home of 4600 people

Fig. 6.48 Old town of Adiri was a transit station and meeting point for commercial as well as pilgrims' caravans

holy places. Among the most important routes, linking Adiri to the Mediterranean Sea and the south, one could mention:

- The route coming from Ghadamis proceeded to Sabha and from there branched out in two directions—the first towards Zawilah, and the second towards Jarmah and Awbari, continuing in the direction of Ghat and West Africa.
- A route also headed out from Adiri towards Jarmah via the sand dunes of Idhan Awbari and from there proceeded to Murzuq, al Qatrun, Tajarhi and thence to central Africa.
- A route left Adiri towards the north via Timzawa and proceeding in the direction of Darj until it reached Tripoli.
- Another route left Adiri towards Ash Shuwayrif. From there, the route branched out in two directions, the first heading towards the cities of the Jufrah depression and then to Sirte and Banghazi, whereas the second proceeded northwards towards al-Qrayyat reaching Mizdah and then Tripoli.
- The desert route which left Tripoli towards Tarhuna and then Bani Walid. At this point, the route intersected with the pilgrims' route at Jufrah. From there, the route proceeded towards Zillah, Awjilah, al al Jaghbub and Siwa, and from there across the Sinai Peninsula towards holy Mecca.

6.7 Al Fugaha—Isolated Oasis in the Libyan Desert

Said Hamid

The al Fugaha oasis is situated to the south of the city of Sirte and north-eastwards of the city of Sabha on the western edge of the great central Haruj volcano and lava field. The oasis is located at latitude 27.50° north and longitude 16.15° east.

The al Fugaha oasis (Figs. 6.49 and 6.50) was considered one of the most isolated oases in the Libyan Desert and among the poorest in superficial natural resources. This was due to its extreme climate and its isolation from the great centres of culture and civilisation. In the past, the Arab travellers and geographers did not pay much attention to the village. Most historical and geographical sources do not make any reference to it. The few sources referring to the village simply stated that halfway between Tmissah and Zillah, there was a campsite belonging to the Waddan people, called al Fugaha.

al Fugaha was considered a transit station for some commercial caravans heading north or south. The oasis was linked to a road which started at Sirte and proceeded towards Tagreft and from there towards al Fugaha. Moreover, another road linked Ajdabiya, Maradah and Zillah to al Fugaha. This oasis was also linked to Sirte via the cities within the Jufrah depression, Waddan, Hun and Soknah. The road which linked the oasis of al Fugaha to the southern regions proceeded towards Tmissah, Zawilah and Murzuq or alternatively headed directly from al Fugaha towards Zawilah. From Zawilah, the road branched out in different directions. The first route headed towards al Kufrah, Waw al Kabir and Waw an Namus and from there to Sudan. The second route proceeded from Zawilah towards the south-west in the direction of Majdul, al Qatrun, Tajarhi, Kuwar and Kuka. A third route left Zawilah towards Umm al-Aranib and Taraghin, reaching Murzuq, Tasawah, Jarmah, al Awaynat and Ghat. From Ghat, the road proceeded towards western African cities, such as Kano and Timbuktu. al Fugaha still retains its old archaeological landmarks. In fact, in its vicinity, one comes across the remains of palaces which used to guard the caravans that crossed that region. Towards the end of the nineteenth century, and as a result of

Fig. 6.49 Oasis of al Fugaha is one of the most isolated desert oases. It is inhabited by Berber-speaking locals

Fig. 6.50 Old town of al Fugaha

the end of caravan trading across the borders, many inhabitants emigrated from the oasis to other regions, particularly to the city of Sabha.[93]

6.8 Zillah—Oasis of the Palm Trees

Mansour El Nayedh

Zillah is one of the oases of the Jufrah depression, which is a range of Libyan Desert depressions that stretch from the Ghadamis depression in the west to the al Jaghbub depression in the east. Zillah is located within a conglomeration of cities and oases, at latitude 29° north and longitude 17.30° east.

Although geographically, Zillah falls within the subsidiary depression of the great Jufrah, and administratively it belongs to the region consisting of the cities of Hun, Waddan, Soknah and al Fugaha. Externally, the depression resembles a triangle, with its base connecting Waddan and Soknah, whereas its peak coincides with Zillah. This depression is renowned for its fertile lands and the abundance of groundwater, which is close to the surface. Its small farms and the date palm groves extend to areas far away from the city centres and the oases of the region.

In the south, the Zillah partial depression borders with the Harudj Mountains as well as the as-Sawda, al-Qarrat and as-Sarir region. Westwards, it borders with the Waddan Mountains and the sand dunes region which almost separate it from the Jufrah oases. Northwards, the depression borders with the Tagreft depression, which also includes some highlands, low grounds and the small rocky tables which extend from the west to the east. Eastwards, it borders the sand dunes which almost separate it completely from the oases of Jalu and Awjilah and from the Maradah depression.

In view of its geographical location, Zillah falls within the transitional sector between the southern dry desert region and the climatically moderate coastal regions in the north. This area, which is situated within the desert region, with its high temperatures in summer and cold weather in winter, is characterised by small amounts of rainfall in winter. In fact, the yearly rainfall does not exceed 50 mm, and therefore, life in the oasis depends completely on the groundwater, which is at a depth quite close to the surface, and on its sandy, limy desert soil which formed as a result of the harsh natural and climatic conditions. At times, the soil manifests itself in different forms such as sand dunes, gravelly ground or desert terrain.

6.8.1 Transit Centre in the Middle Ages

The geographical and historical writers such as al Idrisi, al Himyari or Abu al-Fida refer to the area as Zala, whereas al

Bakri, in his *Kitab al-istibsar,* refers to it as Zalla. Later, the name became Zillah. Many Arab geographers and travellers of the Middle Ages described it as a city with thriving markets, a flourishing trade and a valorous, generous population. In the year 1862, the German traveller Rohlfs, who had visited the oasis several times, estimated the population of Zillah at approximately 1200 persons.

According to popular tradition, the ancestors of the present inhabitants reached Zillah about one thousand years ago. Before the Islamic conquest, the area was inhabited by the Mzata tribes. These were the original inhabitants who had resided in the area and had adapted themselves to the climatic conditions prevalent at that time, making use of the fertile lands of the oasis to secure their agricultural needs.

The Arab Muslim leader Uqba ibn Nafi took over Zillah during his campaign to conquer Zawilah, in the year 643. From then onwards, the Arabic influence prevailed and the inhabitants started speaking the Arabic language. During the twelfth century, the oasis acquired its commercial importance, becoming one of the transit centres of the desert trade. During the Ottoman period, the region administratively formed part of the Fezzan, in view of its dependency upon Jufrah. Subsequently, after the abolition of the federal system, it became dependent upon the province of Jufrah, within the Fezzan region, comprising five districts: Zillah, Hun, Waddan, Soknah and al Fugaha.

6.8.2 Economic Life

For a long time, agriculture was the foundation of the economy of the oasis. Zillah was renowned for rich and arable land as well as for the big variety and excellent quality of its dates. In addition to dates, products such as corn of the *qafuli* type, maize, barley, wheat, tomatoes, onions, melons, cabbages and legumes were cultivated in the oasis. These agricultural products were mostly produced for local consumption. Whenever there was a surplus of produce, it would be given away to the passing caravans. In fact, Zillah owed its importance to its extensive groves of date palms. In

Fig. 6.51 Turkish fort of Zillah towers above the area

1906, Abd al-Qadir Jami estimated that there were 120,000 date palms. This figure was later confirmed by Jamal ad-Din ad Dinasuri in his book *The Geography of the Fezzan*. The date palms spread out beyond the oasis, covering great areas which may reach out as far as 50 km to the surrounding area.

As mentioned before, Zillah was located on the main caravan routes from Sirte and Banghazi. The caravans used to be loaded with grain, barley and traded products to be exchanged with the inhabitants and the merchants of the oasis. Usually, the exchange was done on the basis of barter, especially in the case of dates, which were available in large quantities in addition to locally manufactured products made of palm leaves, wool and fur. But some merchants paid with money. Among the currencies in circulation, one could mention the Austrian Riyal which was minted in 1780 and was widespread across North Africa at that time. Zillah's importance was not restricted to the exchange of products between the local inhabitants and the merchants, but it was also considered a market and a rest area for travellers. It was a transit station for caravans directed south, towards the Fezzan region and from there to central Africa. It was also a halt for caravans in eastern direction, i.e. towards the oases of Banghazi, al Jaghbub and Egypt, and from there on to the Hijaz. This was called the pilgrimage route. It linked the west with the east, and vice versa. Zillah became famous for its commercial markets which used to be frequented by merchants from nearby areas, who traded and exchanged products and merchandise. The oasis was now and then influenced by adverse climatic and economic conditions, leading to a commercial recession in some seasons, especially in summer, due to the high temperatures.

6.8.3 Meeting Point of Trade Caravans and Pilgrims

The oasis of Zillah became very significant due to its strategic location as a passage for pilgrims to and from

Fig. 6.52 Remains of the old town of Zillah

Mecca as well as for the trade caravans (Fig. 6.51). The route used to start off at Marrakesh, Algeria and other regions, proceeding via Ghat and Awbari, then to Sabha and north-eastwards towards al Fugaha, until it reached Zillah. From there, the caravans would resume their journey eastwards, towards Banghazi, or north-westwards, towards Sirte. Following the coastal road, the caravans would then proceed towards Egypt and from there to the holy places. On the return journey towards the Arab West, the caravans stopped at Zillah's market to exchange products and merchandise from Egypt. Moreover, the oasis was bound with strong economic, social and intellectual ties with its neighbouring oases, particularly with the Jufrah oases, Maradah, Jalu and Awjilah, renowned for its commercial, economic and religious activity.

Among the most important commercial routes which passed through this region was the so-called middle route which left Tripoli, passing through Tarhuna towards Bani Walid, Abu Nujaym, reaching the Jufrah oases, Waddan, Hun and Soknah. From there, the route proceeded eastwards, towards Zillah, and from there to Maradah, Awjilah and Jalu. At this point, the route proceeded either southwards, towards al Kufrah, and thereafter towards the Sudan, or eastwards, in the direction of Banghazi. The route continued towards Egypt, through Siwa, reaching the Nile Delta and al-Fayyum, and from, there it progressed towards the holy lands. Zillah benefitted from some pastures, seasonal rivers and wells of potable water. They secured enough water supplies for men and camels along the routes linking the various neighbouring cities and villages. In the east of the oasis, on the route towards Awjilah, the so-called Rakb well was located. This was one of the oldest wells in the region, characterised by the abundance and the sweetness of its potable waters. There was also the so-called Sabil well, situated north of Awjilah. This did not fall under a monopoly and was available to everybody. The caravans had drawn much benefit from this well, satisfying their own needs and those of their animals. Moreover, there were other wells in the middle of the oasis upon which the inhabitants of the oasis officially depended for their own personal use and for their agricultural needs.

The founding and development of the city of Murzuq, in the middle of the sixteenth century, coupled with the change in the desert route which went through Jufrah towards Abu Nujaym and Bani Walid to Tripoli, led to an economic decline in the Zillah oasis and in its commercial significance (Fig. 6.52).[94]

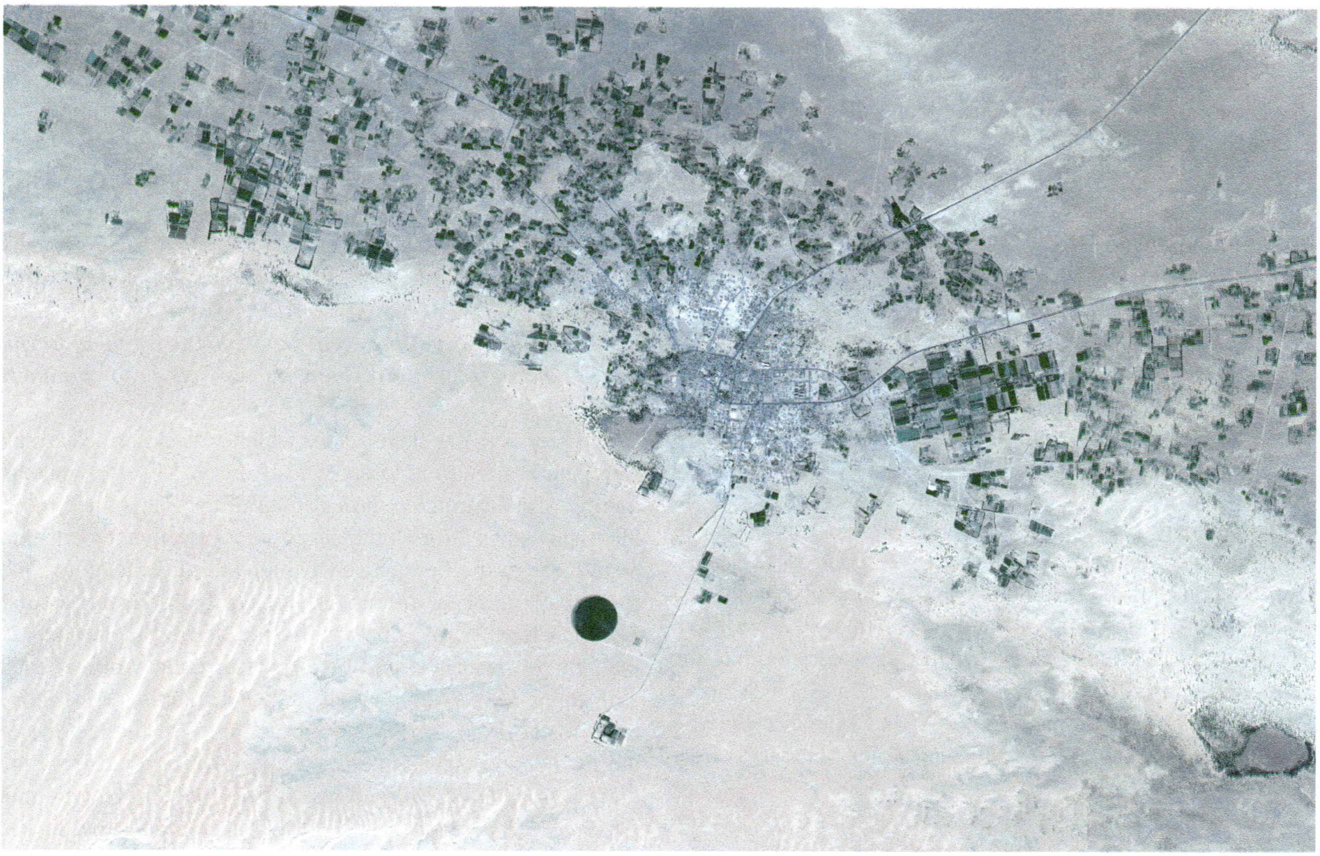

Fig. 6.53 Quickbird image of Murzuq: in the end of the sixteenth century, the oasis gained more importance than Ghadamis or Ghat

6.9 Murzuq—Historic Heart of the Fezzan

Jacqueline Passon, Klaus Braun and Joschua Metzger

Murzuq is an oasis town and the capital of the Murzuq District in the Fezzan region of south-west Libya. It lies on the northern edge of the Murzuq desert, an extremely arid region of ergs or great sand dunes which is part of the greater Sahara Desert.

Murzuq—only a few towns in the Sahara have a name with such a ring to it. As the historical centre of the Fezzan region, trade hub and most significant slave market of the Sahara, it was well-known far beyond the borders of Africa (Fig. 6.53).

Already in the fourteenth century, Murzuq had established itself as the centre of Fezzan and retained this position until well into the nineteenth century. The town hosted the headquarters of the local authorities. Its significance, however, was based on its function as a mid-point of trans-Saharan trade. Many trade routes crossed here, linking together sub-Saharan Africa and the Mediterranean, such as the famous Bornu Road running between Tripoli and Lake Chad. The most important "goods" were black slaves, who made up 90% of sales. The best of times, however, lie well in the past of this legendary caravan town. The ban on slave trading and recurrent malaria epidemics caused by the salt marshes near the town led the way into insignificance. Today, the good times are rarely remembered: the kasbah, situated in the north-west of town, and the remains of the once famous women's market can only bear insufficient witness to the former significance of the place.

Pioneers of African exploration like Friedrich Hornemann, Gustav Nachtigal, Heinrich Barth, Gerhard Rohlfs and James Richardson made stopovers many times at the oasis and left to posterity their reports on the town.

6.9.1 Murzuq—The Old Capital of Fezzan

In antiquity, Murzuq belonged to the territory of the Garamantes, and its centre Garama was 60 km north-east of the oasis. But its roots are lost in the dark past of Fezzan history. Around the year 1310, a pilgrim named Sidi al-Muntasir al-Muhammad is said to have stopped off here on his way to Mecca, where until this time there were only a few huts. Consequently, further settlers came to the place and the stone and mud fort, *kasbah*, and the town wall were built. The history of Murzuq began, which was to develop into the economic and political centre of the Fezzan. That Murzuq took on this role dates back to the Awlad Muhammad, who brought the Fezzan with its fertile and water-rich oases under their rule.[95]

The founder of the Awlad Muhammad was a *sharif* from Morocco or Mauritania named Muhammad al-Fasi. It is said that he arrived first in the Fezzan as the leader of a pilgrimage caravan of the west and that he was requested by the local Fezzani rulers to take over political control. This story, true or not, marks the reinvigoration of pilgrim traffic and the slave trade. A decisive role in the founding of the town was presumably played by its situation at a crossroad of a pilgrimage route from Cairo via Audjila and Ghat to Timbuktu and a long-standing trade route between the Mediterranean coast and the African interior. Particularly significant was the fact that this route from Kuwar, a region in modern-day Niger, through Fezzan and Murzuq to Tripoli or to Banghazi, was the most important route for the slave trade. At the start of the sixteenth century, Murzuq arose as a new centre of the trans-Saharan slave trade.

Situated in the middle of the desert, the town developed over the course of time into a trans-Saharan trade hub. The caravan routes, which led from the coast to the African interior, always passed through Murzuq. The caravans from Tripoli toiled through the desert for up to 30 days to travel a good 900 km. A large part of the journey led through a region called Hamadah al Hamra, a barren rocky desert. Here, it was possible to travel for days without finding water. Fezzan, as the settled area in modern-day south-west Libya is called, appeared to travellers as an oasis landscape. The region, whose capital was Murzuq, lays at 26° of latitude north of the equator and therefore right in the middle of the Sahara. However, fossilised groundwater ensured that the land was fertile enough for meagre agriculture in the gardens of the town. It was sufficient to feed a population, which according to reports in the 1860s was said to have amounted to up to 8000 inhabitants.[96]

At the beginning of the nineteenth century, Murzuq had developed into a significant economic centre in the Sahara. Due to its situation at a hub point of caravan routes, traders from all directions met here. In 1867, Gerhard Rohlfs wrote: "[Fezzan] *is a transport station for goods transports from the north, from Tripolitana and Egypt, to the south, to Bornu and the bordering Negro countries, as well as for products coming from Central Africa.*"[97]

Particularly, the slave trade, which was described by all European African explorers who came to Murzuq, and the trading of gold from countries south of the Sahara had made the trade routes through the desert very lucrative. Goods that went through the Fezzan mostly came through Murzuq. The town was primarily a transport station. In 1857, the scholar Heinrich Barth wrote that the traders offering their wares at the Murzuq bazaar would take most of their earnings back with them to their hometowns, e.g. Ghadamis.[98] The slave trade made up the overwhelming majority of the income of the town—and the dignitaries of the town profited from the duties. From 1812, Murzuq was in the hands of the Ottoman Empire. The potentates of the town understood how to use its unique location to its advantage. Rohlfs described the taxes that were levied on the importing and exporting of individual slaves. This method of making money was not unusual in the oasis towns of the Arab-Islam world.[99] The "*island situation*"[100] of the oasis offered no alternatives for accommodation or marketplace to the caravans passing through. In this way, the greatest slave market in the Sahara grew up in Murzuq. This was continued here until the middle of the nineteenth century. By the time the slaves reached Murzuq, they already had a 2-month crossing of the desert from Sudan behind them. From Murzuq, some went further to Tripoli or Banghazi, from where they were sold on to Europe and particularly to other parts of the Ottoman Empire. For others, the terminating point of their journey was achieved on reaching the oasis. The Fezzan traders who lived there often owned fields themselves and had a considerable need for workers for laborious cultivation. The same held true for the nomads, who also farmed cultivated areas in the oasis. In the mid-nineteenth century, slavery was banned in the Ottoman Empire. However, the trade in humans was too lucrative for it to stop immediately. The significance of the town was only to fall rapidly at the end of the century as the flow of slaves gradually dried up.[101]

The level of economic development achieved by Murzuq in the second half of the nineteenth century received varying evaluations in the travel reports of the time. Already in 1857, Barth emphasised that the consequences of stagnating trade through the desert were "*quite obvious*"[102] and Nachtigal confirmed this estimation in 1879, writing that Murzuq was a place that had "*long forfeited its significance as a trade location*".[103] In 1867, however, i.e. exactly during the decade in between Rohlfs saw that the economic downturn was at least becoming milder. He claimed that under the Ottoman government more stable times had come, whilst plundering had previously been the order of the day.[104]

The heyday of economic development in Murzuq had certainly already passed around 1850. The town depended on products from the north that could not be manufactured in the desert town. Goods that required a mechanical infrastructure for their manufacture were particularly difficult to produce in Murzuq. Nachtigal listed the goods which had been offered by traders from Europe and the Ottoman Empire: matches, cigarette papers, Turkish tobacco, confectionery, coffee sachets, cookware, bowls, Dutch cheese, pipe bowls, razors, needles, hand mirrors, scissors, knives, jewellery, bracelets, necklaces, fabrics, etc.[105] Meanwhile, the demand from the north for slaves and gold decreased throughout the nineteenth century. The demand for gold in Europe was increasingly met by gold from South America. In the course of the nineteenth century, the ban on slave trading gradually prevailed even in the Ottoman regions. However, at the time of the three African explorers, Barth, Rohlfs and Nachtigal, the trade still represented an important factor.[106] Nachtigal offers a good illustration of the connection described here: "*In Fezzan northern goods flowed together from Tunisia, Tripoli and Egypt, going into the landscape of the desert and the Negro lands, where the products of these countries became stockpiled. Throughout centuries so much gold had been brought to Fezzan from Timbuktu from the regions of upper Niger that until the beginning of this century the mitqual of gold here was only a fraction of the current value. Only as the gold supplies ran out were the Austrian and Spanish Taler introduced (…). From the Hausa countries came [various goods and] the most profitable and widespread commodity: slaves. (…) The old people of Fezzan still get lively when they talk of the times of their youth, when throughout the year the great pilgrim caravans came laden with gold from Timbuktu (…) and when several times a year the trade caravans to central Sudan passed through and were a thousand fold stronger on their return. (…) Bad trading relationships (…) the creation of new distribution channels and the weakening of the slave trade had the consequence of a sad regression.*"[107]

The overall turnover of Murzuq in the mid-nineteenth century was estimated by the African explorer Heinrich Barth at 100,000 Austrian Taler, where "*seven eighths (…) came from the slave trade*".[108]

6.9.2 Town View and Inhabitants Since 1850

The travellers of the eighteenth and nineteenth centuries described the oasis in great detail. A lively picture emerges when the reports are taken together. They describe walls and towers surrounding the town and a "*castle*" found in the middle (Figs. 6.54 and 6.55). That Murzuq was "*a very well-formed place with wide streets*" was noted as "*something highly wonderful in an Arab town*".[109] The sometimes very detailed descriptions can be compared with the plans drawn up by Heinrich Barth in 1850 (Fig. 6.56) and Moritz von Beurmann in 1862 (Fig. 6.57). Rohlfs wrote: "*The town of Mursuk makes a square comprising two English miles (…). It is surrounded by in places cracked (…) clay-built walls, which (…) are flanked by square towers. (…) A wide street (…) leads in a relatively straight line from west to east through the town (…)*".[110]

The formation of the town corresponded largely to its economic and political development. Until the time of Italian colonisation, Murzuq was considered to be the "capital of Fezzan" and accordingly housed the seat of the Ottoman Governor, and the *kasbah*, an old fortress and barracks (*kischka*) with its associated garrison. The castle still surmounts a hill in the north-west corner of the town (Map ⑬⑭⑮). It was built by Muhammad al-Fasi and contained the state offices of the Sultan, his residence and that of his sons. A mosque still adjoins the castle. In the various accounts, the central feature of the town was still said to be

Fig. 6.54 View towards the kasbah of Murzuq

Fig. 6.55 Kasbah as seen by Lyon

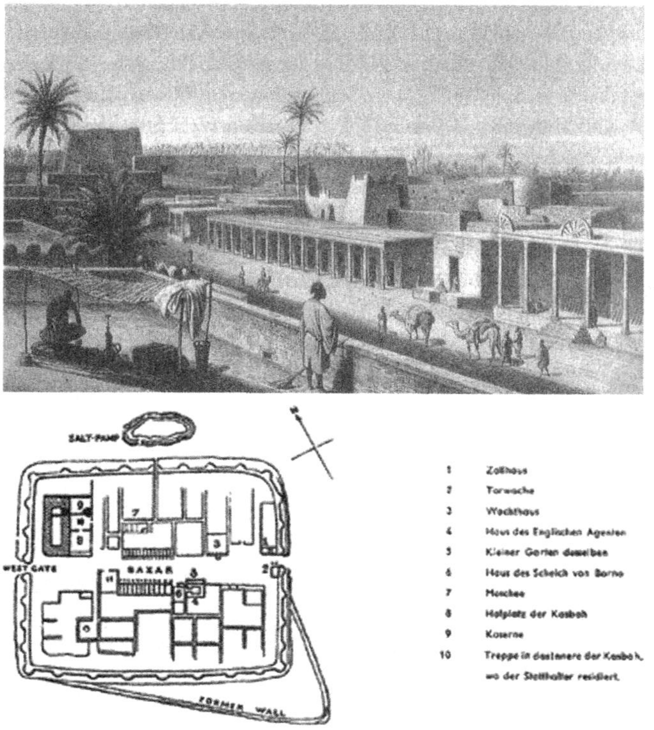

Fig. 6.56 View from the roof of the English consulate of the "Dendal", the main street of Murzuq, which crosses the town from east to west and ends at the square in front of the fortress (Barth 1857)

the wide main street, the "Dendal" (Figs. 6.58 and 6.59), with its small shops, mosques and town gates (Bab al Kebir ⑥, Bab al Garbi ⑦, Bab al Bachri ⑧), the custom house ③, the British Consulate ④ and a cemetery.[111] In the construction of the town, the travellers recognised traces that hinted at the influence of black inhabitants from the Bornu Empire.[112] The course of the roads with the "Dendal" as central axis divided Murzuq into quarters, placing it in clear contrast to the course of twisting Arab streets. Barth and von Beurmann mentioned that this type of construction differed strongly from the usual Islamic-Arabic styled town construction of a twisting *suq*.

Mentioned in the travellers' accounts, there is also a market, a great square in the middle of the settlement. The market was the living and breathing meeting place in the town. According to the Austrian traveller Chavanne, the varied activities of the traders with their goods from all corners of the earth could be observed here. His visit to the market in Murzuq led him to the conclusion that the population of the oasis differed greatly from that of Tripoli and was much more similar to that of the African interior. There was even a coffee house where the (rare) travellers from Europe mixed with the soldiers of the Ottoman garrison. During the day, however, the hot and dry air often drove people into their cool huts, where according to Barth they indulged in sensual pleasures like palm wine.[113]

All historical settlements in the Fezzan were built from materials that were available locally: in Murzuq, these were salty clay bricks. The layout features the typical, Eastern construction style of dry regions. Narrow alleys meander past twisting complexes of houses. The layout of the historical town construction is best described as "block principle". The living areas of a family are grouped around an internal courtyard. A flat roof can serve as a terrace. Besides the fortress complex, the town once consisted of three parts: "En-Nazla", which was situated to the north of the Dendal, contained lodging houses, inns and warehouses with an open area where caravans could rest overnight. South of the Dendal, the area was named "Ez-Zoueiya", so-called from the Sanussi *Zawiya* in the vicinity. Finally, in the south of Murzuq one could find the area "Er-Ras" (Fig. 6.60). The expansion of the town and also its number of inhabitants had presumably long exceeded its peak. Of the three former parts of the town, "Er-Ras", which once had been the principal commercial section and home of the shops and houses of the wealthy merchants, was already abandoned in 1857. Eldblom gathered together the available data on the number of inhabitants in Murzuq in the nineteenth century in the following way: in 1845/46, Richardson estimated the number of inhabitants to be 2000, in 1850–54 Barth and Vogel said 2800, in 1874 Rohlfs said 3000 but for the town and the surrounding area together he named a total figure of 8000, while in 1879 Nachtigal spoke of 3500 inhabitants in the town and a further 3000 in the proximity.[114]

Many travellers described the climate of the town as particularly uncomfortable because it was very dry and hot due to the lack of air movement. The salt pools of stagnant water on the northern edge of the town further added to the negative impression. This may well have made Nachtigal characterise the town as not particularly "*beautiful*". Furthermore, it was very unusual in the desert a malaria area. In 1869, Gustav Nachtigal wrote: "*The oasis of Murzuq lacks the characteristic of beauty, which most of her sisters enjoy; the town is built on a Sabha area (dried up salt lake) and so endows the inhabitants with malaria, otherwise seldom seen in this region. The houses are built from salty clumps of earth and they disintegrate with great delight from time to time, when no more than a tropical rain shower beats down, which luckily seldom happens.*"[115]

In the mid-nineteenth century, Murzuq represented the furthest south-west reaches of the Ottoman Empire. The Ottomans deployed military units to the garrison there.

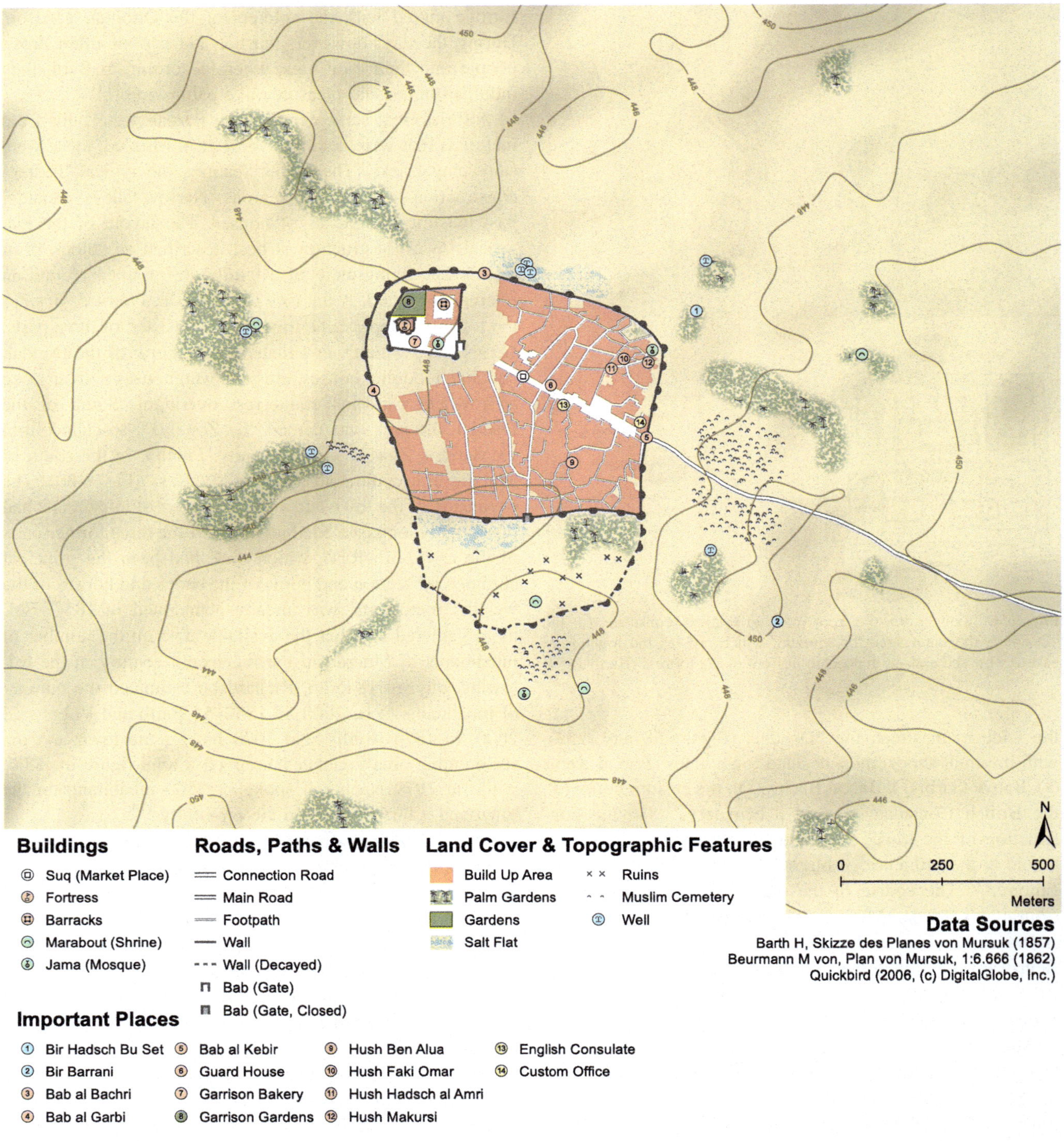

Fig. 6.57 Map of Murzuq in 1862

Outside the town, the influence of the Ottomans fell sharply. Various Tuareg tribes lived in their groups in the surrounding desert and mountain landscape. There was very little contact with Africa south of the Sahara. The way through the desert was too arduous. Only the caravan traders used these routes. Yet the difficulties were worthwhile, particularly when it concerned such valuable goods like gold and slaves.

Although the Italians had occupied Murzuq for the first time in 1914, the town was only definitively conquered in 1930. Under Italian rule, Murzuq finally lost its rank to Sabha as the political and economic centre in the

Fig. 6.58 'Dendal' in a photograph of today

Fig. 6.59 'Dendal' photographed during the Italian colonial period

Fig. 6.60 Traces of the former quarter Er-Ras, which was already abandoned when Beuermann came to Murzuq in 1862

Fezzan.[116] Today, there is a population of 43,732 inhabitants. In contrast to Ghadamis, Murzuq was not able, however, to profit from developments in the tourism business. Until 2010, only few tourists came to the town; it functioned merely as a transport station for desert travellers. The cause of this may have been the urban development of the place during the 1970s. In the course of national modernisation efforts, the historical townscape largely disappeared. The fortress complex with the *kasbah* from Ottoman times is still just about in existence, but measures for sustainable retention of the property must be urgently carried out. The southern part of the Medina with its traditional clay houses is still in existence. The state of the buildings is, however, extremely dilapidated. Currently, this part of the town is still inhabited, a circumstance which has up to now saved it from total collapse. For the most part, Murzuq has been robbed of every property that possesses any appeal to enable it to function as a tourist magnet.[117] Today, it is clear that even less remains of what the nineteenth-century travellers, for example, once described and glorified as to some extent romantic.

6.10 Sharba—Covered by the Sand of the Sahara

Said Hamid

The city of Sharba is situated deep in the Libyan Desert, 30 km to the south-west of the city of Tasawah. It can be reached from the city of Murzuq via Tasawah, a distance of 55 km, and from there to Sharba, which is located at latitude 25.50° north and longitude 13.30° east.

The remains of the ancient city of Sharba are scattered across an area of eight square kilometres within the Sharba valley, which is one of the valleys branching out from the Barjuj valley. Archaeological remains indicate that Sharba was a flourishing city, as a result of its strategic important location on the trade caravan route linking Jarmah with Niger and central Africa. In fact, this route left Jarmah towards the south, passing through Maknusa, Tasawah, Qsar Mara, and from there to Sharba, and then across the valleys leading to Kuwar and Zinder, until it finally reached the banks of the River Niger.[118]

6.10.1 Riddle in the Sand

The remains of this spot pose a big question. For no traveller or any Muslim historian has ever mentioned Sharba. al Idrisi, the author of *The book of pleasant journeys into faraway lands* (*Nuzhat al-Mushtaq*), which was written during the twelfth century, did refer to the land of Fezzan and mentioned Jarmah and Tasawah, praising them for their greatness. He stated: "*In the Sudan they call Tasawah 'Minor Jarmah', and between Jarmah and Tasawah there is a distance of one stage*".[119] Sharba was not mentioned in the early Arab sources, and it was not visited by the European explorers of the eighteenth and nineteenth centuries, nor did they mention it. Perhaps Sharba was one of the cities referred to by the Arab historian Ibn Abd al Hakam, who wrote about the conquest of the land of the Garamantes by Uqba ibn Nafi. He wrote about a big city on a hill in the desert, and whose inhabitants resisted Uqba, compelling him to take the city by force.

It is known that the Arab leader Uqba ibn Nafi led a campaign in 669, starting from the city of Mghamdas, in the vicinity of Sirte, and moved deep inside the Libyan Desert, conquering, one by one, its cities, oases and fortresses. Subsequently, he returned to Zawilah, perhaps via the al Qatrun route. From Zawilah, he returned to his headquarters in Mghamdas, after the absence of five months, which he spent in a jihad, aimed at spreading Islam. Thanks to this campaign, the Arab Muslims came to know the passage to the Ajal valley. Moreover, they could experience the shortest route that linked the Mediterranean Sea to the Sudan, which passed through al Qatrun and Tajarhi.[120]

6.10.2 Prosperous Town Deep in the Sahara

The archaeological remains of the city of Sharba indicate that it was a prosperous city, given its location on the caravan route. Rainfall, as well as underground waters which were close to the ground, secured the city's water supplies. It was located in one of the valleys. A huge wall with numerous towers used to encircle Sharba. There are also remains of two big buildings which might have been two fortresses. Sand covered most of the city's buildings, and nothing remained of them except high structures. In the vicinity of the city, on a low hill, there is a big burial ground that comprises a number of big circular tombs. These may be

Fig. 6.61 Pre-islamic graves: In the vicinity of the city, on a low hill, there is a big burial ground comprising a number of big circular tombs. These may be the tombs of the inhabitants of the ancient city of Sharba, and they resemble the tombs of the Garamantes period

the tombs of the inhabitants of the ancient city of Sharba, and they resemble the tombs of the Garamantian period.[121] British archaeologists tentatively identified the site as an Garamantian town.[122]

The prosperity of the city of Sharba continued even after the appearance of the camel in the Sahara desert. The camel was brought from Western Asia after the Pleistocene period. Rock engravings show that the camel soon appeared. According to Mori's chronology, the camel rock drawings were made during the period between BC 1500 and the end of the first century AD. The first reference to the camel in North Africa goes back to 46 BC. The utilisation of the camel in caravan trading and in agriculture soon increased. Rock inscriptions were also found in the city of Qarza and in other areas.

Following a change in the climatic conditions in the Libyan southern region, the temperature rose and the quantity of rainfall and humidity gradually decreased. As a result, horses could not penetrate the desert, whereas the camel ventured on new tracks crossing the desert. New desert commercial cities emerged during the sixteenth century, among them Murzuq in 1550. Soon after, it became the seat of the Awlad Muhammad tribe. As regards Sharba, one may say that it continued to be active for some time after that. In fact, it was connected with Murzuq via a number of fortresses known as the Umm al-Hammam fortresses, of which the Tqalqalat, al-Manashi and Tamira fortresses were the most important.[123] What is clear until now is that the series of three dates from different contexts confirms that occupation continued into the post-Garamantian period. The latest of the three dates belongs to the eleventh to early thirteenth century AD.[124]

After that, Sharba started to lose its importance and its inhabitants abandoned it. Nothing is left of it except its remains which point to its prosperity for a number of centuries. Sharba has fallen into a state of slumber.

A number of factors led to Sharba losing its commercial importance. Among them are the depletion of its water resources, and their penetration deep inside the ground, which made it impossible for the inhabitants to reach them. Another reason was the inexorable advance of sand, and the deviation from it of the trade caravans, which opted for new cities like Murzuq. These factors prompted Sharba's inhabitants to abandon their city. Sharba was covered by sand and became yet another old city which fell into oblivion (Fig. 6.61).[125]

6.11 Jarmah—Capital of the Garamantes

Said Hamid

The city of Jarmah is situated in the so-called Valley of Life, previously known as Ajal valley, about 160 km south-west of the city of Sabha. Jarmah is about 30 km east of Awbari. It is located at the intersection of latitude 26.40° north with longitude 13.00° east.

6.11.1 The City of the Garamantes

The Greek historian Herodotus, who lived in the fifth century BC, was the first to write about Jarmah. In the fourth book of his *Histories,* he wrote that after a journey of ten days from Awjilah, one reached another hill of salt, water and many date palms, as in other places, and this place was inhabited by the Garamantes. He added that they were in great number and that they planted the soil after having spread salt over it. These Garamantes, mounted on their chariots driven by four horses, and chased after the Ethiopian cave dwellers. Historically the Garamantes initially built their dwellings on the high summit of Zenkekra Mountain. Later, during the third century AD, they came down and settled at the foot of the same mountain. They left many inscriptions on rock, as well as drawings on Zenkekra Mountain, depicting cows, giraffes, horses and ostriches.

Historical sources refer to Roman raids against Jarmah led by Cornelius Balbus in the year 19 BC. During the first century AD, the city of Oea (Tripoli) sought the help of the Garamantes in its dispute with the city of Leptis regarding the ownership of some lands. This led to the Garamantes' siege of Leptis which lasted for many days, until the Romans intervened and put an end to it. Later, in the year 69 AD, the Roman leader Valerius Festus sought vengeance against the Garamantes and raided their territories. By doing so, he demonstrated the Romans' ability to penetrate the desert shield of the Garamantes. A third Roman campaign was led by Suilius Flaccus in 87 AD. Apparently, this campaign was launched for economic reasons, namely to reach the desert's commercial centres. The Romans succeeded in paralysing the Garamantes' movement towards these commercial centres, which were situated north of Jarmah, thus limiting this city's influence over the surrounding regions (Fig. 6.62).

During the second and third centuries AD, the region enjoyed a period of peace and stability. This enabled the Garamantes to build a new city in the innermost sector of the valley. They built many buildings with stone and mud, and their structures included columns with Ionian, Doric and Corinthian pillars. The new city of Jarmah was enclosed within defensive walls which could be penetrated by three gates.

6.11.2 Wind of Change

The Arab Muslims, under the leadership of Uqba ibn Nafi, took control of Jarmah in 663. The people of Jarmah were invited to accept Islam, and they did not fail to accept the

Fig. 6.62 Old Garama has been a domicile for the Garamantes, an ancient nation dwelling in the Fezzan. From the fifth century BC to the fifth century AD, the Fezzan was home to the Garamantian Empire, a city state which operated the trans-Saharan trade routes between north and south

Fig. 6.63 Garamantes' traces are still there among the ruins of the old city. The different types of cemeteries are scattered around the region. On the left, a Roman grave, and on the right the kings' graves of the Garamantes near Jarmah

invitation. The Arabs then directed themselves towards the desert castles (*qsur*) of the Fezzan and took control of them as well.

The Arab historians and geographers made very few references to Jarmah; however, it seems that during the eleventh and twelfth centuries, the city still retained its importance. According to the great traveller al Idrisi, Jarmah was located at a distance of less than one day from Tasawah, situated in the Ataba valley. These two cities were of the same size and had the same number of inhabitants. They cultivated date palms, white corn and barley, and they used to irrigate their fields and gardens with particular equipment which drew water from wells.

Peaceful times, particularly during the seventeenth and eighteenth centuries, led to the flourishing of the caravan trade in Jarmah. However, at the end of the nineteenth century, the oasis witnessed the decline of this desert commercial activity due to the fact that in Europe the demand for desert products, such as ostrich feathers and ivory decreased dramatically. Moreover, this period coincided with the opening of new, alternative and cheaper sea and land routes. In view of the drastic fall in their income, the caravan traders had no other choice but to give up their business. The caravan trade came to an end at the beginning of the twentieth century.[126]

6.11.3 Connecting Routes

Jarmah was connected to numerous caravan routes, either those coming from the north, from the south or from the west. The following are among the most important ones:

Linking Jarmah to the Mediterranean coast:

- The Ghadamis, Adiri, Awbari and Jarmah route.
- The Tripoli, Gharyan, Hamadah al Hamra, Adhan lakes, Awbari, the Valley of Life and Jarmah route.
- The Sirte, Jufrah oases (Zillah, Waddan, Hun, Soknah), Sabha and Jarmah route.

Linking Jarmah to the south:

- The Jarmah, Murzuq, Taraghin, Zawilah, Tazerbu, al Kufrah, al Awaynat mountains and Marwa route (in the Sudan).
- The Jarmah, Tasawah, Qsar Mara, Sharaba, Murzuq, al Qatrun, Fayalarju, and Kanem route.
- The Jarmah, Tasawah, Qsar Mara, Sharaba, Idhan valleys, Murzuq and Kuwar (in Niger) route.
- The Jarmah, Mathendous valley, Amsak Staft heights and Kuwar (in Niger) route.
- The Jarmah, al Awaynat Sardalas (near Acacus Mountains), Ghat, Wargla, Tamanrasset as far as the banks of the river Niger (Fig. 6.63).

6.12 Ghat—a Picturesque *qser* on the Frontier Between the Fezzan and the Algerian Sahara

Mansour El Nayedh

Ghat is situated in the south-west of Libya at a distance of about 1360 km from Tripoli, 582 km from Ghadamis and about 15 km from the Algerian border. It is located at the intersection of longitude 57.7° east with latitude 24.37° north.

Ghat was a *qsar* of the Sahara among the Tuareg Adjer and a major terminal point on the trans-Saharan trade routes linking the Mediterranean ports with the main commercial centres in Kano, Zinder and Agades. Moreover, Ghat is situated in an area which is ideal for human settlement, given the availability of underground water which is quite close to the surface. Its valleys are vibrant with vegetation and small trees of the acacia type, the Alhagi maurorum, the Ampelodesma tenax and the Branbah plant, in addition to the herb bushes which cover extensive areas of the valleys. Its desert climate is characterised by great heat in summer, cold weather and lack of rain in winter.

6.12.1 Important Cultural and Religious Centre

The city's origin is still shrouded in obscurity. It may go back to ancient times, because near the valley of Tanezzouft, whose valley lies in a north–south direction between the bank of primary sandstone on the side of the Tadrart in the east and the similar plateaus of the Tassilin'Adjer in the west, the land is fertile and the waters are abundant. It has been a transit area for commercial caravans plying the desert. Furthermore, numerous rock engravings and some necropolises found, for example, in al Barkat or Tin Alkoun, and indicate that the region was inhabited in ancient times. The origin of modern Ghat probably goes back to the fourteenth century. The name Ghat was first confirmed by the great Arab traveller Ibn Battuta who travelled the area as far back as the second half of the fourteenth century.[127]

Islam and the Arabic language probably reached Ghat during the fourteenth century, and it developed into an important religious and cultural centre of the region. For travellers of the desert, the oasis must have appeared as a big fortified city. Indeed, Ghat's Medina is fortified in an irregular rectangle of 700 by 500 m in area and surrounded by a wall made of stones and clay. Part of it is also the two suburbs, quite close to it, namely Tunein and Tedremit (Fig. 6.64). It has four gates leading to the city:

- The northern gate, known as Bab al Cher ⑤.
- The eastern gate, known as the Bab al Kalala ⑥.
- The southern gate, known as Bab Tafachachat ⑦. There once had been the Bab Eschef en Kena in the south ③.
- The western gate, known as Bab Tamalchat ④.

Ghat (Figs. 6.65, 6.66, 6.67 and 6.68) has been famous for its markets and these were the destination of many merchants. Dozens of them from different places lined the *aswaq*. The nineteenth-century traveller Muhammad al Otsman al Hachaichi remarks that most of the traders of Ghat were people from Ghadamis and Tripoli. Those groups

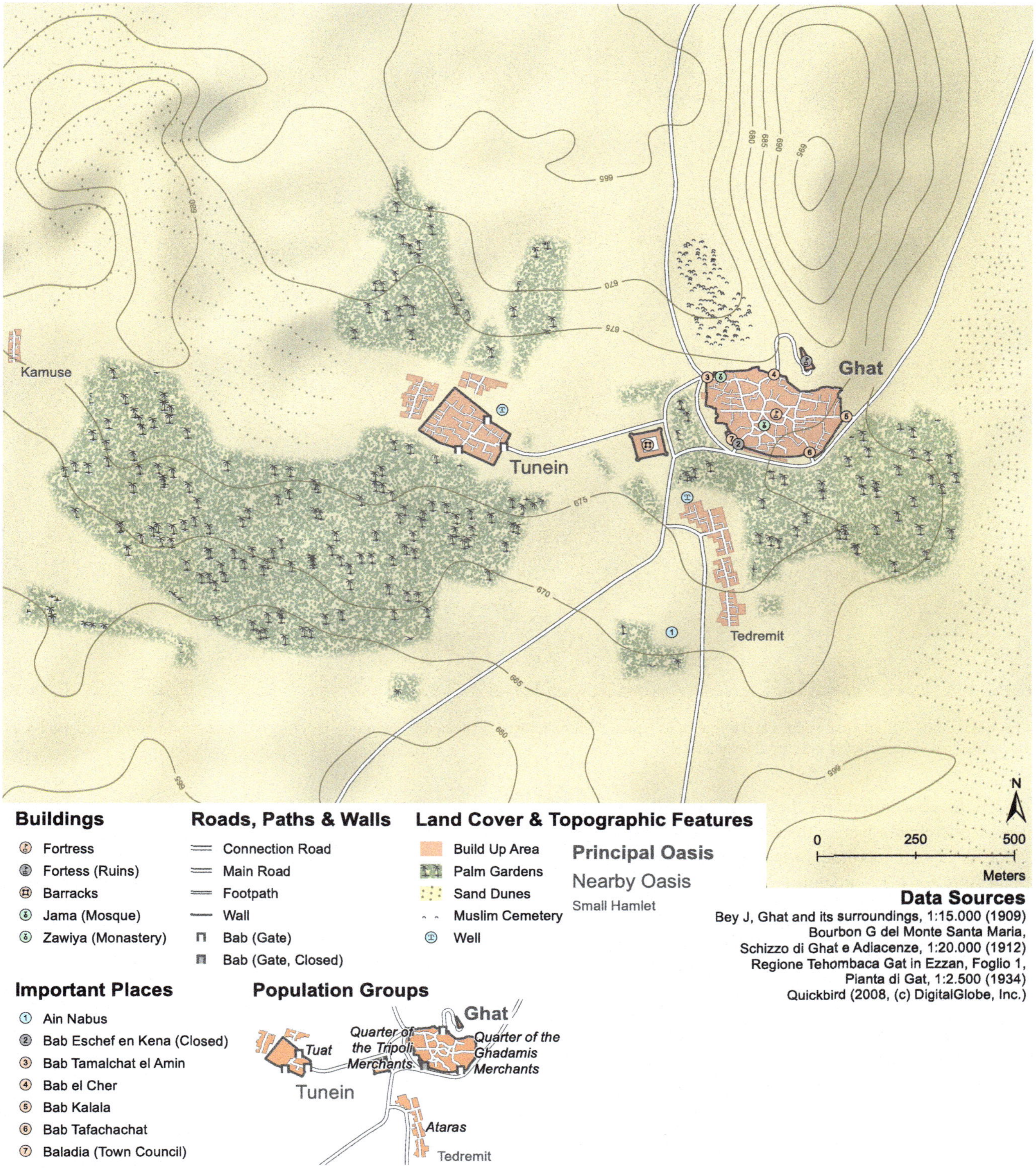

Fig. 6.64 Map of Ghat at the beginning of the twentieth century

Fig. 6.65 Old town of Ghat is built at the foot of Cocaman Mountain, which situated to the north, holding the fortress, which was initiated by the Ottomans and reconstructed later by the Italians

tended to concentrate on their own neighbourhoods. The Tripoli merchants were situated in the west of the Medina, while the Ghadamis traders had their quarter in the east of town (Fig. 6.64: Population groups). The Tuareg hired their camels to them. At the Tafachachat gate, which overlooked a spacious square, merchants from Ghat met their counterparts from many other regions.

There were also local markets which used to be set up in the city as well as outside. Two important markets used to convene twice yearly, one in summer and another in winter. They usually lasted at least for a period of about three months. It was the time when the large trans-Saharan caravans came to Ghat. There were also markets dedicated to women's needs, selling perfume, incense, herbs, clothing, etc. The traveller Richard Dasson described the arrival of trade caravans from Kano, Bornu and Twat in the year 1845, carrying all kinds of goods for the city's markets. One should mention that the caravan merchants did not arrive in Ghat exclusively from the desert regions, the Fezzan, Ghadamis and Tripoli, but also from Tunisia and Algeria. These traders would come to Ghat to trade and exchange merchandise, particularly during the month of October. The most important merchandise transported by the caravans to Ghat included slaves from the various Sudanic areas, ivory from Bornu, ostrich feathers, tanned leather, leopard hides and musk from Mali, rhinoceros horns from Wadai, Bajarmi textiles, Sudanese traditional crafts as well as products manufactured by the inhabitants of Ghat. The oasis acquired considerable revenues from passage dues paid by transiting caravans. The city and its traders benefitted from the exchange of products and from the city's vigorous markets.

In addition to trading activities, agriculture was considered one of the main economic resources of the oasis. The inhabitants cultivated grain, barley, white corn and various types of good quality dates which were at the centre of their diet. Moreover, vegetables and fruits such as pepper, onions, tomatoes, watermelons, peaches, figs and lemons were also grown in the valleys. Here, there was plenty of underground water, rather near to the surface. For the irrigation of the agricultural products, water was drawn from wells and springs. The inhabitants of Ghat were also engaged in animal husbandry, including the breeding of camels, particularly of

Fig. 6.66 Ghat was a major terminal point on the trans-Saharan trade route. It was a stronghold for the Kel-Adjer Tuareg federation, whose territory covered most of south-western Libya, until 1913, when the city was occupied by Italy as a colony. Many goods were traded in the old town. The most important merchandise transported by the caravans to Ghat included slaves from the various Sudanic areas, ivory from Bornu, ostrich feathers, tanned leather, leopard hides and musk from Mali, rhinoceros horns from Waddai, Bajarmi textiles, Sudanese traditional crafts as well as products manufactured by the inhabitants of Ghat

Fig. 6.67 Ghat photographed during the Italian colonial period

Fig. 6.68 Ghat photographed during the Italian colonial period (Castigliola 1930, 84–85)

the Mahdiya type, as well as goats and sheeps. Moreover, people from Ghat were engaged in such trades like tanning, the manufacture of traditional weapons, drums, leather shoes, products from palm leaves, woollen and fur textiles and tents. They also trained camels of the Mahri type, used in camel races, especially on social occasions.

For a long time, Ghat remained independent, governed by a hereditary *amghar* and was under the somewhat heavy protection of the Tuaregs. In 1875, the Ottomans installed a garrison there, which is situated on top of a hill besides the old town, and remained until 1914. As a major terminal point of trans-Saharan trade, Ghat was located at strategic crossroads linking four routes:

1. The western route towards Timbuktu, passing through Twat and Ain Salih. Merchandise would come from Timbuktu passing through Arwawan in the direction of the north-east through Algerian desert routes and reaching Ain Salih in Algeria about 900 km west of Ghat. Caravans would stop in Ghat on their way towards the east having traversed desert routes dotted with springs and wells of potable water.

2. The southern route, or the Bornu and Zinder route, passing through Tamanrasset. This was where the roads branching out from Bornu in Niger and Zinder in Mali. Both roads met near the Libyan–Algerian border, passing through the regions of Tamanrasset which offered some pasture and wells of drinking water. The route then reached the region of Ain az-Zan situated on the Libyan southern border. The caravan route moved north in the vicinity of Tadrart Mountain and the Janit area just 40 km away from Ghat. From Ghat, the route proceeded northwards towards the Mediterranean harbours.

3. The eastern route, or the Bornu–Ghadamis route, passed through the regions of Kuwar, al Qatrun, Tajarhi and Murzuq until it reached Ghat. The route started at Bornu and Agadez, and it proceeded northwards, passing through Atqiqmi and the regions of Kuwar. It followed some southern desert routes until it reached the western parts of Tibesti, known as Tammu. It moved north for about 180 km, towards Tajarhi and al Wigh, renowned for its spring of drinking water. The route then led further north through Mara territory, in the region of Madrusah and al Qatrun, west of the Bin Ghunaymah Mountain, with its rich vegetation. From there, it moved slightly towards the north-west until it reached Ghwat and al Baydan, about 200 km east of Murzuq. It then led to Murzuq and from there other roads branched out. The western road was the one which led to Ghat. It passed through Umm al-Hammam and Tasawah and then moved south-west through the Idhan areas of Murzuq, turning towards the south-west, in a semicircle, passing through the paths of Milt, close to the Akakus Mountains, towards Tin alKoun and al-Birkat, in the southern neighbourhoods of Ghat, which were rich in drinking water. In the vicinity of the paths of the Milt and Akakus Mountains, some vegetation, such as the ampelodesma tenax, and pasture for camels grew. Finally, the route reached the commercial city of Ghat. The distance between Murzuq and Ghat was about 546 km.

4. The northern route started from Ghat, at the northern gate, known also as Bab al-Khayr, and proceeded northwards through Tanezzouft valley and the al Awaynat region. Here, a number of wells were located, among them the Bilaj well. The route stretched further until it reached the western parts of Awbari which were well known for the abundance of wells and springs, such as Ain Taqhiri and Ain Azwa, and also for their rich vegetation. This route was quite easy for caravans. They reached the rocky desert of Tatghart, leaving behind them the Hasawinah Mountains to their east. Here, the caravans took supplies of water. This region was a sandy area of about 70 km, rich in wells and springs, among them the Tawil well, Adamir well, Yur Sida well and the Azar spring. The caravans then came to a mountain range in the limits of Ghadamis, that reached an altitude of 500 m above sea level, west of the red rocky desert. Moving northwards, along the Algerian border, the caravans arrived at the city of Ghadamis. From there, roads branched out, according to their importance, towards the north-east until ending in Tripoli.[128]

6.13 Al Qatrun—a Pearl in the Midst of Sand

Najmiya as-Sadeq at-Tellisi

Al Qatrun is located at the south-west of Libya in a depression known by the Italian writers as the depression of wisdom. Today, it is known as al Qatrun depression which is at about 510 above the sea level and at coordinates longitude 25° north and latitude 14.45° east.

6.13.1 Most Severe Arid Part of the Fezzan

The area of al Qatrun includes four small different villages called al Qatrun, Tajarhi, al-Bakhi and Madrusah. This region is the poorest area of the Fezzan region in its natural, agricultural and animal sources. It is located at the most severe arid part of the Fezzan due to its location at the heart of the Sahara which is famous for its aridity and high temperature. Generally, the depression is a number of small, sparsely populated villages within an area of about 60 km northbound. The area is bordered by the southern edge of Jabal bin Ghunaymah, the highlands of Debassa Mountain, the depression of Alweg al Kabir and the Idhan of Murzuq. The depression itself includes a number of surfaced wells of not high-quality water, which were used by the caravan travellers crossing the area towards Murzuq, Ghat and Tripoli or towards the south to the central African countries

Fig. 6.69 Al Qatrun is a flat oasis, where palm trees are plenty and surface water is nearby

through areas of Tania al Kabir, Ghiadar, Tania Saghira and Bir al Waar. The wells at the northern part of the depression are located in immediate proximity to each other, but the distance between the water holes increases as the route continues southwards to Tummo Well. Towards Murzuq, the next well is situated at Bir um al-Adam, and another well is located near the castle of Hatiyat Mastuta half way between al Qatrun and Murzuq.

The scenery on the southern route is characterised by hills as well as the picturesque highlands especially between Tajarhi and Tummo Well. These specific geographical features were used as landmarks to guide caravans and Sahara travellers. Regarding the western direction, the surrounding area is mostly characterised by high sand dunes (Idhan of Murzuq) that crouch eastwards to cover a large part of the depression.

6.13.2 Al Qatrun and Its Neighbouring Villages

The people of al Qatrun and the surrounding villages are mainly of Tubu tribes. They once came from Bornu and the area of the Tibesti Mountains. The other inhabitants are called the al Qatruns, who descend from the mix of Garamantes and the Arabic tribes which settled in the area. Popular indigenous literature recounts that Oulad Jaber (Sons of Jaber), the Marabouts who form the majority of al Qatrun village, descended from their grandfather Jaber, one of Saguia el-Hamra Marabouts, who came from the Maghreb to the area in the fifteenth century. Jaber is considered the founder of al Qatrun. The local people of this town are divided into four tribes: the Aoulad (Sons of) al-Haj Hamed, Aoulad Abel Rasoul, Aoulad Mohammed and Aoulad Abdullah. The next town is al-Bakhi, which was founded by a Marabout called Yahya, who came from Saguia al Hmara in the same period. In addition to the locals, the Tubu tribes formed the society of this town, and some of them moved into the nearby village of Madrusah. These Tubu tribes settled the region since ancient times and were governed by a strict tribal system (Figs. 6.69 and 6.70).

The economy of the area depended mainly on the field products for local consumption and on the palm trees, which formed the main food source of the area. In 1911, there were about 102 farms including 112 surfaced wells with 31,500 productive palm trees in addition to 17 camels, 124 donkeys and 228 goats. Furthermore, the wells of al Qatrun are relatively salty, while their depth ranges from one metre to three metres. Due to the vicinity of the water table from the surface, several kinds of grass and plants, which bear the weather aridity and the salty soil, grow in the area. In addition, palm tree forests extend for long distances from Hatiyat ad-Dakeer to Qsar Mariana and Bir Shibou near south Tajarhi and from the eastern side at Bir Mastuta, Hatiyat Umm al Edam, Majdul area and the surrounding villages of the sand dunes of Idhan Murzuq. Desert plants such as Acacia and Tamarisk as well as different local

Fig. 6.70 Old town of al Qatrun: The people, who live there belong mainly to the Tubu tribes

Fig. 6.71 Traditional objects are exhibited in the old town of al Qatrun

grasses also grow there and are grazed by camels and other domestic animals.

Dates were the most-traded product. They were produced locally and transferred to the Murzuq market to be sold to the passing caravans towards Tibesti, Zinder, Bardai, Timbuktu, Kanem and southwards to Lake Chad and the upper regions of the river Niger. By the beginning of autumn, large caravans used to come from the north to al Qatrun to exchange different kinds of local dates with the needed products in the area such as grains, spices and garments.

According to the census of 1954, al Qatrun's population was about 1654 persons distributed among four villages: al Qatrun (820 persons), al-Bakhi (510 persons), Madrusah (200 persons) and Tajarhi (124 persons). Most of the people depended on the traditional ways of field cultivation or grazing animals in a number of vicinal wadis. The people of the area also practised traditional handicrafts, which depended on local raw materials such as palm trees wood or fronds, pottery, wool and spinning and skin tannery (Fig. 6.71). Moreover, some of the inhabitants rent their camels to the traders. They also worked as Sahara route experts and as guides for the caravans. It is no surprise that a famous caravan leader, called Mohammed al Qatruni, has originated from that geographical area. He accompanied a large number of Arab and European travellers, and he was highly recommended by the German explorer Heinrich Barth.

The features of the people of al Qatrun and its neighbouring villages are similar to the people of Gabr Oun and some other villages of Wadi al Hayat, who inhabit the area of the sand dunes to the south of Wadi ash-Shati and north of Wadi al Hayat. There is a similarity between them and the people of Bornu; their tongue is Arabic with a lot of Berber and Kanori words.

Due to the rocky terrain, the rarity of water and the insecurity of the region, it was not easy to pass the area of al Gatrun. Even though the distance between Tajarhi and Majdul was guarded by several castles (*qsur*), caravans usually were attacked by Saharan raiders. To provide security and food to the caravans, many castles and wells were developed. Until now, some of the relevant evidences and remains are still visible in the area such as small mosques (usually of a row of stones with an apse at the eastern side showing the direction of holy Mecca), large stones which were used for fixing tents, sand-filled wells and damaged watching towers and forts which were used for guarding the caravan routes.[129]

6.13.3 Tajarhi—the Most Remote Spot

The oasis of Tajarhi is considered to be one of the most distant centres of culture in the Fezzan area. The small oasis is situated about 82 km south of al Qatrun at the intersection of latitude 14.20° north and longitude 14.25° east.

Tajarhi came into being in a small basin, on the edge of Idhan Murzuq in the depression of what the Italians called al-Hikma ('wisdom') valley. The locals attach this name to a small hill situated east of Qasrawa, located between al Qatrun and Tajarhi. The al-Hikma valley was considered among the poorest valleys of the Fezzan with regard to sweet water. Nevertheless, there was a water table at a depth ranging from one metre to three metres underground level from which the locals drew water using a counterpoised sweep for the irrigation of their limited cultivations. This water table made it possible for some grass to grow, serving as pasture for cattle. On the other hand, date palm trees were available in great numbers.

The leader Uqba ibn Nafi took control of the region in the year 664 in a campaign which led him as far as the fortresses of Kuwar. Probably, he chose the Tasawah road and the Taraghin regions in his quest to take over the fortresses of al Qatrun and Tajarhi. On this road, one finds easy paths and inhabited cities. In this manner, the Arabs came to learn the shortest way linking the southern Mediterranean cities to the sub-Saharan territories. Tajarhi's extreme location in the Libyan Desert rendered the oasis very important as a halting place for caravans coming from al Qatrun. In fact, Tajarhi was the last oasis of the Fezzan. One would leave it to proceed towards the city of Sajdin, in Kuwar, 580 km away. Caravans covered this distance in about fifteen days. When travelling in this region, the caravans had to exercise great caution. For two whole days after leaving Tajarhi, no water resources were available. After that, one could replenish one's water reserves from the Sharu well.

The Tubu communities called the southern Fezzan region "Zila" or "Zilan", particularly the area of al Qatrun, Tajarhi and Taraghin. This was the area from which one could reach Jadu, from Tajarhi, having travelled 550 km. On this journey, one followed the Tummo and Madama road and crossed the Jadu hill, in a straight line. Caravans were able to cover this distance in 15 days. Some Tuaregs used a path from Ghat which went through Idhan Murzuq, reaching Tajarhi in a straight line. Some sections of this road were quite safe, but others were rather dangerous. In fact, large caravans did not take this road, and only those with few camels ventured through it.

Caravan trade, which linked the Mediterranean coastal cities to al Qatrun, Tajarhi and Kuwar, flourished during a period of stability and security in the region, particularly during the eighteenth and the end of the nineteenth centuries. However, by the end of the nineteenth century and the beginning of the twentieth century, the situation had changed, especially after the French occupation of Chad and the opening of alternative roads, which saved traders both time and money. Moreover, the drop in the European demand for African products affected the region economically, until it became partially isolated.[130]

6.14 Sirte (Sultan)—Custodian of the Gulf of Sirte

Said Hamid

The old city of Sirte (nowadays called Sultan) is situated about 55 km east of the modern city of Sirte and about 520 km away from Tripoli in the west. It lies approximately halfway between Tripoli and Banghazi.

The old city of Sirte is situated in the great Gulf of Sirte, which is considered among the biggest land indentations that can be clearly seen in the African continent's coastline. It is suggested that this gulf was created in the wake of the collapse of land due to tectonic movements. These movements were linked with similar tectonic activity which gave rise to the Cyrenaican plateau in the East and the Tripolitanian plateau in the west.

6.14.1 Important Commercial Spot by the Sea

Sirte was built in the Middle Ages in the vicinity of the Roman city of Acena. It seems that this Roman city was built on the remains of the Phoenician port of Korax, which became famous after the establishment of the Phoenician city of Carthage in 814 BC, serving as a market for the commercial exchange between the Cyrenaican Greeks and the Carthaginian Phoenicians. The old city of Sirte was taken over by the leader Amr ibn al-As shortly before he took control of the city of Tripoli in 643. Probably, it was just a small city sustaining a limited population. The city's importance goes back to the tenth century.[131] Back then the Arab geographer Ibn Hawqal noted: "*[...] It possess more property, revenue, freewill offerings, grazing camels and cattle than the city of Ijdabya* [Ajdabiya] *in our time. It is rich in date palms from which ripe dates are picked [...] It has enough date palms, and it also has vines and fruits of reasonable price [...] Their imam undertakes the collection of their alms, taxes, and the dues payable by the passing caravans [...] Its revenue is greater than that of Ijdabya [...] Merchandise-carrying vessels also reach Sirte, and from it, Sirtian alum is exported, which is abundant in this region as well as wool. As regards goats are more abundant than sheep, and more useful. Its inhabitants drink rain water (which they gather) in cisterns.*"[132]

The old city of Sirte witnessed considerable development during the tenth century, and it was very important commercially due to the region's considerable economic resources such as Sirtian alum, meat and wool. Moreover, its imam used to collect the legally prescribed alms tax and the land tax, apart from collecting the dues from the caravans passing through the city. Its harbour, notwithstanding its small size, contributed towards the flourishing of commercial activity. It was the port of small vessels which carried exported goods for exchange with local products.

The Fatimids recognised the importance of the old city of Sirte, especially at the time of al-Muizzli-Din Allah. That was because the city was located on the way leading to Egypt, against which he led his military campaign. In fact, Sirte fell on the shortest route between the city of al-Mahdiya (in Tunisia) and Egypt. For that reason, some fortresses, wells and cisterns were built, and Sirte's mosque was reconstructed. It became one of the most important commercial cities in the Gulf of Sirte because of its location, halfway on the route that linked Misurata and Ajdabiya on the one hand and the route linking the cities and oases of al Jufrah and the Fezzan territory to the northern coastal route on the other. In view of its vicinity to the desert, the travelling distance between Sirte and the Sub-Saharan

Fig. 6.72 Remains of the old city of Sirte (above left) and scenes from modern Sirte taken during the Italian occupation

commercial centres was about one month less than the distance between those centres and other southern Mediterranean coastal cities. Moreover, it was situated on the northern coastal pilgrimage route. It had abundant water reserves stored in wells and cisterns, and the surrounding area was also rich in pasture for the caravans' camels. Among the most important commercial routes which used to pass through Sirte was the coastal route adopted by the North African and Moroccan pilgrimage caravans and travelling parties. It contributed towards the flourishing of commercial exchange between the inhabitants and the pilgrims' caravans. Bartering between the two sides was the predominant way of doing business. Some travellers describe how the people of Sirte resorted to some tricks in their commercial dealings with the pilgrims. For example, they would inflate water skins and hang them in front of their shops, giving the impression that oil was available in abundance. In this way, the price of oil would become cheaper.

The old city of Sirte was influenced by the events which took place during mid-eleventh century. The city and its region were exposed to the migration of the Banu Hilal and Banu Sulaym. Moreover, during the thirteenth century, the city as well as the routes leading to it suffered from chaos and insecurity that had resulted from the raids of the Qaraqosh and Bani Ghaniya armies. This brought about the loss of Sirte's commercial and economic importance. Furthermore, Sirte was linked to two main southern routes:

- Sirte, al Jufrah oases (Zillah, Waddan, Hun, Soknah), Tmissah and Zawilah. From here, two routes branched out: the first one towards Murzuq, Tasawah, Ghat, Zinder and Kano and the second one left al Qatrun towards Madrusah, Tajarhi, Kuwar and Kuka on the Lake Chad.
- Sirte to Tagreft, and from here two routes branched out: the first one towards Zillah, Tazerbu, al Kufrah, proceeded southwards towards al Fasher and the western cities of the Sudan. The second route proceeded towards Tagreft, al Fugaha, Tmissah and Zawilah. From here, one proceeded either to Murzuq, Ghat or towards al Qatrun and from there to Kuka.

6.14.2 Modern City of Sirte

As for the modern city of Sirte, which is situated at latitude 31.14° north, and longitude 16.06° east and located about 55 km west of the old city of Sirte, it was established during the second Ottoman period (Fig. 6.72). In 1842, the Ottomans built a fortress at Marsa az-Zafran (modern Sirte). This fortress was originally called az-Zafran palace, and later it was referred to as Sirte palace. In 1912, the Italians restored the old fortress, and the modern city of Sirte developed around it. Later, it became the most important resting and provisioning terminal for the westward and eastward routes. These routes followed either the northern coastal road towards Cyrenaica, Egypt and the Hijaz or westwards towards the Maghreb Arab countries and southwards towards the Fezzan and Central, West and East Africa.[133]

6.15 Banghazi—Port and Trading Centre in Eastern Libya

Najmiya as-Sadeq at-Tellisi

The city of Banghazi is situated about 1050 km east of Tripoli. It overlooks the Mediterranean coast and is located at latitude 32° north and longitude 20° east.

It is considered one of the most important cities in the north-east of Libya and the biggest among them, both in area and in the number of inhabitants. Banghazi is also considered the gate to the Green Mountains, which is characterised by its enchanting natural environment and by the various Greek and Roman archaeological remains.

6.15.1 Port and Trading Centre

The old city was one of the famous five Cyrenaican cities established by the Greeks in the year 446 BC. They named it Euesperides. The Libyan Pentapolis included Cyrene, Apollonia, Ptolemais (or Barca), Arsinoe (or Taucheira) and Euesperides (or Berenice). The settlement was located on the edge of a lagoon which opened from the sea. The name Euesperides is thought to refer to the mythological gardens of Hesperides. After Ptolemy III married Berenice, the daughter of the governor of Cyrene in 246 BC, the city was named Berenice after his wife. It was under the rule of Ptolemy that the entire city was moved to the present location of Banghazi. The move may have been triggered by the silting up of the lagoon. It later became a Roman city and prospered for 600 years. In the third-century AD, the city became a Christian bishopric. The first of its bishops, whose name is recorded in extant documents, is Ammon, to whom Dionysius of Alexandria wrote in about 260. The base of all wealth and prestige of the city was the trade. Networks of trade linked ancient coastal Libya to the oases of the Sahara.[134] The Arab conquest from 641 and 642 brought an end to the Roman world in Northern Africa. But the city ceased to exist. For several centuries, an Arab settlement prospered. Cattle, wool, honey, and olive oil were exported through the port of Ptolemais and overland by caravan. Sometime after 1050, the invasions of the Bedouin tribesmen of the Banu Hilal and Bani Suleim tribes ended sedentary life in the area.[135]

Fig. 6.73 Italian quarter was traditionally home to many people, before many families left the old town to resettle elsewhere in the city

Subsequently, the city was given other names, among them Koyat al-Milh. Ultimately, in 1450, when the governor as-Salih (Sidi Ghazi) settled here, the city was called Banghazi. It became one of the areas, which were submitted to Ottoman rule. In 1517, Cyrenaica became part of the Ottoman Empire, but at that time exercised little local control for the most part. It was in the beginning of the sixteenth century, that many Jews from Tripoli helped to repopulate Banghazi, earning their livelihood by trade with North Africa and the Mediterranean area. In 1635, at the time of Saqizli Pasha, a military campaign under the leadership of Othman Saqizli was sent against Banghazi with the aim of occupying it. Othman established a fortress (Fig. 6.74 ①) overlooking the harbour, fully equipped with cannons, in order to control the commercial activity in the port. Later on between 1711 and 1835, the control of Banghazi passed on to the Karamanlis.

Time and again, Banghazi flourished both economically and culturally. This was due to the eminent role it played in commercial exchange, involving agricultural and animal produce as well as European and African products that reached its harbour. The city was at the crossroads of both maritime and caravan routes, the latter coming from the Sudan and Fezzan, loaded with different kinds of merchandise. This phenomenon rendered Banghazi a centre of commercial exchange and a base for the export of African commodities on the one hand and the importation of European products through its harbour on the other hand. In this respect, European states like England or France had signed commercial treaties to open consulates inside the Medina of Banghazi (Fig. 6.74 ③④).

During the nineteenth century, it became an important outlet for the Saharan slave trade. There had been an ancient trade route between the eastern Sudan and Cyrenaica up to the Middle Ages. This route had been abandoned and came back into use in the early nineteenth century. Back then, the Wadai traders needed more outlets to Egypt and the Mediterranean than the route trough Fezzan offered. For this, they reopened this dangerous but direct road northwards via Kufrah and Awjilah. Slaves in the nineteenth century

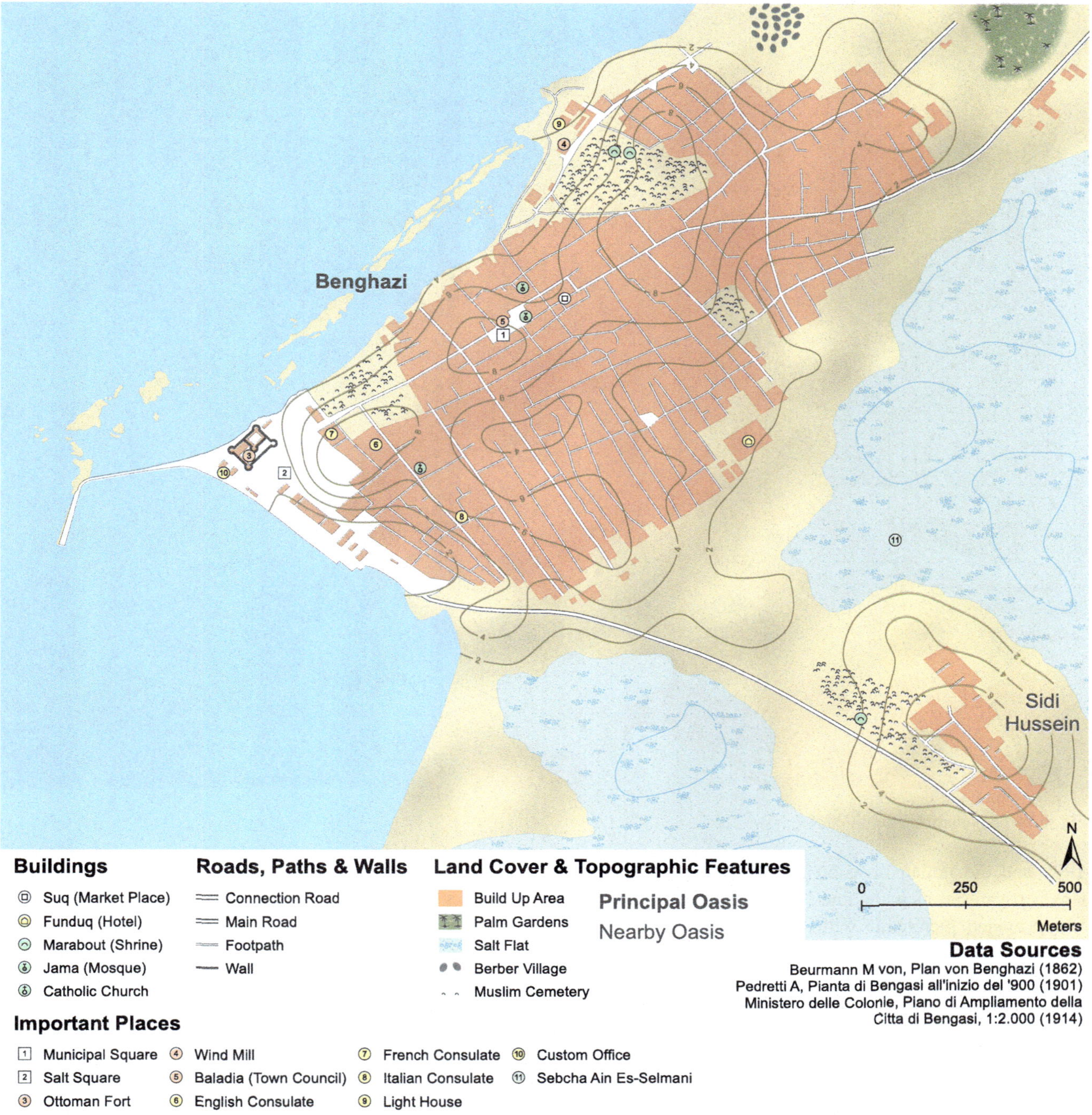

Fig. 6.74 Map of Banghazi

reached Banghazi from two main sources: from Bornu and Hausaland through Fezzan, and from Wadai through Kufrah. During the second half of the century, the Wadai road became even more important. Under pressure from the European states in terms to the abolition of slave trade, the Turkish measures in the 1850s affected the slave trade through Tripoli massively. On average, about 700 slaves were shipped from Banghazi every year to the ports of Crete and Constantinople. The local demand in the city itself is estimated at 200–300 slaves a year.

The Wadai road remained active later than any other route connecting the Mediterranean littoral with the interior of Africa. It lasted well into the 1920s. What was important here was the fact that, until the fall of Wadai to the French in

Fig. 6.75 Banghazi market in 2006

1909 and the Italian invasion of Cyrenaica in 1912–14, a single organisation maintained the Cyrenaican-Wadai road and provided merchants with legal stability, maintenance, security and shelter. That organisation was the Sanussiya brotherhood.[136]

The city was also the destination of numerous European travellers, who documented their impressions about the historical and cultural sites they had visited. Grangier and Gerard (1675–1668) were among the most important travellers. Other visitors, who described Banghazi's main places of interest, were the Picci brothers (1821), Pascio (1825) or Ludwig Salvator (1873).

6.15.2 Banghazi Markets

The Banghazi markets were characterised by intense commercial activity during the passing of the pilgrims' caravans on their outgoing and incoming journeys to and from the holy lands (Figs. 6.75, 6.76 and 6.77).

Goods like cows, wool, cattle hides, barley and small quantities of grain were exported along these routes. Some central African goods carried by the caravans were exported to Malta. Other commodities such as butter, honey, wax, salt, for which Banghazi was famous, and some goods coming from the caravan trade, such as elephant tusks and ostrich feathers as well as slaves were exported to the eastern Arab countries. Moreover, Banghazi was also involved in bartering with the island of Crete and Tripoli. Cattle constituted an important segment of Banghazi's exports. Many Maltese and other ships used to call at the port of Banghazi to load cattle and cattle products for the European markets, particularly during the summer months. On the whole, Banghazi's traders considered trading in cattle a very profiting business, whenever suitable climatic conditions allowed the rearing of such livestock.

The events in sub-Saharan Africa, including its occupation and the opening of alternative routes as well as the Italian occupation (Fig. 6.78), influenced caravan trading negatively. As a result of these developments, Banghazi lost

Fig. 6.76 Market scenes (photograph on the left Laronde 1997, photograph on top right Mathuisieulx 1912) and a street in Banghazi during the Italian occupation (Laronde 1997)

Fig. 6.77 A small caravan in camps near Banghazi on the 12 March 1881, as seen by the painter G. Haimann

Fig. 6.78 Banghazi municipality is the former town hall of Banghazi (Fig. 6.74 ⑥), which was built in 1924. It is located in the centre of the Italian quarter in Maydan al-Baladia (municipality square). The building is the creation of several architects and designers. The portico of the facade and the interiors were designed by Marcello Piacentini, while the facades were designed by the architect Ivo Lebboroni. The interior frescoes are the work of Guido Cadorin, while the chandeliers are by Umberto Bellotto. The furniture inside the town hall was designed by the company Ducrot[138]

its economic importance. The commercial activity, which was the source of the previous affluence of the Banghazi traders, came to an end.

Banghazi was connected to numerous caravan routes directed towards Africa; the most important of which were the following:

- The Wadai route: This route departed from Banghazi and crossed the Awjilah, Jalu, al Kufrah and Tibesti oases. It passed by a number of wells, such as the Abu Tifl and the Zighan wells. The outgoing and incoming journey, including the various stops which were made, lasted about eight to ten months. Along this route, there were no inhabited areas, where one could find inhabitants renting camels or desert guides as was the case with the Ghadamis–Ghat–Murzuq route. Trade along the Wadai-Banghazi route flourished in the wake of the takeover of the Sudan by the followers of the Mahdi. Later, after the British occupation of the Sudan in 1889, this commercial activity declined, due to the transportation of exported products both on rivers and on railways built by the British in 1900.
- Banghazi-al Jaghbub route: This route branched out in two directions: the western route, which was approximately 870 km long, was the preferred route. It presented no water problems, especially between Banghazi and Tilmun given to the existence of numerous wells. Among these wells, one could mention al-Khawabi, al-Gasiba, Umm Mabruka, Abu Idris or Saqiyat al-Baba. Those were shallow wells, ranging in depth between two and three metres. The Tilmun well had the sweetest waters, but it was rather deep. After Umm ar-Raha, one entered level ground, after which one would come across wavy sand dunes. This was a difficult tract, extending as far as

the Matar valley. It was a low land with a deep well of sweet water. Beyond the valley, the routes came across the Hanaqiyya, Shaqiq and Hasi Hussein wells, which provided water to camels and horses. From there, the caravans used to reach the Asila oasis, situated in a low land with quicksands. Here, one found some wells such as the Tarfawi and the Abu Salama wells, whose waters caused diarrhoea. For this reason, the water was consumed, neither by men nor by animals, unless the caravan suffered from severe thirst. The eastern route comprised two wells, the Fastus and the Mazriq wells. In the tract between Banghazi and Fastus, there were some reservoirs in which rainwater was collected, supplying caravans with water. The most important were the reservoirs of the Hmarayn wells, situated deep in the desert, and the reservoirs of the al-Hakim wells, which collected rainwater mixed with mud. These waters were consumed by the caravans' camels.

- The Banghazi–Sudan–Chad route: This route was used to pass through Ajdabiya, Awjilah, Jalu, Zighan and al Kufrah. Caravan routes came together in this region, turning the oases into what looked like cities that offered various possibilities of rest, with markets fully supplied with different goods. In fact, this oasis region owed its rise to importance, as well as its later decline, to the caravan route.[137]

6.16 The Islamic City of Barqa

Said Hamid

The Islamic city of Barqa (modern al Marj) is situated in the area of the Green Mountains, about 100 km to the north-east of the city of Banghazi. It is located on longitude 20.48° east and latitude 32.63° north.

6.16.1 A Place Full of History

The old city of Barqa was founded at the time of Arcesilaos II, the fourth of the kings of Cyrene who ruled between 554 and 544 BC. The city witnessed various historical periods which characterised the area of Barqa including the Greeks, the Romans and the Byzantines.

Having conquered the city of Alexandria in the year 642, the Arab leader Amr ibn al-As wanted to continue his conquests in the region west of Egypt. The objective was to enable him to secure his position against the Byzantine threat as well as to spread Islam in new areas. After the Arab conquest, the region of Cyrene became known as 'Barqa' after its provincial capital. It enjoyed both security and stability at that time. During the early Islamic period, the inhabitants of Barqa were prosperous because the place was not controlled by any tax collector. The people of Barqa used to send the amount of the land tax at the right time. Abdallah ibn Amr ibn al-As is reported to have said: "*Had it not been for my property in the Hijaz, I would have settled in Barqa. For I do not know of any place more secure and peaceful than this.*"

The city of Barqa (Fig. 6.79) is renowned for its fertile coast and it produced grain which used to be exported to Greece. In fact, the Greek historian Herodotus, who lived during the fifth century BC, praised the city's abundant produce.

During the early years of the Islamic period, Barqa became the capital of the region as well as an important commercial centre. Many Arab geographers and travellers described the city. One of them, al Yaqubi, mentioned the oasis in the ninth century in his *Book of the Countries (al-Buldan)*: "*The city of Barqa is situated in a wide meadow. Its soil is of a very reddish colour. It has a wall, iron gates, and a ditch. al-Mutawakkil al Allah had ordered the building of the wall. Its inhabitants drink rain-water coming from the mountain, through the valleys, which is then collected in big cisterns commissioned by the caliphs and emirs for the benefit of the inhabitants of Barqa. The outskirts around the city are inhabited by different peoples. [...] Six miles separate the city of Barqa from the sea coast*".[139]

Barqa is famous for being the first city which one reaches coming from Egypt. In his book *The shape of the earth (Surat al-ard)*, Ibn Hawqal reports about Barqa: "*[...] (in size) it is neither large, nor small. It has flourishing small towns. It is the first base which the traveller from Egypt reaches on his way to Qayrawan. It is constantly inhabited by strangers, because of the commerce which goes on in it, as well as by western and eastern visitors who pass through it on account of the tar commerce for which the city is famous, as well as for hides, which are then tanned in Egypt, and for dates from Awjilah. It has bustling markets selling wool, pepper, honey, wax, oil, and various products from the east and the west. Very often its food is abundant and cheap.*"[140] The inscriptions and coins found in Barqa underline the city's significance during the early Islamic centuries.

Trade in Barqa started to shrink during the twelfth century, leading to a decline in its role and importance. In his book *The pleasure of him who longs to cross the horizons (Nuzhat al-Mushtaq)*, written during the mid-twelfth century, al Idrisi remarked that the city had a small population at that time, its markets were depressed, and that in the past, the situation had been different.[141] On his part, Ibn Said reported in 1250 in his *Book of the Extension of the Land on Longitudes and Latitudes (Bast al-ard)* that Barqa "*[...] was the base of the Cyrene region which was much destroyed by the Arabs and is today called al Marj.*" The city, and the whole region in general, was exposed to the migration of the Banu Hilal and

Fig. 6.79 Old town of al Marj (old Barqa) built in Italian style with churches and markets

Banu Sulaym tribes on the initiative of the Fatimid caliph al-Mustanir Billah, who, during the mid-eleventh century, allowed these tribes to cross the Nile on their way towards the Maghreb. This migration led to the destruction of Barqa, an event which was recorded by the historian Ibn Khaldun. A glorious phase in the history of the city had come to an end.

Barqa's position as a meeting place along the northern route of the pilgrimage to Mecca as well as its caravan routes was important factors that affected the city's prosperity. The desert caravans used to depart from Barqa on their way to Ajdabiya and from there to al Kufrah going south to Central and West Africa. It was connected with the Siwa oasis via the towns of al Mkheyli and the port of Marsa Lak near Tubruq, apart from being connected with Egypt via Derna and Alexandria. Barqa was linked to the Maghreb via Banghazi and Tripoli along the main road adjacent to the Mediterranean coast, apart from its connections by sea with Tripoli and other maritime cities via the port of Tolmeitha, which is situated about 20 km from the city of Barqa. Small ships could make use of this port during some periods of the year.[142]

6.17 Ajdabiya—Important Trading Centre During the Early Islamic Period

Said Hamid

The city of Ajdabiya is situated 160 km south-west of Banghazi, and it is 28 m above sea level and located at latitude 30.47° north and longitude 20.12° east.

Ajdabiya relies on underground wells situated in the sand dunes region for its water supplies. The city derived its importance from its distinct geographical location, enabling it to control the caravan trade routes directed south towards the Jalu and Awjilah oases and from there to al Kufrah and the Sudan as well as the coastal routes directed towards Sirte and from there to Tripoli and the Arab Maghreb. This was considered one of the most important caravan routes. It was also adopted by the pilgrims' caravans coming from Morocco. These used to stop at Ajdabiya both on their outgoing and their incoming journeys from the holy lands after performing the pilgrimage.

6.17.1 Eventful History

The origins of Ajdabiya are not very clear. The city seems to have arisen in Antiquity before the Roman occupation during mid-first century BC. Probably, its Roman name was Corniculanum, and it was included in Roman maps. During the first century AD, Syrian warriors established themselves in Ajdabiya. They used to defend the city and guaranteed the passage along the road from Cyrenaica to the Gulf of Sirte. Roman inscriptions on rocks have been found near old water wells, which testify to the Syrian presence. The Arab geographer al Yaqut al Hamawi refers in his book *Mujam al-Buldan*, which he finished in 1228, to the name of the city of Ajdabiya, stating that Ajdabiya may be an Arabic name, being the plural of *jadb* that is "aridity". Moreover, it is probable that Ajdabiya is the Arabic name given to the city by the Arab conquerors upon capturing it. They had done the same with other cities which they had taken.

The city was captured by the Arab leader Amr ibn al-As in 642 AD. Upon entering into a peace agreement, the inhabitants of the city agreed to pay a tribute of 5000 dinars. Most of them embraced Islam. It was from this city that the Arab armies led by Amr ibn al-As departed on their way towards Tripoli and under the leadership of Uqba ibn Nafi headed towards the south in the direction of Awjilah and Zawilah. During the eighth century, Ajdabiya fell under the rule of Egypt. After the establishment of the Fatimid state, during the ninth century, the Fatimids started to focus their attention on the east, hoping to control Egypt. However, before doing so, they did their best to control the cities situated on the way leading to Egypt. The Fatimids were interested in the city of Ajdabiya because they soon became aware of its important commercial potential, as it was situated on the road linking North Africa to Egypt. In 966, the Fatimid caliph al-Muizzli-Din Allah started preparing his departure to Egypt. He arrived in Sirte on the "4th of Jumada al-Awwal" and then he left it, and stayed at the palace which was built for him in Ajdabiya. During his stay in the city, the caliph ordered the construction of reservoirs for the collection of rain water. It seems that the city prospered to some degree after the Fatimids started showing interest in it. Indeed, caravan trade flourished, and Ajdabiya represented the most important city in the Sirte region. It was linked to the Sudan via a route passing through Awjilah. After the death of al-Muizz in 976, the caliphate passed to his son al-Aziz Nizar. The latter confirmed Yusuf Bilkin as governor of North Africa and entrusted to him the territories of Tripoli, Sirte and Ajdabiya. Yusuf put into service his own agents in these cities, and his position was greatly enhanced. Later, the Zirid–Fatimid relations deteriorated, leading al-Muizz ibn Badis in 1049 to sever ties with the Fatimids and recognise the Abbasids. Consequently, the Fatimids supported the Arab tribes of the Banu Hilal and Banu Sulaym across the Nile and ordered them to move towards North Africa, promising them that the cities they conquered would become theirs.

Ajdabiya lost the prosperity it had witnessed during the early Islamic and the Fatimid periods. At the beginning of

Fig. 6.80 There are archaeological remains of an old Fatimid mosque, which has been built on Roman ruins

the Arab conquest, it was the place to which armies retreated and from which they departed on their campaigns. During the Fatimid period, it was an important commercial centre, and there were regions which belonged to it. In Ajdabiya, there was a fortress, a big mosque (Fig. 6.80) and a permanent market. It also had a harbour which was situated not far from the centre of the city.

6.17.2 Written Accounts on Life in Ajdabiya

Many Arab travellers and geographers have made reference to the city of Ajdabiya. At the end of the ninth century, the Arab geographer al Yaqubi described Ajdabiya in his *Book of the Countries (al-Buldan)* as "*[...] a city above which there is a fortress. In it, there is a mosque and markets. From Barniq to Ajdabiya there are two stages, and it is four stages away from Barqa. It has regions and six miles of coast on the salty sea, in which ships cast anchor.*"[143] The tenth century Muslim writer, geographer and chronicler Ibn Hawqal gives important details about Ajdabiya in his book *The shape of the earth (Suratu l-Ard)*. From his description, one gets the impression that the place was a prosperous and heavily populated city. Its inhabitants were self-sufficient because of their orchards and date palms. Its governor assumed the collection of the alms, tributes and tithes as well as the fees from the trade caravans and sent them to the Fatimid caliph in Egypt.[144] Abu Ubayd Allah al Bakri informs us about the condition of the city of Ajdabiya during the eleventh century. He states that "*[...] it is a big city in the desert, its land is smooth, its wells are dug in the smooth rock, and their waters are good. It has a spring of sweet water, as well as beautiful orchards and abundant date palms. It has only Arak trees. In it there is a beautifully constructed mosque built by Abu l-Qasim ibn Ubayd Allah. The mosque has an octagonal minaret built with great skill. The city has many baths, inns and well-supplied markets which are well-frequented by customers. Its inhabitants are well-off. It has a harbour known as al-Mahur [...] Prices are not expensive, and its abundant dates, of different kinds, come from the city of Awjilah.*"[145] Whereas al Idrisi describes Ajdabiya during mid-twelfth century as "*A city in plain land, made of even stone. It used to have a wall. Now, only two fortresses remain in the desert. The sea is four miles away from it. There are no girls in it, nor around it, and most of its inhabitants are Jewish and Muslim traders [...] In Ajdabiya and Barqa there is no running water. For water, they rely on cisterns and waterwheels, enabling them to cultivate small quantities of wheat, and especially barley, some types of corn and other grains.*" The traveller al Abdari confirms the level of desperation that Ajdabiya must have been enduring his trip in 1289 and states that "*... it has an old fortress, the size of a high house. Some historians mentioned that it had running water and date palms. But now, there is only a fortress in the open country, and there is neither running water, nor a single tree.*" There is much historical evidence attesting to the decline of the area and to its loss of its commercial importance. Many of its inhabitants abandoned Ajdabiya because of the lack of security and stability. Its farms and orchards were vandalised by adventurers like Karakosh and Yahya ibn Ghaniya. Their actions had indirect repercussions which were even more dangerous than their actual vandalism. This picture about Ajdabiya presented by al Abdari is the last one we have from the Arab geographers and travellers.

During the nineteenth century, the city of Ajdabiya was completely abandoned. It was not until the twentieth century, in 1921, that the place started to be repopulated and became again one of the most important cities of the Gulf of Sirte. Among the most important archaeological remains, which are still to be found in the city, are the Fatimid palace that was built prior to the departure of the Fatimid al-Muizzli-Din Allah to Cairo. There is also a mosque attributed to the Fatimid caliph Abu l-Qasim ibn Ubayd Allah (1528–1539). Since 1971, the archaeological department in collaboration with the Society of Libyan Studies in London has carried out digs at the Sidi Hasan cemetery in Ajdabiya, which brought to light most of the remains of the old mosque (Fig. 6.80).[146]

6.18 Awjilah—One of the Old Libyan Cities

Salih al-Mahdi Khalifa

The city of Awjilah is situated in the south-east of Libya deep in the desert. It is situated 250 km south of the Mediterranean coast in a small, semi-closed basin within the northern section of the Wahat Great Depression. Awjilah is located at latitude 29.20° north and longitude 21° east.

6.18.1 Economic and Scientific Centre in the Midst of the Desert

Awjilah is considered one of the old Libyan cities (Fig. 6.81). Its name is featured in ancient Egyptian, Greek and Roman sources. In fact, Egyptian hieroglyphs going back to about 1200 BC indicate that the inhabitants of the region, including the tribe of the Nasamoneans, extended from Banghazi and the Gulf of Sirte in the north and even penetrated southwards as far as the oasis of Awjilah. Herodotus, who lived in the fifth century BC states, that in summer the Nasamoneans used to leave their herds on the coast and travelled towards the interior to a place called Awjilah in order to gather dates. Date trees of the best quality abounded in that region. Moreover, Herodotus also referred to desert routes and geographical discoveries made

Fig. 6.81 Great Mosque of Atiq is the oldest mosque in the region. The present building dates back to the twelfth century

by some Libyans in the desert, in a country which nowadays is called Chad. In 643, Uqba Ibn Nafi conquered the Awjilah oasis and spread Islam in the surrounding regions.

This oasis drew the attention of a number of Arab and foreign travellers among them Ibn Hawqal, a famous geographic author of the tenth century. He described the oasis as "*[…] an area with great date-palms and dates*". al Bakri wrote in the eleventh century that the city was opulent and prosperous, with plenty of date palms. al Idrisi, who lived during the twelfth century, was of the same idea. In his book, *A description of Africa (Wasf Ifriqlya),* al Hasan al Wazzan refers to the cultivation of date palms in Awjilah. Some historians regard the city as having three distinct agricultural areas, namely Mazuz, as-Swani and Awjilah. Marmol, the Spanish traveller, who visited Libya during the last third of the sixteenth century, stated that Awjilah was situated on the main pilgrimage route, linking Mauritania, Shengit and Egypt with the Hijaz. It was only natural that, with the flourishing of agriculture, caravan commerce was bound to thrive. In fact, in the first third of the seventeenth century Ibn Mlih stated that "… *its welfare is abundant, because sustenance comes from far-away countries*".[147] On his part, the German traveller Rohlfs visited Awjilah twice, in 1869 and in 1879.

He tried to monitor the commercial activity during the ten-year period which had elapsed since his first visit to the city. During the second visit, Rohlfs registered the changes witnessed in Awjilah. Regarding the number of mosques, for example, in 1869, he noticed that the oasis housed one big mosque and four small ones, whereas in 1879 there were thirteen mosques. Rohlfs confirmed Bashu's, Hamilton's and Hornemann's description of Awjilah and stated that he had nothing to add. He also observed the arrival of a caravan from Wadai, led by the brother of Ali Gurgi from Tripoli and stated that the arrival of caravans brought about a change in the way of life of the city.[148]

Awjilah's location, south-east of Ajdabiya and Banghazi, west of al Jaghbub and Jalu and east of Maradah, put the inhabitants of this oasis in an advantageous situation. This location enabled them to establish economic and commercial links with the internal regions of Libya as well as with the sub-Saharan African countries. Moreover, the city's location in the midst of the desert, faraway from the coast, protected it from the centres of conflict and from the great naval powers, such as the Phoenicians, the Romans and the Byzantines. This contributed towards the political, economic and social stability, growth and prosperity of the oasis. Furthermore, its distance from the main cities, such as Banghazi, Tripoli, Sabha, Darnah, Misurata and Sirte, which had witnessed a number of armed uprisings, was not quelled before the Ottoman Turks suffered great losses.

In view of Awjilah's location, at the crossroads between the north and the south, the east and the west, the city became a meeting place for traders and men of science. Its markets were full of African products, including gold, silver, ivory, hides, ostrich feathers and other products. Products from the north, including textiles, utensils and agricultural products were also available in these markets (Fig. 6.82). The oasis attracted traders, people of science and pilgrim caravans, transforming Awjilah into an economic and scientific centre.

6.18.2 Commercial Activity of the Oasis

Since Antiquity, Awjilah has been also renowned for its groves of date palms and the availability of good quality dates of different types. This has enabled Awjilah to supply different areas with dates. The people of Awjilah were good-mannered, honest and truthful in their dealings, as well as multilingual. In fact, most of them spoke more than one language or dialect such as Arabic, the Kano language, Hausa and dialects of central Sudan and of the Tuareg regions. These linguistic abilities facilitated their understanding of African peoples. Moreover, traders from Tripoli and Banghazi relied on the people of the oases, including Awjilah, to guide and defend their commercial caravans through the deserts. Apart from agriculture, the inhabitants of Awjilah engaged in other professions, the most important of which being trade, camel breeding and lending as well as guiding caravans due to their knowledge of the desert and its paths.[149] They taught their children the art of trade, as confirmed by the traveller Hornemann who stated that: "*From a tender age, men dedicate themselves to trade. I have witnessed children of thirteen or fourteen years accompanying, on foot, the caravan on its long and strenuous journey from Awjilah to the Fezzan, hardly ever riding horses or camels*".[150]

The commercial activity, including the long yearly journeys to and from Awjilah, left its social and cultural marks on the city. On the social level, commerce influenced the social make-up of the inhabitants of Awjilah, as a result of intermarriages between the locals and people from other regions, who settled in the city. Culturally, the inhabitants of Awjilah, who came into direct contact with foreign people, were able to observe these ways of life and culture and were influenced by them. Some of its people became poets, indigenous scholars and judges.

The inhabitants of Awjilah relied upon the waters from the many wells in their region. In 1934, there were 270 wells of which 140 were fit for drinking. The other remaining wells were reserved for consumption by the livestock. Among the most famous wells, one comes across Zaydan well, Hayawan well, Tughut well, Awslam well, Ahmad Abu Silsla well, Diryana well, Sabil well, ar-Rasm well and ar-Rakb well. The latter is considered among the oldest wells, and according to tradition of the people of the oasis, it was dug out at the time of the beginning of the Islamic conquest. It is considered an important stage on the caravan route because here, in the middle of the desert, one can supply oneself with fresh water. The pilgrims' caravans coming from Morocco on their way to the Hijaz adopted this well as a reference point. It has been called ar-Rakb because of the great number of caravans which frequented it.

6.18.3 Decline of the Oasis

News about the control of the Awjilah inhabitants over the caravan routes and their prosperity reached the governor of Tripoli, Muhammad as Saqizli, who entered into agreements with the rulers of the Fezzan, Bornu, Timbuktu and Wadai. These, in turn, had commercial relations with the people of Awjilah. With the aim of monopolising the caravan trade, the governor instructed Othman as Saqizli, the leader of his army, to carry out a military campaign and occupy Awjilah in 1633. Through deception, he managed to arrest Ahmad Abd al Hadi and his family, tracked the merchants, robbed them of their money and threw them in jail. Moreover, he plundered Awjilah's gold and silver and imposed heavy taxation on the population. With the great amounts of gold, silver and taxation money, Muhammad Pasha as Saqizli was able to mint the coin known as the Qarmil, each Qarmil being equivalent to half a Dirham. This coin continued to be in circulation in Tripoli until the minting of another coin by Khalil Pasha. So much wealth was amassed, that Tripoli witnessed a boom in construction. It was at this time that the Othman Pasha as Saqizli school, the Grand Hammam (public baths) and other construction projects were carried out. The Ottoman campaign against Awjilah, the imposition of heavy taxation and Ottoman injustice and oppression resulted in most inhabitants emigrating from the oasis. This led to the oasis losing its identity as a commercial and administrative centre. Consequently, Jalu assumed the role of the main commercial centre in the region. In addition, the distinctive geographic location of Awjilah led its people to establish economic and commercial ties with various cities of the desert, making Awjilah a transit centre for the caravan trade. Among the most important routes which cross the region and link Awjilah to the Mediterranean coast are:

- The Hijaz route, otherwise known as the pilgrims' route: This route was taken by people coming from Algeria, Morocco and Mauritania. It passed through places as Ghat, Awbari, Sabha, al Fugaha, Zillah and Maradah until it reached the Awjilah oasis. From there, it

Fig. 6.82 Market scene from Awjila

progressed towards al Jaghbub, Egypt and finally the holy lands. What made this route distinctive is that it abounded in water sources and in fertile lands, thus providing the caravan camels with pasture. Moreover, there were many desert animals ideal for hunting, such as the gazelle, the Waddan goat and rabbits. This solved the problem of supplying oneself with food supplies. However, the fact that this route very often went through rocky areas made travel time-consuming.

- The Cairo route through al Fayyum, Siwa to the Awjilah oasis, and from there to Tripoli, Ghadamis and westwards across Twat towards the city of Timbuktu.
- The Awjilah-al Kufrah route reaching Wadai through Tmissah, Zawilah and Kuwar.
- The route which left Banghazi towards Awjilah, Jalu, al Kufrah, Tibesti, Anjuanja, Iski until it reached Wadai.
- The desert route left Tripoli towards Tarhuna and then Bani Walid. At this point, the route intersected with the pilgrims' route at Jufrah. From there, the route proceeded towards Zillah, Awjilah, al Jaghbub and Siwa, and from there across the Sinai Peninsula towards holy Mecca (Fig. 6.82).[151]

6.19 Jalu—Important Crossroads in Eastern Libya

Najmiya as-Sadeq at-Tellisi

The oasis of Jalu is located south-east of the city of Ajdabiya which is about 250 km ahead. Moreover 30 km north-west of Jalu, one finds the famous oasis of Awjilah. Jalu is located at latitude 29.02° north and longitude 21.33° east.

Geographically, the city of Jalu is situated in the middle of the north-east region of the great oasis depression. The Jalu basin is about 30 metres lower than the surrounding land. In its centre, there is the so-called Nakhfush valley. The southern section of this valley is called Dweylith-Thalab that is *"the small tail of the fox"*. Here, the valley becomes somewhat wider. Extensive parts of the valley bed are covered with stones and pebbles which, in some areas, are mixed with sand formations. In other areas, sand dunes encircle the basin, particularly on its eastern and southern part.[152]

Fig. 6.83 Areal view of Jalu

Fig. 6.84 Scenes from Jalu taken from the Archive of the CLAR.HS (above) and the book written by Hassanein Bey in (1924, 242). The geographic explorer Hassanein Bey noted: "Jalo is one of the most important oases in Cyrenaica, partly because of the dates which it produces, but more especially because it is the destination of the caravans coming north from al Kufrah. Ivory and ostrich feathers from Wadai and Darfur come to Jalo to be forwarded either eastward to Egypt or northward to Bengazi. This trade is chiefly in the hands of the Majabra tribe, whose headmen are the—merchant princes of the Libyan Desert. A Majabri (singular of Majabra) boasts that his father died on the basur (camels saddle) in the same way that a soldier boasts that his father died on the held of battle."[157]

6.19.1 Jalu—a Spot with a Vibrant Life in the Nineteenth Century

In the beginning of the nineteenth century, Jalu became one of the most important commercial centres (Figs. 6.83 and 6.84). Sadiq Muayyid al-Azam described the city of Jalu on his way to al Kufrah in 1895 as follows: '*It resembles a small island, at the centre of wide expanses of sand dunes and stony deserts. In it, one finds about two thousand houses and eleven mosques and two zawiyas. It is surrounded on every side by about 76.000 palm trees. Sand dunes and desert vegetation encircle Jalu on all sides*'.[153] But the history of caravan trading in the area goes back much further. Jalu consists of four small villages, the biggest of which being al-Araq, in which most economic and social activities took place. It represented the centre of the oasis region which, apart from Jalu, comprises the two oases of Awjilah and Jakhra. Next in importance after al-Araq, and to the east of it, is the village of al-Libba, which is actually an extension of al-Araq. The two remaining villages are ash Sharaf and Rashida. Ash Sharaf is situated west of al-Araq, whereas Rashida is to the east of it. al-Araq is considered the principal centre from which all caravan routes branched out, linking the oasis with other areas like Banghazi, al Kufrah, al Jaghbub, Maradah, Awjilah, Jakhra and the southern commercial centres in the Fezzan region and in the Sudan. Since antiquity, it was renowned as a main caravan trading centre, enabling its inhabitants to establish strong ties with African countries. These inhabitants not only managed caravans, but also opened and established a number of tracks, such as the "track of the forty" or the Majabira track. The inhabitants of Jalu, or the Majabira, as they are called, soon controlled the trade paths leading to the Sudan, especially after the prohibition of the slave trade, which constituted important revenue to caravan traders. On their way to the Sudan, the caravans of the Majabira used to pass through the areas of Adiri, Anbadi and the Lake Chad region, where they were engaged in commercial activity. The Majabira possessed three commercial houses inhabited by families from Jalu. One of these houses was situated in the Majabira valley, the second one in Kardasa, near Cairo, and the third one in Zawilah or Murzuq, in Fezzan. Hides constituted the most important merchandise sold in this region.[154]

The Jalu oasis derives its water reserves from the water table which is close to the ground. This water tends to be salty, and the salt level in some wells renders their waters unpalatable. Nevertheless, the inhabitants drank this water. Some of them fetched sweet water from the Abu at-Tifl wells, situated about eight hours away from the city. The waters of these wells are sweeter than those of the al-Qimmi and Masliywa wells situated within the oasis. The wells of the oasis are distributed among Jalu's four villages. Moreover, there are other wells in the al-Gibli, Banus and Ghuratla areas, and their water is similar to those of the other wells of the oasis.

Jalu has a number of historical monuments representing different historical periods. Among them one finds the ancient circular burial grounds in the Nasha area. These cemeteries are in the form of well-ordered high circular wells. Moreover, in various locations one comes across fossilised palm trees. Among Jalu's most renowned historical buildings are the so-called four fortresses, called Hiri, Yanbus, Ghardafat and ash Sharaf, respectively. A number of inscriptions have been found on them.

The people of Jalu belong to the Majabira Arab tribes, who colonised the oasis since the first phase of the Islamic conquests. However, in view of long-lasting and intense commercial activity between Jalu and the Sudan, and the inhabitants' intermarriages with Sudanese women, their skin became darker and their body features changed.

Most of the inhabitants of the oasis used to engage in agriculture. Dates were considered the most important commodity, apart from other agricultural produces such as grain, barley, white and yellow corn and some vegetables. Nevertheless, the agricultural production, especially the cereal production, did not suffice for the local consumption. Hence, the oasis depended on the importation of additional quantities of products from the nearby coastal cities. Moreover, they inhabitants of the oasis practised cultivation on dry land in the southern regions of the Green Mountains which registered considerable rainfall. For this reason, at the end of summer or the beginning of autumn, a number of inhabitants from the oasis moved to the northern regions, near the coast, to plough and sow cereals, particularly grain and barley, relying on the rainy seasons. They then returned to their fields during harvest time to collect their produce. With regard to agriculture within the oasis, they relied on tomatoes and some other products of limited productivity, such as pepper, turnips and other products for local consumption.[155]

6.19.2 Declining Significance of the Trade

Of course the inhabitants were engaged in trade, transport of merchandise, and they were also acting as caravan guides directed southwards or eastwards. That was because the city of Jalu was considered a main caravan trading station. The inhabitants of the oasis were considered among the first people, who drove trade caravans along the longest distance linking the Mediterranean coasts with central and eastern Africa. The traders of the region established trade and marriage relationships with the inhabitants of Bornu, Timbuktu, Kano, Murzuq, Zawilah, Soknah, Ghat and al Kufrah. Within the local market of the oasis, as well as between the trade caravans reaching the oasis from all directions, the traders exchanged such Western commodities as textiles and woven products and domestic appliances on the one hand, and slaves, ivory, feathers, ostrich eggs, incense and perfumes on the other hand. The people of Jalu had become famous for their trade with the Sudan and central Africa. Trade exchange brought great revenue to the African continent. With the prohibition of slave trade and with the imposition of restrictions on the trading of some commodities, trade activity declined. Similar to the fate of other oases, Jalu lost its importance as a commercial centre linking the north and the south. Moreover, the inhabitants' agricultural activity became very restricted due to the limited fertility of the land and inadequate water reserves. Nevertheless, date palms still constitute an important source of revenue for the inhabitants of the oasis.[156]

6.20 Al Kufrah—the Travellers' Dream

Said Hamid

The city of al Kufrah is located in the south-eastern region of Libya, south-east of the city of Banghazi. Both cities are linked by an road of about 1000 km which goes through Ajdabiya, Awjilah and Jalu. al Kufrah lies at the intersection of latitude 24.10° north and longitude 23.05° east.

Deeply located as it is in the desert, al Kufrah did not attract the attention of the Arab travellers and geographers of the early Islamic centuries or of the Middle Ages. Nonetheless, it was ranked among the biggest stations on the caravan roads in the south-east of Libya. al Kufrah is made up of a number of oases in the very heart of the Sahara desert and is considered to be the most important of them all. The other oases are al-Hawari, al-Hwawiri, al-Tullab, at-Tlaylib and al-Juf. The closest oasis to al Kufrah is the Bazima oasis, which is located 120 km in the north-west. al Kufrah is also linked to Tazerbu in the north-west via a road of 240 km long and which passes through the Bazima oasis. The distance between al Awaynat, in the south-east, and al Kufrah is approximately 350 km.

6.20.1 Hard to Get to

Many foreign travellers dreamed of reaching al Kufrah. Hassanein Bey, an Oxford-educated Egyptian explorer who accompanied the travel writer Rosita Forbes to the al Kufrah oases in 1920, has pointed out in his book *The lost oases*: "*(...) that it has been said by explorers that the tribes of Cyrenaica, who are all under Sanussi influence, are actuated by religious fanaticism when they refuse foreigners permission to enter their territory. To my great interest, however, I discovered that not religious, but patriotic, fervor is the impelling principle. Their reasoning is simple. They do not want any aliens to come into the country, for they say foreigners mean domination, domination means paying taxes, and they do not want to pay taxes! Therefore, the best way to avoid taxes is to prevent any stranger from entering the region.*"[158] Therefore, no outsiders, but especially no Christians, were permitted to enter the area.

The first European who succeeded in reaching the remote oases was the German traveller Rohlfs who arrived there in 1879. But he was an exception. After long negotiations with the Zuwaya *shaykhs* (Fig. 6.85), they agreed to provide guides for Rohlfs'expedition. His small caravan managed to cross the waterless desert between Jalu and al Kufrah in a

Fig. 6.85 Zuwaya chiefs of al Kufrah as photographed by Hassanein Bey (1924). The Zuwaya are the conquerors of al Kufrah and the inhabitants of it now. They are the tribesmen, who destroyed all the notes and scientific results of the German explorer Rohlfs, when he visited them in 1879

record five days. But when the suspicious tribesmen, who inhabited al Kufrah, tried to attack him, he escaped with little more than his life. All his notes and scientific equipment were destroyed, and only the arrival of a messenger from the *shaykh* of the Sanussis in Jaghbub prevented him from being killed. Nevertheless, he managed to describe the al Kufrah oasis, the roads leading to it from different directions, and the oasis' economic and social conditions.[159] Hassanein Bey noted: "*But when I, a Moslem, the son of a religious man and a friend of their head man, arrived, they scratched their heads. Here was a foreigner whom they did not want, but they could not convict him on religious grounds. I had to be tolerated, temporarily at any rate*".[160]

The traveller Sadiq Muayyid al-Azam visited al Kufrah in 1895. He then published a book about his journey, highlighting the economic and social conditions of the oasis. The author stated that the al Kufrah oasis was situated halfway between Banghazi and the Wadai kingdom. For this reason, al Kufrah was an important station for caravans coming and going to Wadai and other African kingdoms in its vicinity. All caravans would meet et al Kufrah. As a result thereof, it was very important commercially because it connected the north with the south and vice versa.[161]

6.20.2 The Importance of the Sanussi Brotherhood

The al Kufrah oasis derived its importance from its distinctive location and also from the availability of groundwater, which was found close to the surface. In fact, in some places, lakes of sweet water were formed (Figs. 6.86 and 6.87). The various types of plants growing in the oasis provided ideal pastures for camels. The road that linked Ajdabiya, Awjilah, Jalu, Bazima, al Kufrah and al Awaynat could have been a main artery leading deep into the desert, but the long distances between the watering holes limited its regular use during the Middle Ages.[162] Another significant obstacle for trade was raids committed by tribes east and west of the Cyrenaica. However, with the institution of the Sanussi brotherhood, al Kufrah became an important trading centre, where traders exchanged different goods and where several trade routes crossed. According to Bey, it has been estimated that between a million and a half and three million people owed allegiance to the Sanussi brotherhood when el Mahdi became its active head. He saw that there was more scope for the influence of the brotherhood to the southward, and in the year 1894, he removed his headquarters from Jaghbub to al Kufrah. This move marked not only the beginning of an important era in the history of the Sanussi, but also in the development of trade between the Sudan and the Mediterranean coast by way of al Kufrah. Under the rule of the Zuwaya tribe of Badawi, who had conquered al Kufrah from the black Tubus, the group of oases was the chief centre of brigandage in the Libyan Desert.[163]

During the first half of the nineteenth century, the al Kufrah road was hit by an earthquake. Subsequently, less caravans used it due to the lack of security along its track. This situation lasted for 15 years, after which the road recovered its importance as security and stability were restored during the middle part of the same century.

6.20.3 Caravan Roads that Linked al Kufrah to the Cities of the Mediterranean Coast and the South:

Linking al Kufrah to the Mediterranean coast:	Linking al Kufrah to the south:
Sirte, Tagreft, Zillah, Tazirbu, Bazima, al Kufrah Banghazi, Ajdabiya, Awjilah, Jalu, Tazerbu, Bazima, al Kufrah Marsa Lak, Siwa Santriyya, al Kufrah	al Kufrah, al Awaynat, al-Athrun, Waddai or the Sudanese cities al Kufrah, Tibesti, Uzu, Kawar, Ghadāmis

There were also two roads, one going towards the east, to Egypt, and the other to the west, towards Rebyana, to Waw an-Namus, Waw al-Kabir, Zuweyla, Sabha. Here, it connected to the northern road, or the road coming from

Fig. 6.86 Date palms in the valley of al Kufrah. In the middle distance, the light streak is the lake of al Kufrah. In the foreground are the dwellings of the natives (Hassanein Bey 1924, 244)

Fig. 6.87 Scenes from al Kufrah taken from the book written by Hassanein Bey (1924). The saltwater lake at al Kufrah covers an area of two square miles, and it is very deep in the centre. The oasis surrounds the lake. In the foreground, it is one of the notable Bedouin chiefs of al Kufrah and one of the Sanussi soldiers. When Hassanein Bey asked this chief if there were fish in the lake, he replied: "What is a fish? Does it walk? Does it fly?" He found it to be almost impossible to describe a fish to a person who had never seen one (Hassanein Bey 1924, 246). Hassanein Bey stated: "The approach of al Kufrah from the north has an element of surprise in it that makes it doubly interesting. We marched through gelly rolling country with an irregular ridge of no great height forming the horizon. Suddenly the top of the ridge resolved itself into the outlines of a group of buildings, their walls hard to distinguish at any distance from the rocks and sands they match so weil in colour and in form. This is al Taj, the headquarters of the Sanussi family in al Kufrah. As we enter the town, we see that the ground drops abruptly away beyond it, down to the valley of the al Kufrah. This pleasant valley is a shallow, roughly oval bowl, 40 km on its long diameter in extent and up to 20 km on the short one. It is dotted with palm trees, and across it in an irregular line from north-east to south-west are strung the six settlements of Boema, Buma, Jof, al Zurruk, Tallalib, and Tollab. Near Jof lie the blue shimmering waters of a lake of large extent. At this point in the sand waste of the desert, this expanse of water is both a boon and an aggravation. Unfortunately, its waters are salty. However, one can swim in it with a pleasure unknown in the land of sea and river and bath-tubs." Hassanein Bey started exploring the valley of al Kufrah: "I visited the villages, the Suk where the weekly market is held, and the Zawya at Jof, which is the oldest Sanussi school and the first building erected in al Kufrah. It was interesting to find side by side in the Suk rille cartridges whose marking shows them to be thirty years old, Italian tomato sauce in tins from Banghazi, blue and white calico from Egypt, and leather, ivory, and ostrich feathers from Wadai. These products of the south, however, are not plentiful now in al Kufrah, except when a merchant who has brought them from Wadai is prevented for some reason from going on to the north to sell them in Egypt or Cyrenaica".[166]

Zuweyla to al Qatrun, Madrusah, Tajarhi, Sajdin, Kuwar and Kuka.

Until the beginning of the twentieth century, al Kufrah used to receive one caravan every week. The number of camels making up such caravans varied between one hundred and three hundred, carrying hides, ivory and ostrich feathers from central and western Africa towards the Mediterranean coasts. The use of this road came to an end with the Italian occupation of Libya in 1911.[164]

6.20.4 Al Kufrah at the Dawn of the Twentieth Century

At the beginning of the 1920s when Hassanein Bey was visiting the oasis, he noticed that "*in addition to the trade that passes through al Kufrah, most of the big Zuwaya chiefs go in for agriculture. They raise barley and maize. The Sanussi are more progressive, and grow melons, grapes, bananas, marrows, and other vegetables of the more delicate kinds, all of which are a great treat after the monotonous fare of the desert. They also raise mint and roses, from which they make rose-water and mint essence, so essential in their ceremonies of hospitality. From a few olive trees olive oil is produced in primitive presses. But dates are the main item in the Badawi's diet, and there are many date trees all through the al Kufrah valley. Dates are the only article of exportation from the oases, but for other commodities, whether for food or clothing, al Kufrah is mainly dependent on the outside world. Tea, sugar, rice, flour, and white calico are the main imports of al Kufrah. The Badawi live a simple and primitive life, sheltered in houses built of stone, plastered and whitewashed inside. A roughly made divan, covered with cushions and Badawi rugs, made in the north, is the only furniture one comes across in these small houses. In a rich man's house the reception room is covered with Persian rugs and expensive silk cushions. A gramophone with discs of Arab songs by Egyptian singers is the rich man's luxury.*" But what is very interesting to read is that "*manual work is done mostly by slaves. The price of slaves, learnt* at al *Kufrah, has risen a great deal during the last few years because there are no more slaves coming up from Wadai on account of the vigilance of the French authorities in that province.*" Hassanein Bey reports that the Badawi sometimes marry their slave-girls, "*and if one of these bears a male child she automatically becomes free. The Badawi have no prejudice against colour, that is, if a slave bears the head of a tribe his eldest male child, that child ipso facto becomes in his turn the head of the tribe, however black he may be. On the whole slaves are weil treated by the Badawi. They are weil dressed, for an ill-dressed slave reflects badlyon his master.*"[165]

Towards the end of the twentieth century, al Kufrah saw an increase in camel trading. Herds of camels without any loads assembled at al Kufrah, and from here, they were then transported on trucks to other Libyan cities and sold for their meat (see *Fezzan as a Transit Region in the Late twentieth and at the Dawn of the twenty-first Century*).

Notes

1. Di Vita et al. (1999), Di Vita (1996, 173–190), Mattingly (1995, 122–125), Kenrick (2009b, 16–34).
2. Mattingly (1995, 123).
3. Micara (2008, 387), Ciranna (2017, 67).
4. Rossi (1913–1936), Oman (2010, 2012).
5. Micara (2008, 385), Al-Tijani, Rihlah, in Rossi (1968, 76), al Tijani (1852).
6. Al-Tijani (1852).
7. Halm (1992, 129–157), Khalilieh (1999, 212–225), Salama (1995, 72f), al Khouja (2007, 32, 130).
8. Rossi (1968, 112), Micara (2008, 387).
9. al Khouja (2007, 6, 63).
10. Wright (2005, 20).
11. Micara (2008, 388).
12. Micara (2008, 391), Administration of the Historical Cities (2010).
13. Micara (2008, 391f).
14. al Khouja (2007, 35, 42, 48), Administration of the Historical Cities (2010).
15. Micara (2008, 394–395).
16. al Khouja (2007, 35, 42, 48), Administration of the Historical Cities (2010).
17. al Khouja (2007, 20, 42, 75), Administration of the Historical Cities (2010).
18. Salama (1995, 31f, 44f).
19. Micara (2008, 396).
20. Wright (2005, 45f), Tully (2009, 6).
21. Wright (2008, 153).
22. Wright (1989, 48f).
23. Wright (1989, 48), Dearden (1957, 160).
24. Lyon (1821, 12).
25. Piazza (1996, 139–173), Micara (2008, 397–398).
26. Cowper (1897, 11).
27. Rae (1877, 19–21).
28. Piazza (1996, 139–173), Micara (2008, 397–398).
29. Micara (2008, 402).
30. al Khouja (2007, 33, 43).
31. al Khouja (2007, 104).
32. Salama (1995, 47f), al Khouja (2007, 132).
33. Greenhow (1835, 52).
34. Greenhow (835, 3), Bosworth (2007, 533f).
35. Rohlfs (1871, 84).
36. Rohlfs (1881, 77).
37. Banse (1912).
38. Herzog (1961, 255–262).
39. Krause (2014, 34).
40. Krause (2014, 82).
41. Krause (2014, 71).
42. Krause (2014, 71).
43. Krause (2014, 82).
44. McLaren (2006, 223).
45. Christides et al. (2007, 534), Altekamp (2000, 1998).
46. McLaren (2006, 98), Altekamp (2000).
47. McLaren (2006, 98), Schieder (2006, 707).
48. McLaren (2006, 98), Altekamp (2000).
49. Barth (1849, 295).
50. Rohlfs (1871, 32).
51. Rohlfs (1871, 82).
52. al Khouja (2007, 34).
53. Altekamp (2000, 249).
54. Hamed (2008, 24–37).
55. Altekamp (2000, 191–200, 249–253), see also Altekamp (1998, 159–175), Altekamp (1999, 80–103), and McLaren (2006, 94–103).
56. Altekamp (2000, 194f) and McLaren (2006, 94–103).
57. Braun and Passon (2010), Bianca (2000).
58. Hawqal, A Picture of the Earth, al Bakri, al Mughrib in the mention of the countries of Ifriqiah and Maghreb, a Part of the Book of al Masalik Wal Mamalik, Baghdad, Ash Shammakhi (1995), Goodchild (1999), Hamid (1978), Hamid (2005).
59. Haynes (1955, 36).
60. Goodchild (1954, 76).
61. Despois (2012).
62. Khaldun (1867), Band 7, 58f.
63. Leo Africanus (1984, 191).
64. Helfensteller (1998, 87).
65. Richardson (1848, 88).
66. Chavanne (1879, 221).
67. Rohlfs (1984, 62).

68. Evans (1976, 31).
69. Chavanne (1879, 222f).
70. Rohlfs (1984, 63).
71. Chavanne (1879, 232).
72. Schliephake (1982, 286).
73. Ahmad (1969, 145f), Despois (2012).
74. Richardson (1848, 92).
75. Ahmad (1969, 148f).
76. Rohlfs (1984, 69f).
77. Richardson (1848, 90).
78. Evans (1976, 31ff).
79. Rohlfs (1984, 66f).
80. Eldblom (1968, 1ff).
81. Despois (2012).
82. Chavanne (1879, 234).
83. Rohlfs (1984, 68ff).
84. Despois (2012).
85. Schliephake (1982, 289).
86. Eldblom (1969, 3–8).
87. Passon (2008).
88. McLachlan (2012), Az Zawi (1968), Diary of a Tripoli caravan to Wadday, Rohlfs (1856–1867), Transportation routes in Tripoli, Mattingly (translated by Center for the Libyan Archives and Historical Studies), Nachtigal (translated by Center for the Libyan Archives and Historical Studies), al Arabi, Terry (2004, 2008).
89. Miludi.
90. Barth (1890, 45ff), Az Zawi (1968), Terry (2004, 2008).
91. Terry (2004, 2008), Goodchild (1999); al Barghuthi (1971), Az Zawi (1968), Abbas (1968), An Nims (1978), Haynes (1981).
92. Ad Dinasuri (2004), Terry (2004, 2008), Sharaf (1962).
93. al Bakri; Terry (2004, 2008), Ad Dinasuri (2004), Sharaf (1962), Ayyub (1963).
94. al Idrisi; Rohlfs; Ad Dinasuri (2004), Az Zawi (1968), al Mahdawi; al Mawsili.
95. Weis (1996).
96. Martin (1985, 61–76).
97. Rohlfs (1874), Chap. 6.
98. Barth (1857).
99. Rohlfs (1874), Chap. 7.
100. Wirth (2000, 97f).
101. Weis (1996).
102. Barth (1857, 180).
103. Nachtigal (1879, 92).
104. Rohlfs (1874), Chap. 7.
105. Nachtigal (1879, 93f and 133f).
106. Barth (1857), Nachtigal (1874), Rohlfs (1874).
107. Nachtigal (1879, 130).
108. Barth (1857, 180).
109. Weis (1996).
110. Rohlfs (1874).
111. Barth, 177f, Nachtigal (1879, 73).
112. Rohlfs (1874), Chap. 7; Wirth (2000, 103ff).
113. Chavanne (1879).
114. Eldblom (1968, 314f), Martin (1985, 63).
115. Nachtigal (1879).
116. Weis (1996).
117. Passon (2008, 222–226).
118. Terry (2004, 2008), Hamid (2006).
119. al Idrisi (1981, 120).
120. Hamid (2006), Sharaf (1962).
121. Hamid (2006).
122. Mattingly et al. (2009, 151).
123. Hamid (2006), Ayyub (1993).
124. Mattingly et al. (2009, 151).
125. Hamid (2006), The Tarikhuna encyclopedia supervised by Sadeq an Naihoum.
126. Hakam (1968), al Ansari et al. (1967), Ayyub (1993), al Barghuthi (1972), Terry (2004, 2008).
127. Despois (2012).
128. Krause (1998); Hasan: Studies in the Geography of Libya and the Arab World; Az Zawi (1968); A map from the book of A. G. Hopkins: The Economic History of Arab Africa (translated by Ahmad Fuad Balba) 1998; Terry (2004, 2008); Despois (2012).
129. Terry (2004, 2008), Az Zawi (1968), Sharaf (1962), al Azm (1998).
130. Terry (2004, 2008), Sharaf (1962), Ad Dinasuri (1967).
131. Terry (2004, 2008).
132. Hawqal, A Picture of the Earth, Beirut.
133. Khaldun (1284); Hakam (1983); al Bakri in the Mention of the countries of Ifriqiah and Maghreb; Terry (2004, 2008), Goodchild (1999), Hamid (1978), Hamdani; Abbas and Najm (1968), Fehr Fari (1977–1981).
134. Goodchild (1970), Wright (2007, 104), Lequien (1740, 623–626).
135. Wright (2007, 103–105), Cordell (1977, 21–22), Baier (1978, 27–28).
136. al Buri (2004); Brothers Biatci and the Libyan Coast: translated by al Hadi Abu Luqma; al Hashaishi (1965); Mansur: The Arab Travellers and their role in the writing of Libya's Political and Economic History in the Seventeenth and Eighteenth Centuries; Curo: Libya During the Second Ottoman Era; al Azm (1998); al Fituri: Libya and the Caravan Trade; Bazama: Banghazi through History; Rohlfs (translated by Center for the Libyan Archives and Historical Studies); Hamid (2008).
137. Hawqal: A Picture of the Earth; Hakam (1968); al Bakri in the Mention of the countries of Ifriqiah and

Maghreb; Terry (2004, 2008), al Ayyashi (1977), Abbas and Najm (1968), Hamid (1991), Belhaj (1968).
138. Prestopino (1999, 123).
139. Hawqal: A Picture of the Earth. Hakam (1968); al Bakri in the Mention of the countries of Ifriqiah and Maghreb; Terry (2004, 2008), al Ayyashi (1977), Abbas and Najm (1968), Hamid (1991), Belhaj (1968).
140. Hawqal: A Picture of the Earth.
141. Levtzion and Hopkins (2000, 104).
142. Hawqal: A Picture of the Earth; Hakam: The conquest of Egypt and the Maghreb; al Bakri in the Mention of the countries of Ifriqiah and Maghreb; Terry (2004, 2008), al Ayyashi (1977), Abbas and Najm (1968), Hamid (1991), Belhaj (1968).
143. Hakam: The conquest of Egypt and the Maghreb; Terry (2004, 2008), al Ayyashi (1977), Rohlfs (2002): Across Africa (translated by the Center for the Libyan Archives and Historical Studies); Abbas and Najm (1968), al Sayyad and Hamid (1976); Department of Antiquities (1976).
144. Hawqal: A Picture of the Earth.
145. al Bakri in the Mention of the countries of Ifriqiah and Maghreb.
146. Hakam: The conquest of Egypt and the Maghreb; Terry (2004, 2008), al Ayyashi (1977), Rohlfs (2002): Across Africa (translated by the Center for the Libyan Archives and Historical Studies); Abbas and Najm (1968), al Sayyad and Hamid (1976); Department of Antiquities (1976).
147. Assuwais (2000).
148. Rohlfs: A journey to al Kufrah (translated by The Center for the Libyan Archives and Historical Studies).
149. The Islamic Encyclopedia; Assuwais (2000), al Aqil (2000); Rohlfs: A journey to al Kufrah (translated by Center for the Libyan Archives and Historical Studies); Pashu: Narration of a trip to Marmara, Corina, and the oases of Aujal and Marada; Ash Sharief (1988); Hamid: The ancient history of the Aujala oasis up to the Ottoman era; Sharib: The caravan trade and its relation to the Aujala oasis.
150. Hornemann (1802).
151. The Islamic Encyclopedia; Assuwais (2000), al Aqil (2000); Pashu: Narration of a trip to Marmara, Corina, and the oases of Aujala and Marada. Ash Sharief (1988); Hamid: The ancient history of the Aujala oasis up to the Ottoman era; Sharib: The caravan trade and its relation to the Aujala oasis.
152. Sharaf (1962).
153. al Azm: A Journey in the Great Desert in Africa (translated by Center for the Libyan Archives and Historical Studies).
154. Terry (2004, 2008); Rohlfs: A Journey from Tripoli to Alexandria (translated by Center for the Libyan Archives and Historical Studies); Mansur: The Arab Travellers and their role in the writing of Libya's political and economic history.
155. Terry (2004, 2008).
156. Mansur: The Arab Travellers and their role in the writing of Libya's political and economic history.
157. Hassanein Bey (1924).
158. Hassanein Bey (1924).
159. Rohlfs: Across Africa (translated by Center for the Libyan Archives and Historical Studies).
160. Hassanein Bey (1924).
161. al Azm (1998), Pasha (2006).
162. Sharaf (1962).
163. Terry (2004, 2008), Hassanein Bey (1924, 278).
164. Sharaf (1962).
165. Hassanein Bey (1924, 289).
166. Hassanein Bey (1924, 287–288).

Bibliography

Oral Interviews

Al Thani N, Al Turki M (2006) Collection of source material from the family archive. Interview with Haj Muktar Achmed Taher, Al Bacchi

Al Thani N, Al Turki (2006) Desert castles. Interview with archaeologist of the district of al Qatrun

Al Thani N, Al Turki (2006) Collection of source material from the family archive. Interview with Mohammed Abdel Aziz, Dougal

Al Thani N, Turki Al (2006) Regional development. Interview with representatives of the People‘s Congress from Murzuq, Murzuq

Braun K, Passon J, Ghanem I (2006) Caravan trade, desert castles. Interview with head of the Sahara Research Centre, Murzuq

Braun K, Passon J, Ghanem I (2006) Urban development. Interview with responsibilities from the municipality of Murzuq, Murzuq

Braun K, Passon J, Ghanem I (2006, 2009) Orientation in the desert. Interview with desert guides and field experiments, Sahara Desert

Passon J (2014) Caravan Trade. Interview with Historians from the CLAR.HS, Tripoli

Passon J, Ghanem I (2006) Caravan trade in Ghat. Interview with elders from the town council, Ghat

Passon J, Ghanem I (2006) Caravan Trade, urban development of al Qatrun. Interview with Ali, member of the village council of elders, Al Qatrun

Passon J, Ghanem I (2006) Caravan trade in al Fugah. Interview with elders from the village council. al Fuqah

Passon J, Ghanem I (2006) Caravan Trade and urban development in al Qatrun. Interview with elders of the village council, Al Qatrun

Passon J, Ghanem I (2006) Caravan trade, regional development. Interview with representatives of the People‘s Congress from Tajahri, Tajahri

Passon J, Ghanem I (2006) Caravan Trade, regional development. Interview with representatives of the People's Congress from Trajen, Trajen

Passon J, Ghanem I (2006, 2008, 2009) Orientation in the desert, caravan trade. Interview with desert guides and observations, Sahara Desert

Swifka S, Ghanem I (2006) Trade between Niger and Libya, trade in camels. Interview with merchants from Niger at the camel market in Fungul, Fungul

Literature of the Historical Archive of the CNAR.HS

Abu Sharib M. The Caravan Trade and its relation to the Aujala Oasis, Tripoli

Al Baladhuri Futuh Al Buldan, Cairo

Al Fituri AS Libya and the Caravan Trade, Tripoli

Al Hashaishi M (1965) Trip to Libya, Tripoli

Al Mabruk al Mahdawi M Libya's Human Geography, Benghazi

Al Mawsili M The transportation routes, Tripoli

Bazama MM Benghazi through History, Benghazi

Centre for Libyan Archives and Historical Studies (ed) Diary of a Tripoli Caravan to Wadday, Tripoli

Centre for Libyan Archives and Historical Studies (ed) The Arab Travelers and their role in the writing of Libya's political and economic history, Tripoli

Curo F Libya during the second Ottoman Era (Translated by Khalifa at Tillisi). Tripoli

Department of Antiquities (ed) (1976) Some Islamic Sightseeing's in Libya. Tripoli

Fehr Fari G The Sirte Excavations (1977–1981) (Translated by Mustafa at Tarjum). Edited by Department of the Libyan Antiquities, Tripoli

Hamdani A History of the City of Sirte (Translated by Mustafa at Tarjum). Tripoli

Hamid S The ancient history of the Aujala oasis up to the Ottoman era. Tripoli

Hasan MI Studies in the Geography of Libya and the Arab World. Benghazi

Khushaim AF (1967) Libyan Texts. Tripoli

Mansur AMI The Arab travelers and their role in the writing of Libya's political and economic history in the Seventeenth and Eighteeth Centuries. Tripoli

Naihoum S (ed) The Tarikhuna encyclopedia. Tripoli

Najm M History of Tripoli (Translated by Abdussalam Adham). Tripoli

Pall Service Company (ed) (1980) The general plan for the city of Ghadames

Pashu JR Narration of a Trip to Marmara, Corina and the Oases of Aujala and Marada (Translated into Arabic by Dr. Muftah Abdulla al Meisuri). Tripoli

University of Birmingham: A map from the book of A. G. Hopkins: The Economic History of Arab Africa (Translated by Ahmad Fuad Balba)

K. Braun and J. Passon (eds.), *Across the Sahara*,
https://doi.org/10.1007/978-3-030-00145-2

Archive of the CNAR.HS, District Archives & Family Archives

Collection of source material compiled by al Haj Muktar Achmed Taher (family archive, al Bacchi), collected 19.11.2006
Collection of source material compiled by Mohammed Abdel Aziz (family archive, Dougal), collected 22.11.2006
Collection of source material from the Museum of the Medina of al Qatrun, collected 18.11.2006
Collection of source material from the Sahara Research Centre (Murzuq), collected 23.11.2006
Consular records from Germany, Great Britain, Italian source material collected April 2007, February 2012, November 2013. Centre for Libyan Archives and Historical Studies, Tripoli

Unpublished Master Thesis

Al Arabi M Master's thesis on Gharian

Presentations

Al Zubaidi MS (2002) The efficiency of thermal performance of the desert buidings—the traditional house of ghadames. Montréal, Québec, Canada, 6/5/2002.
Assuwais B (2000) Aujala between the past and the present. Aujala, Libya, 9/17/2000
Assuwais B (2000) Aujala in the eyes of the classical writers (the Greek, the Romans, and the Byzantine). Aujala, Libya, 9/17/2000
Di Lernia S (2003) (Re)Setting the Agenda. Development and cultural heritage in the Libyan Sahara, Uppsala
Fritsch K and Voigt I (2008) Local knowledge is wonderfully good, but …—African Knowledge in European Maps—In: Symposium, ICA commission on the history of cartography. Portsmouth, UK, 9/10/2008
Kohl I (2004) "Kontinuität und Wandel". Transsaharische Handels- und Reisewege durch den Fezzan/Libyen. „Handels- und Reiserouten im Nahen und Fernen Osten". Grazer Morgenländische Gesellschaft. Graz
Masonen P (2001) Leo Africanus: the man with many names. In: Finnish Institute in Rome. Rome (Villa Lante), 11/7/2001
Underdown R, Redfern J, Lisker F (2006) Hydrocarbon maturation history of the Ghadames Basin, North Africa: the influence of exhumation and erosion. In: Saharan Conference, April 2006.

Primary Sources: Literature in Arabic, German, Romanic Languages and English

Africanus L (1984) Beschreibung Afrikas [1526]. In: Karl S-E (ed). Leipzig
Al Bakri A Al Maghrib in the Mention of the countries of Ifriqiah and Maghreb. A Part of the Book of al Masalik Wal Mamalik. Baghdad
Al Bakri A (1913) Description de l'Afrique Septentrionale. Paris
Al Hachaichi M (1912) Voyage au pays Senoussiya. Paris
Al Hakam I Futuh misr wa-l-magrib. Baghdad
Al Hakam I (1922) The history of the conquest of Egypt, North Africa and Spain. In: Torrey CC (ed) Known as the Futuh misr of Ibn Abd al-Hakam. New Haven
Al Hamawi Y Mu'jam al Buldan, vol 1. Beirut
Al Idrisi (1983) From the nuzhat al mushtaq book. Scrutinized by Muhammad Haj Sadeq, Paris
Al Idrisi (1983) The Arab Maghreb. Scrutinized by Muhammad Haj Sadeq, Paris
Al Idrisi M (1592) Kitab nuzhat al-mushtaq. Rom
Al Masudi (1989) The meadows of gold: the Abbasids (Translated and edited by Paul Lunde and Caroline Stone). London
Al Qaysi A (1968) Anis as sari was sareb. Scrutinized by Muhammad Al Fasi, Fez
Al Tijani A (1968) Rihlah. In: Rossi E (ed) Storia di Tripoli e della Tripolitania dalla Conquista Araba al 1911. Rom.
Al Tijani A (1994) Voyage dans la Régence de Tunis pendant les années 706, 707 et 708 de l'Hégire (1306–1309) (Translated by A. Rousseau). In *Islamic Geography* 186
Al Yaqubi A (1892) Kitab al Buldan. In: de Goeje MJ (ed). Leiden.
Al Yaqubi A (1981) Kitab al Buldan. In: Levtzion N and Hopkins JFP (eds) Corpus of early arabic sources for West African History. Cambridge.
Barth H (1849) Wanderungen durch das Punische und Kyrenäische Küstenland oder Magreb. Afrikia und Barka, Berlin
Barth H (1855–1858) Reisen und Entdeckungen in Nord- und Zentralafrika, vol 5. Gotha
Barth H (1857) Reisen und Entdeckungen in Nord- und Central-Afrika in den Jahren 1849 bis 1855, 2nd edn, vol 5 Gotha (1). Available online at www.BioLib.de
Barth H (1857) Reisen und Entdeckungen in Nord- und Central-Afrika in den Jahren 1849 bis 1855, vol 5. Gotha (2). Available online at www.BioLib.de
Barth H (1857) Reisen und Entdeckungen in Nord- und Central-Afrika in den Jahren 1849 bis 1855, vol 5. Gotha (3). Available online at www.BioLib.de
Barth H (1858) Reisen und Entdeckungen in Nord- und Central-Afrika in den Jahren 1849 bis 1855, vol 5. Gotha (5). Available online at www.BioLib.de
Barth H (1863) Die Bestätigung der Todesnachricht des Herrn Moritz v Beurmann. Zeitschrift für allgemeine Erdkunde 15:538–543
Barth H (1863) Neueste Nachrichten über Dr. Vogel und Herrn v. Beurmann. Zeitschrift für allgemeine Erdkunde 14
Barth H (1967) Travels and discoveries in North and Central Africa. In: Kirk-Greene AHM, vol 3. London
Battuta I (1983) Travels in Asia and Africa 1325–1354. London
Battuta I (1985) Reisen bis ans Ende der Welt 1325–1353. Stuttgart
Beechey FW and Beechey HW (1828) Proceedings of the Expedition to explore the Northern Coast of Africa, from Tripoly eastward in 1821 and 1822. Comprehending an account of the Greater Syrtis and Cyrenaica and of the ancient cities composing the Pentapolis. London.
Bernet E (1912) En Tripolitaine. Voyage a Ghadamès, Paris
Bovill EW (ed) (1964) Missions to the Niger. Cambridge
Braun E (1914) The new Tripoli and what I saw in the hinterland. London
Brothers Biatci The Libyan Coast (Translated by Al Hadi Abu Luqma)
Bruce-Lockhart J (1996) Clapperton in Bornu. Journals of the Travels in Bornu of Lieutenant Hugh Clapperton RN From January 1823 to September 1824. Cologne
Cacace P (1924) In Tripolitania: Rapporto di Paolo Cacace delegato della Camera di Commercio di Napoli al viaggio d'istruzione Commerciale in Tripolitania (novembre-dicembre 1924). Napoli

Caillié R (1830) Journal d'un voyage à Tembouctou et à Jenné dans l' Afique centrale. Paris
Chavanne J (1879) Die Sahara. Oder: Von Oase zu Oase. Wien
Clapperton H (1985) Difficult and dangerous roads: Hugh Clapperton's travels in Sahara and Fezzan. In: Narrative of travels and discoveries in Northern and Central Africa. London
Clapperton H, Denham D, Oudney W (1828) Narrative of travels and discoveries in Northern and Central Africa in the years 1822, 1823 and 1824. 3rd edn, vol 2. London (1)
Cowper HS (1897) Further Notes on the Tripoli Hill Range. Geogr J 9 (6):620–638
Cowper HS (1897) The Hill of the Graces. In: A record of investigation among the Trilithons and Megalithic sites of Tripoli. London
Dearden S (1957) Ten years' residence at the court of Tripoli. London.
Dearden S (1976) A Nest of Corsairs. The Fighting Karamanlis of the Barbary Coast. London
Della Cella P (1822) Narrative of an expedition from Tripoli in Barbary to the western frontier of Egypt in 1817 (Translated by Anthony Aufrere). London
Della Cella P (1912) Viaggio da Tripoli di Barbaria alle frontiere occidentali dell'Egitto fatto nel 1817. Genova
de Mathuisieulx H-M (1912) La Tripolitaine d'hier et de demain. Paris
de Mathuisieulx HM (1912) Attraverso la Libia. Milano
de Polignac L, Mircher H (ed) (1863) Mission de Ghadamès (septembre, octobre, novembre & décembre 1862): Rapports officiels et documents à l'appui. Alger
Di Giovanni M (1938) Tripoli. Bologna
Dickson CH (1852) Extract from Vice-Consul C. H. Dickson's report of his journey from Tripoli to Ghadamis. J R Geogr Soc Lond 22:131–136
Dickson CH (1860) Account of Ghadamis. J R Geogr Soc Lond 30:255–260
Duveyrier H (1860) Nachrichten von Herrn H. Duveyrier. Aus einem Briefe Duveyrier's an Herrn Dr. Barth, Ghadames 14. Aug. 1860. In: Zeitschrift für allgemeine Erdkunde vol 9, pp 234–236
Duveyrier H (1864) Exploration du Sahara. Les Touareg du Nord, Paris
El Hachaichi MBO (1903) Voyage au Pays des Senoussia à travers la Tripolitaine et les pays Touareg. Paris
Forbes R (1921) Across the Libyan Desert to Kufara. Geogr J 58 (2):81–101
Forbes R (1921) The Secret of the Sahara: Kufara. London
Furlong CW (1909) The gateway to the Sahara: observations and experiences in Tripoli. New York
Ghisleri A (1912) Tripolitania e Cyrenaica dal Mediterraneo al Sahara. Milano
Goutier EF (1987) Sahara. The Great Desert, New York
Greenhow R (1835) The history and present condition of Tripoli. In : With some accounts of the other barbary states (1835). Richmond
Hamilton J (1856) Wanderings in North Africa. London
Hassanein Bey AM (1909) Ghat and its surroundings. Geogr J 34 (2):171–173
Hassanein Bey AM (1924) Through Kufra to Darfur. Geogr J 64:2743291
Hassanein Bey AM (1925) The Lost Oases. London
Hawqal I (1938) Opus Geographicum. Leiden
Hawqal I (1964) Configuration de la terre (Translated by G. Wiet). Paris-Beyrouth
Herodot (2004) Neun Bücher zur Geschichte. Wiesbaden
Hornemann FK (1802) Fr. Hornemanns Tagebuch seiner Reise von Cairo nach Murzouk, der Hauptstadt des Königreichs Fessan in Afrika, in den Jahren 1797 und 1798. In: König C (ed), Weimar
Ibn Said al Maghribi (1958) Bast al-ard fi al-tul wa-al-'ard. Titwan
Instituto Agricolo Coloniale (ed) (1946) Tripolitania. Some photographic representations of Italy's Action. Firence
Instituto Geografico Militare (1914) Rilievi in Libia ed in Somalia. With assistance of Ministero delle Colonie. Rom
Khaldun I (1892) Kitab al Ibar. In: de Goeje MJ (ed). Leiden
Khaldun I (1951) Ausgewählte Abschnitte aus der Muqaddima. In: Schimmel A (ed). Tübingen
Khaldun I (1958, 1986) The Muqaddimah. In: Rosenthal F (ed) An Introduction to History, vol 3. New York
Khaldun I (1980, 1995) Le Voyage d'Occident et d'Orient. In: Cheddadi A (ed). Paris
Khaldun, I. (1981): Kitab al Ibar. In Levtzion, N. & Hopkins, J. F. P. (Ed.): Corpus of early Arabic sources for West African History. Cambridge
Khaldun I (1992) Buch der Beispiele. Die Einführung. In: Pätzold M (ed). Leipzig
Khaldun I (2002) Le Livre des Exemples. In: Cheddadi A (ed) Tome I. Paris
Khaldun I (2006–2013) Kitab al-'Ibar wa-diwan al-mubtada' wa-'l-habar fi aiyam al-'arab wa-'l-'agam wa-'l-barbar wa-man 'asar-ahum min dawi 's-sultan al-akbar. Tunis
Krause GA (1878) Zur Geschichte von Fessan und Tripoli in Afrika. In: Zeitschrift der Gesellschaft für Erdkunde zu Berlin Band, vol 13, pp 356–373
Krause GA (1882) Aufzeichnungen über die Stadt Ghat in der Sahara. In: Zeitschrift der Gesellschaft für Erdkunde zu Berlin Band, vol 17, pp 266–337
Krause GA (1888) Mit einer Sklavenkarawane in der Sahara, geschrieben 1875 von Musa. In: Kreuzzeitung Berlin, pp. 364, 366, 368, 370, 372, 374, 376, 378.
Krause GA (1998) The Libya Hinterlands (Translated by Dr. Imad Ghanem). Tripoli
Krause GA (2014) Tripolitanisches Kriegstagebuch. In: Seebald P (ed). Bremen
Lenz O (1887) Timbouctou: Voyage au Maroc, au Sahara et au Soudan, vol 2. Paris
Lequien M (1740) Oriens christianus in quatuor Patriarchatus digestus, vol. II. Paris
Levtzion N and Hopkins JFP (eds) (1981) Corpus of early Arabic sources for West African history. Cambridge
Lyon GF (1821) Narrative of travels in Northern Africa in the Years 1818, 19 and 20. Accompanied by Geographical Notices of Soudan and of the Course of the Niger. London
Denham D, Clapperton H, Oudney W (1826) Narrative and discoveries in Northern and Central Africa in the years 1822, 1823 and 1824. London@@
Nachtigal G (1879–1889) Sahara und Sudan, vol 3. Leipzig, Berlin
Nachtigal G (1971–1984) Sahara and Sudan (Translated by Fisher A, Fisher GB, Humphrey J). vol 4. London
Nachtigal G The Sahara and the Sudan Country (Translated by Imad Ghanem). Tripoli
Pollacchi P (1929) Atlas Colonial Français. Colonies, Protectoratset Pays sous mandat
Primaudaie ME (1805) Le Littoral de la Tripolitaine. Commerce, Navigation, Géographie Comparée. Paris
Reynolds JM, Ward Perkins JB (1952) The inscriptions of Roman Tripolitania, London
Richardson J (1849) Travels in the Great Desert of Sahara in the years of 1845 and 1846, vol 1. London
Richardson J (1849) Travels in the Great Desert of Sahara in the years of 1845 and 1846, vol 2. London
Richardson J (1853) Narrative of a mission to Central Africa, vol 2. In: Bayle S-J (ed). London
Richardson J (1853) Narrative of a mission to Central Africa performed in the years 1850–51, vol 2. London

Rohlfs G A Journey across Africa: Observations of German traveller Rolfs in Libya, Bornu and the Gulf of Guinea, 1856–1867 (Translated by Dr. Imad Ghanem). Tripoli
Rohlfs G A Journey from Tripoli to Alexandria (Translated by Dr. Imad Ghanem). Tripoli
Rohlfs G A Journey to al Kufra (Translated by Dr. Imad Ghanem). Tripoli
Rohlfs G A Trip from Tripoli to Alexandria (Translated by Dr. Imad Ghanem). Tripoli
Rohlfs G Across Africa: A journey from the Mediterranean to the Lake of Chad to the Gulf of Guinea (Translated by Dr. Imad Ghanem). Tripoli
Rohlfs G (1868) Reise durch Marokko, Uebersteigung des grossen Atlas, Exploration der Oasen von Tafilet, Tuat und Tidikelt und Reise druch die grosse Wüste über Rhadames nach Tripoli. Bremen
Rohlfs G (1871) Von Tripolis nach Alexandrien, vol 2. Bremen
Rohlfs G (1874) Quer durch Afrika. Reise vom Mittelmeer nach dem Tschad-See und zum Golf von Guinea, vol 2. Leipzig
Rohlfs G (1875) Drei Monate in der libyschen Wüste. Kassel
Rohlfs G (1876) Beiträge zur Entdeckung und Erforschung Afrikas. Berichte aus den Jahren 1870–1875. Leipzig
Rohlfs G (1881) Kufra. Reise von Tripolis nach der Oase Kufra. Ausgeführt im Auftrage der afrikanischen Gesellschaft in Deutschland. Leipzig
Rohlfs G (1883) Expedition zur Erforschung der Libyschen Wüste unter den Auspizien Sr. H. d. Chedive von Ägypten im Winter 1874/1875. Kassel
Rohlfs G (1885) Von Tripolis nach Alexandrien, vol 2, 3rd edn. Norden
Rohlfs G (1984) Quer durch Afrika: Die Erstdurchquerung der Sahara vom Mittelmeer zum Golf von Guinea 1865–1867. Stuttgart
Rohlfs G (2002) Across Africa: A journey from the Mediterranean to the Lake of Chad (Translated by Dr. Imad Ghanem). Tripoli
Schweinfurth G (1863) Bericht über die von M. v. Beurmann 1862 aus dem mittleren Sudan eingesandten Pflanzenproben. In: Zeitschrift für Allgemeine Erdkunde, vol 5
Schweinfurth G (1874) Im Herzen von Afrika. Leipzig
Strabo (2005) Geographica. In der Übersetzung und mit Anmerkungen von A. Forbiger. Wiesbaden
Todd ML (1912) Tripoli the mysterious. London
Tully M (2009) Letters written during a ten years′ residence at the Court of Tripoli, 1783–1795. In: Stone C (ed). Kilkerran
Tully R (1816) Narrative of a ten years' residence at Tripoli in Africa, 2nd edn. London
Vinassa de Regny P (1913) Libya italica: Terreni ed acque, vita e colture della nuova colonia. Con appendice di ricerche ed osservazioni originali ed analisi di terreni fatte dall'autore durante la su. Milano
Vischer A (1911) Tripoli. Geogr J 38(5):487–494
Vischer H (1909) A Journey from Tripoli across the Sahara to Lake Chad. Geogr J 33(3):241
Vischer H (1910) Across the Sahara from Tripoli to Bornu. London
Vogel E (1855) Mission to Central Africa. In: JRGS 15, pp 237–245
Vogel E (1860) Schilderung der Reisen und Entdeckungen des Dr. Eduard Vogel in Central-Afrika in der großen Wüste, in den Ländern des Sudan. Leipzig
von Beurmann M (1862) Brief des Herrn Moritz v. Beurmann an Herrn Dr. H. Barth. In: Zeitschrift für allgemeine Erdkunde, vol 12, pp 404–413
von Beurmann M (1862) Brief des Herrn Moritz v. Beurmann an Herrn Dr. H. Barth. In: Zeitschrift für allgemeine Erdkunde, vol 13, pp 347–352
von Beurmann M (1862) Brief des Herrn Moritz v. Beurmann an Herrn Dr. H. Barth. In: Zeitschrift für allgemeine Erdkunde, vol 13, pp 44–50
von Beurmann M (1863) Briefliche Mittheilung des Herrn v. Beurmann an Prof. Ehrenberg. In: Zeitschrift für allgemeine Erdkunde, vol 15, pp 287–293
von Beurmann M (1863) Briefliche Mittheilungen des Herrn Moritz v. Beurmann an Herrn Dr. H. Barth. In Zeitschrift für allgemeine Erdkunde, vol 15, pp 273–286
von Maltzan H (1870) Reise in den Regentschaften Tunis und Tripolis, vol 3. Leipzig
von Maltzan H (2008) Tripoli and Tunis (Translated by Dr. Imad Ghanem). Tripoli
Vuillot P (1895) L'exploration du Sahara: Etude historique et géographique. Paris
Warrington GH (1844) A short account of Tripoli in the West. In: JRGS 14, pp 104–107
Yusha BQ (1986) Fihris makhtutat Ghadamis. Tripoli

Secondary Literature

Abbas I (1970) Tarih libya mundu al-fath al-arabi hatta ma tla al-qarn at-tasi al-hijri. Benghazi
Abbas I, Najm M (1968) Libya in the history and biography books. Benghazi
Abou-El-Haj R (1983) An agenda for research in history: The history of Libya between the sixteenth and nineteenth centuries. Int J Middle East Stud 15(3):305–319
Abu Halaq M, An Nims M (1978) The Museum of Antiquities in the Saraya Al Hamra. Tripoli
Ackermann-Liebermann PI (2012) Contractual partnerships in the Geniza and the relationship between Islamic law and practice J Econ Soc Hist Orient 54(2011):646–676
Ad Dinasuri J (1967) The Geography of Fezzan. Benghazi
Ad Dinasuri J (2004) The Geography of Fezzan. Benghazi
Administration of the Historical Cities (ed) (2010) Guide of the historical landmarks of the old city of Tripoli. Tripoli
Ahmad NA (1969) Die ländlichen Lebensformen und die Agrarentwicklung in Tripolitanien. Heidelberg
Ahmida AA (1994) The making of modern Libya. State Formation, Colonization, and Resistance, 1830–1932. New York
Ahmida AA (ed) (2009) Bridges across The Sahara. In: Social, economic and cultural impact of the trans-Sahara Trade during the 19th and 20th Centuries. Newcastle
Ahmida AA (2009) The making of modern Libya. In: State formation, colonization, and resistance, 2nd edn. Albany, [N.Y.]
Al Anies D (ed) (2010) Civilization of the Sahara. Benghazi
Al Ansari AN (1967) Al manhal al adb fi tarih Tarabuls al-Garb. Tripoli
Al Aqil AM (2000) Aujala in the eyes of traveller Rosita Forbes and researcher Jean Mason. Aujala
Al Ayyashi AS (1977) Ma'ul Mawai'd. Indexed by Muhammad Higgi. Rabat
Al Azm SM (1998) A journey in the great Desert in Africa (Translated by Abdul Kareem Abu Shweireb). Tripoli
Al Barghuthi A (1971) The Ancient Libyan History. Beirut
Al Barghuthi A (1972) The Islamic History of Libya. Beirut
Al Barghuthi A (1973) Tarih libya al-islami. Beirut
Al Buri WA (2004) Benghazi in the period of Italian Colonialism
Al Khouja YK (2007) Tripoli. The fragrance of ancient times. Tripoli

Al Miludi A A picture of castles, forts, palaces and schools on the Libyan soil during various eras. Tripoli

Al Sayyad A, Al Hamid A (1976) The old lslamic city of Ajdabiyah. Tripoli

Alföldy G (1987) Römisches Städtereisen auf der neukastilischen Hochebene. Ein Testfall der Romanisierung. - In: Abhandlungen der Heidelberger Akademie der Wissenschaften, Heidelberg, pp 7–140

Alhasanat MB, Kabir S, Hussin MA, Addison E (2012) Spatial analysis of a historical phenomenon: Using GIS to demonstrate the strategic placement of Umayyad desert palaces. GeoJournal 77:343–359

Allaghi FA (1981) Rural women and decision making: A case study in the Kufra settlement project. Fort Collins, Libya

Allan J (1980) Management of agricultural resources in coastal Libya. Maghreb Rev 5–6:104–114

Almabruk A (1997) Estimation of evapotranspiration in Libya under the impact of plausible global climate change. Geogr Polonica 70:25–42

Altekamp S (1999) Ein Libyen-Bild für Reisende. Tourismusplanung in Libyen unter italienischer Herrschaft 1911-1943. In: Voyage. Jahrbuch für Reise-und Tourismusforschung, vol 3, pp 80–103

Altekamp S (2000) Rückkehr nach Afrika: Italienische Kolonialarchäologie in Libyen 1911–1943. Köln

Altekamp S, Marcks-Jacobs C, Seiler P (eds) (2017) Perspektiven der Spolienforschung 2. Zentren und Konjunkturen der Spoliierung. Berlin

Altekamp S (1998) Italienische Kolonialarchäologie in Libyen zur Zeit des Faschismus. In: Petersen J (ed) Faschismus und Gesellschaft in Italien. Köln, pp 159–175

Altekamp S (1999) Ein Libyen-Bild für Reisende. In: Voyage, pp 70–103

Altekamp S (2000) Rückkehr nach Afrika – Italienische Kolonialarchäologie in Libyen 1911–1943. Köln

Anderson L (1984) Nineteenth-Century Reform in Ottoman Libya. Int J Middle East Stud 16(3):325–348

Arditi C (1993) Commerce, Islam et État au Tchad (1900–1990). In: Grégoire E and Labazée P (eds) Grands commerçants d'Afrique de l'Ouest. Logiques et pratiques d'un groupe d'hommes d'affaires contemporains. Paris, pp 174–220

Arditi C (1995) Le Commerce des Dattes du Borku (Tchad). Cahiers des Sciences Humaines 31(4):849–882

Arditi C (2005) Le Rôle des Étrangers dans la Constitution d'une Classe de Commerçants au Tchad au XX. Siècle. In: Bulletin Méga-Tchad, pp 21–26

Arditi Claude (2003) Le Tchad et le Monde Arabe: Essai d'Analyse des Relations Commerciales de la Période Précoloniale à aujourd'hui. Afrique Contemporaine 207(3):185–198

Ash Shammakhi M (1995) The Book of Biographies. The part dedicated to the biographies of the Maghreb scholars. Scrutinized by Muhammad Hasan. Tunisia

Ash Sharief KBA (1988) The evaluation of commodities and goods by the Ghadamsi caravan trade. Hist Res Mag 1(10)

Attir M (1983) Libya`s pattern of urbanization. Ekistics 50:157–162

Auskunftsamt für das Ausland (ed) (1981) Die Arabisch Libysche Sozialistiche Volksjamahiriya. Horizonte und Perspektiven

Austen R (1979) Trans-Saharan trade: a tentative census. In: Gemery H and Hogendorn JS (eds) The uncommon market: essays in the economic history of the Atlantic Slave Trade. New York, pp 23–79

Austen R (1990) Marginalization, Stagnation and Growth: Trans-Saharan Caravan Trade, 1500–1900. In: Tracy J (ed) The Rise of Merchant Empires: long-distance trade in the early modern world, 1350–1750. Cambridge, pp 311–350

Austen R (1992) The Mediterranean Islamic slave trade out of Africa: a tentative census. In: Savage E (ed) The human commodity, pp 214–248

Austen R (2012) Sahara. Tausend Jahre Austausch von Ideen und Waren. Berlin

Austen R and Cordell D (2002) Trade, transportation, and expanding economic networks: Sahara caravan commerce in the era of European expansion 1500–1900. In: Jalloh A and Falola T (eds) Black business and economic power. Rochester, pp 80–113

Austen R and Derrick J (1999) Middlemen of the Cameroons rivers: the duala and their Hinterland c. 1600–c. 1960. Cambridge

Awotona A (1990) Housing in Libya: 1950–1980. HABIAT INTL 14 (1):55–85

Ayad MK (1930) Die Geschichts- und Gesellschaftslehre Ibn Khalduns. Forschungen zur Geschichts- und Gesellschaftslehre. Stuttgart, Berlin

Ayyub MS (1963) Antiquity Sites from the south of the Jamahiriya. Tripoli

Ayyub MS (1964) Abridged history of Fezzan from ancient times to the year 1811. Tripoli

Ayyub MS (1993) Antiquity sites from the south of the Jamahiriya. Tripoli

Az Zawi TA (1963) Tarih al-fath al-arabi fi libya. Cairo

Az Zawi TA (1968) Dictionary of Libyan Cities. Tripoli

Baali FG (2005) The science of human social organization. In: Conflicting views on Ibn Khaldun's (1332–1406) Ilm al-Umran. Lewiston, N.Y.

Badry R (1985) Die Entwicklung der dritten Universaltheorie (DUT) Mu'ammar al-Qaddafis in Theorie und Praxis: Aus ideengeschichtl. u. historischer Sicht. Frankfurt a. M

Badry R (1987) Libyen. In: Operschall C (ed) Libyen (die verkannte Revolution?). Wien, pp 146–163

Bagnold RA (1931) Journeys in the Libyan Desert 1929 and 1930. Geogr J 78:13–39

Bagnold RA (1933) A further journey through the Libyan Desert. Geogr J 82:103–129

Bagnold RA (1936) The Libyan desert. Afr Aff 35(140):294–305

Bagnold RA (1941) Libyan Sands. Travel in a Dead World. London

Baier S (1977) Trans-Saharan trade and the Sahel: Damergu 1870–1930. J Afr Hist (JAH) 18(1):37–60

Baier S (1980) An Economic History of Central Niger. Oxford

Baier SA (1978) History of the Sahara in the Nineteenth Century. Boston

Banse E (1912) Tripolis. Weimar

Bantle S (1994) Schattenhandel als sozialpolitischer Kompromiss: Die „libyschen Märkte" in Tunesien. Münster

Barker G, Gilbertson D, Jones G, Mattingly D (1996) Farming the Desert: the UNESCO Libyan Valleys Archaeological Survey, vol 1. London

Barnes JR (1987) An introduction to religious foundations in the Ottoman Empire. 2nd imprint. Leiden

Baume O, Sager I (1992) Naturräumliche Gliederung und agrarische Nutzungseignung der libyschen Küstenregion – Gedanken zum wirtschaftlich bedeutendsten und am dichtesten besiedelten Gebiet Libyens. In: Zeitschrift für den Erdkundeunterricht (44/6), pp 208–213

Bayer E, Ritter H (2006) Die Pflanzen der Wüste. In: Müller CC, de Castro I (eds) Die Wüste. Begleitbuch zur Landesausstellung im Ausstellungszentrum Lokschuppen Rosenheim 2006. Germering/München, pp 32–45

Belhaj SM (1968) Islamic Tombstones from Cyrenaica. Ancient Libya Mag 5

Bennafla K (2000) Tchad: L'Appel des Sirènes Arabo-Islamiques. Autrepart 16:67–86

Bennafla K (2004) La Réactivation des Échange Transsahariens: L'Exemple Tchadolibyen. In: Marfaing L, Wippel S (eds) Les Relations Transsahariennes à l'Epoque Contemporaine. Un Espace en Constante Mutation. Paris, Berlin (Hommes et sociétés), pp 89–110

Bergs R (1987) Das libysche Entwicklungsmodell und die wirtschaftliche Krise seit 1981. In: Operschall C (ed) Libyen (die verkannte Revolution?). Wien, pp 78–113
Bergs R, Schliephake K (eds) (1999) Libyen und nordwestliches Ägypten. Aktuelle Elemente der räumlichen Entwicklung, Würzburg
Bernia C (1969) Tarablus min 1510 ila 1850 (Translated by Khalifa Muhammad at-Tillisi). Tripoli
Besler H (2002) Complex barchans in the Libyan Desert: Dune traps or overtaking solitons? In: Zeitschrift für Geomorphologie, vol 126, pp 59–74
Bethencourt F, Egmond F (ed) (2007) Correspondence and Cultural Exchange in Europe 1400–1700. Cambridge
Blair S, Katz JG, Hamès C (2010) Zawiya. In: Bearman P, Bianquis T, Bosworth CE, van Donzel E, Heinrichs WP (eds) Encyclopaedia of Islam, 2nd edn. Leiden: BrillOnline, checked on 6/8/2018 http://dx.doi.org/10.1163/1573-3912_islam_COM_1384
Blake G (1983) Tradition and change in a Libyan Market Town: Misratah 1966–1974. Tijdschrift voor Economische en Sociale Geografie 74(3):175–184
Boahen AA (1962) The Caravan Trade in the nineteenth century. J Afr Hist 3(2):349–359
Boahen AA (1964) Britain, the Sahara and the Western Sudan, 1788–1861. Oxford
Bosworth CE (ed) (2007) Historic Cities of the Islamic World. Leiden
Bovill EW (1928) Saharan explorers of the fifteenth century. JRAS 28 (113):19–27
Bovill EW (1929) The silent trade of Wangara. JRAS 29(109):27–38
Bovill EW (1933) Caravans of the Old Sahara: An introduction to the history of the Western Sudan. London
Bovill EW (1958) The golden trade of the moors. London
Braun K, Passon J (2005) Von der Karawanserei zu internationalen Standards. Afrika Wirtschaft 5:36–38
Braun K, Passon J (2010) Vom Berberzelt in die Moderne. In: Glaser R, Kremb K (eds) Afrika. Darmstadt, pp 56–66
Braun K, Eisele T, Passon J (2008) Libya—a new tourist destination. GR Int 2:16–20
Brenner L (1971) The North African trading community in the nineteenth century central Sudan. In: McCall DF, Bennett NR (eds) Aspects of West African Islam. Boston, pp 137–150
Brett M (1983) Ifriqiya as a Market of SaharanTrade from the Tenth to the Twelfth Century A.D. JAH 24:347–364
Brett M (1983) Islam and Trade in Bilad al-Sudan: tenth-eleventh century. JAH 24:431–440
Brett M, Fentress E (2010) The Berbers. Malden
Briggs LC (1960) Tribes of the Sahara. Cambridge
Brockelmann C (1949) Geschichte der arabischen Literatur. Leiden (2)
Brooks N, Chiapello I, Di Lernia S, Drake N, Legrand M, Moulin C, Prospero C (2005) The environment-society nexus in the sahara from prehistoric times to the present day. JNAS 10:253–295
Brooks N, Di Lernia, Drake N (2004) The prehistory of Western Sahara in a regional context. In: Leroy S, Costa P (eds) Environmental catastrophes in Mauritania, the desert and the coast. Mauretanien, pp 35–39
Brooks N, Di Lernia S, Drake N, Raffin M, Savage M, Savage T (2003) The geoarchaeology of Western Sahara: preliminary results of the First Anglo-Italian expedition in the "free zone". Sahara 14:63–80
Brooks N, Drake N, McLaren S, White K (2003) Studies in geography, geomorphology, environment and climate. In: Mattingly D (ed) The Archaeology of Fezzan. London, pp 37–74
Bruce-Lockhart J, Wright J (2000) Difficult and dangerous roads: Hugh Clapperton's travels in Sahara and Fezzan. London
Brusberg F (1985) Production and Exchange in the Saharan Air. Curr Anthropol 26(3):394–395
Bubenzer O, Bolten A, Darius F (eds) (2007) Atlas of cultural and environmental change in Arid Africa. Köln
Busche D (2004) Nasse und trockene Reliefgeschichte der Zentralen Sahara. In: Dreikluft R, Busche D (eds) Die Sahara: Natur und Geschichte. Darmstadt, pp 16–109
Cachia AJ (1975) Libya under the Second Ottoman Occupation (1835–1911). Tripoli
Cana FR (1915) The Sahara in 1915. Geogr J 46(5):333–357
Capresi V (2007) I Centri rurali Libici. L'architettura dei centri rurali di fondatione costruiti in Libia - colonia italiana - durante il fascismo (1934–1940). Wien
Chapelle J (1982) Nomades noirs du Sahara: Les Toubous. Paris
Charles-Picard G (1962) Nordafrika und die Römer, Stuttgart
Ciammaichella G (1987) Libyens et Français au Tchad, 1897–1914. La Confrérie Senoussie et le Commerce Transsaharien. Paris
Ciranna S (2017) Pulcherrima Spolia in the Architecture and Urban Space at Tripoli. In: Altekamp S, Marcks-Jacobs C, Seiler P (eds) Perspektiven der Spolienforschung 2. Zentren und Konjunkturen der Spoliierung. Berlin, pp 67–94
Clam JJ, Hubel H (1987) Die Krise um Libyen. Bonn
Cleaveland T (1998) Islam and the construction of social identity in the nineteenth-century Sahara. J Afr Hist 39(3):365–388
Cohen A (1971) Cultural strategies in the organization of trading diasporas. In: Meillassoux C (ed) The development of indigenous trade and markets in West Africa. London, pp 266–281
Cohen A (2017) Custom and politics in Urban Africa. Berkeley
Collins C (1974) Imperialism and revolution in Libya. In: MERIP Reports, vol 27, pp 3–22
Coquery-Vidrovitch C, Lovejoy PE (ed) (1985) The workers of African Trade. Beverly Hills
Cordell DD (1977) Eastern Libya, Wadai and the Sanusiya: A Tariqa and a trade route. J Afr Hist 18(1):21–36
Cordell DD (1985) The Awlad Sulayman of Libya and Chad: power and adaption in the Sahara and Sahel. Can J Afr Stud 19(2):319–343
Corradini L (2014) Die Rolle libyscher Stammesgesellschaften in der Revolution von 2011. Köln
Curtin P (1984) Cross-cultural trade in the world history. Cambridge
Curtin PD (1975) Economic change in Precolonial Africa. Senegambia in the Era of the Slave Trade. Madison
Daniels C (1970) The Garamantes of Southern Libya. Stoughton
Daniels C (1974) The Garamantes (Translated by Ahmad al Yazuri). Tripoli
Davis RC (2001) Counting European slaves on the Barbary coast. Past Present 172:87–124
Despois J (1946) Mission scientifique du Fezzan (1944–1945). Alger
Despois J (2010) Fazzan. In: Bearman P, Bianquis T, Bosworth CE, van Donzel E, Heinrichs WP (eds) Encyclopaedia of Islam, 2nd edn. BrillOnline, Leiden. Checked on 6/8/2018 http://dx.doi.org/10.1163/1573-3912_islam_SIM_2344
Despois J (2010) Ghadamès. In: Bearman P, Bianquis T, Bosworth CE, van Donzel E, Heinrichs WP (ed) Encyclopaedia of Islam, 2nd edn. Leiden: BrillOnline, checked on 6/9/2018 http://dx.doi.org/10.1163/1573-3912_islam_SIM_2438
Despois J (2012) Encyclopaedia of Islam. In: Bearman P, Bianquis Th, Bosworth CE, van Donzel E, Heinrichs WP (eds), 2nd edn. Brill Online, Ghadames
Deutsches Orientinstitut (ed) (1995) Libyen im 20. Jahrhundert: Zwischen Fremdherrschaft und Selbstbestimmung. Hamburg
de Rubeis A (1929) Ghat. La città sahariana. Rivista delle Colonie Italiane 3(8):735–768
de Vries B (1979) Research at Umm El-Jimal, Jordan, 1972–1977. Biblical Archaeol 42(1):49–55

Di Vita A (1996) Leptis Magna, die Heimatstadt des Septimius Severus in Nordafrika. Antike Welt 27(3):173–190

Di Vita A, Di Vita G, Bacchielli L (1999) Das antike Libyen. Köln

Dittrich P (1983) Biologie der Sahara. Ein Führer durch die Tier- u. Pflanzenwelt d. Sahara mit Bestimmungstab. 2., verb. u. erw. Aufl. München

Djian G (1991) Étude sur les Senoussistes et leur Action dans le Centre Africain. Islam et Sociétés au Sud du Sahara 5:109–138

Dowler A, Galvin ER (eds) (2011) Money, trade and trade routes in Pre-Islamic North Africa. British Museum Research Publication no. 176

Dreikluft R, Busche D (eds) (2004) Die Sahara: Natur und Geschichte. Darmstadt

Edwars J, Oglivie S (2012) Contract enforcement, institutions, and social capital: The Maghribi traders reappraised. Econ Hist Rev 65 (2):421–444

Eldblom L (1968) Structure foncière. Organisation et Structure Sociale. Une étude comparative sur la vie socioéconomique dans les trois oasis libyennes de Ghat, Mourzouk et particulièrement Ghadames. Lund

Eldblom L (1971) Land tenure—social organization and structure. In: A comparative sample study of the socio-economic life in the three Libyan oases of Ghat, Mourzouk and Ghadames, 2nd edn. Edited by The Royal University of Lund, Department of Geography. Lund

El-Hesnawi HW (1960) Encyclopaedia of Islam. The New Edition, Leiden

Elmahmudi A, Reineke U (1999) The islamic cities in Libya. In: Europäische Hochschulschriften, vol 66.

Englert W (1999) Der aktuelle Stand der libyschen Wirtschaftsentwicklung und die Perspektiven für die Zeit nach dem Embargo. Würzburger Geographische Manuskripte 51:201–212

Ergenzinger P (1969) Die Siedlungen des mittleren Fezzan (Libyen). In: Hövermann J, Jensch G, Valentin H, Wöhlke W, Hagedorn H (eds) Berliner Geographische Abhandlungen, vol 8. Arbeitsberichte aus der Forschungsstation Bardai/Tibesti. Berlin

Evans M (1976) The traditional house in the Oasis of Ghadames. Libyan Stud 7:31–40

Evans-Pritchard EE (1940) The Place of the Sanusiya Order in the History of Islam. Tripoli

Evans-Pritchard EE (1945) The Distribution of Sanusi Lodges. J Int Inst Afr Lang Cult 15:183–187

Evans-Pritchard EE (1949) The Sanusi of Cyrenaica. London

Evans-Pritchard EE (1963) The Sanusi of Cyrenaica. London

Exenberger A (2002) Außenseiter im Weltsystem. Die Sonderwege von Kuba, Libyen und Iran. Frankfurt a M

Eydoux H (1938) Die Erforschung der Sahara. Paris

Faath S, Mattes H-P (ed) (1989) wuqûf 3. Hamburg

Fall YK (1982) L'Afrique a La Naissance De La Cartographie Moderne: Les Cartes Majorquines. XIVe-XVe Siecles, Karthala

Fallers L (1967) Immigrations and Associations. The Hague

Faloos MF Studio storico-archeologico dei siti di epoca romana e bizantina nel territorio di Tripoli (Oea). Macerata

Fergiani MG (1971) Beauty of Libya. A pictural Book, Tripoli

Fischel WJ (1967) Ibn Khaldun in Egypt. In: His Public Functions and His Historical Research, 1382–1406. A Study in Islamic Historiography. Berkeley

Fisher HJ (2010) Kukawa, or Kuka. In: Bearman P, Bianquis T, Bosworth CE, van Donzel E, Heinrichs WP (eds) Encyclopaedia of Islam, 2nd edn. Leiden: BrillOnline, checked on 6/8/2018 http://dx.doi.org/10.1163/1573-3912_islam_SIM_4489

Folayan K (1971) Tripoli-Bornu political relation, 1817–25. JHSN 5 (4):463–476

Förster F, Riemer H, Bolten A, Bubenzer O, Hendrickx S, Darius F (2010) Tracing linear structures. remote sensing, landscape classification and the archaeology of Desert Roads in the Eastern Sahara. In: Möhlig JG, Bubenzer O, Menz G (eds) Towards interdisciplinarity. Experiences of the Long-term ACACIA project. Köln, pp 49–78

Frank H (ed) (1985) Faschistische Architekturen. Hamburg

Frank S, Kamp M (1995) Libyen im 20. Jahrhundert. Zwischen Fremdherrschaft und nationaler Selbstbestimmung. Hamburg

Fritsch K (2009) "You Have Everything Confused and Mixed Up…!" Georg Schweinfurth, Knowledge and Cartography of Africa in the 19th Century. Hist Afr 36:87–101

Garrard TF (1982) Myth and metrology: the early trans-Saharan gold trade. J Afr Hist 23(4):443–461

Gatti F (2005) Die Verdammten der Sahara. Flüchtlingsrat - Zeitschrift für Flüchtlingspolitik in Niedersachsen 110:107–110

Gemery H, Hogendorn JS (eds) (1979) The uncommon market: essays in the economic history of the Atlantic Slave Trade. New York

Ghalbun I (1960) At tidkar fi man malaka Tarabuls wa-ma kana biha min ahbar, 2nd edn. Tripoli

Gilbertson D, Hunt C, Fieller N, Barker GWW (1994) The environmental consequences and context of ancient floodwater farming in the tripolitanian Pre-Desert. In: Millington AC (ed) The environmental change in drylands. Wiley, Chichester, pp 229–252

Gladstone P (1970) Travels of Alexine: Alexine Tinne, 1835–1869. London

Glaser R, Kremb K (eds) (2010) Afrika. Darmstadt

Goodchild R (1954) Oasis forts of Legio III Augusta on the routes to Fezzan. Papers of the British School, Rome, pp 56–68

Goodchild R (1999) Libyan studies (Translated into Arabic by Ahmad al Yazuri and Abdul Hafiz al Mayyar). Tripoli

Goodchild RG (1951) Boreum of Cyrenaica. J Roman Stud 41(1–2):11–16

Goodchild RG (1952) Mapping roman Libya. Geogr J 118(2):142–152

Goodchild RG (1970) History of the City of Benghazi (Translated into Arabic by Saleh Gibreel). Tripoli

Goodchild RG (1971) Kyrene und Apollonia. Zürich

Grégoire E, Labazée P (eds) (1993) Grands commerçants d'Afrique de l'Ouest. Logiques et pratiques d'un groupe d'hommes d'affaires contemporains, Paris

Greif A (1989) Reputation and coalitions in medieval trade: evidence on the Maghribi traders. JEH 64:857–882

Greif A (1993) Contact enforceability and economic institutions in early trade: the Maghribi Traders' coalition. Am Econ Rev 83:525–548

Greif A (1994) Cultural beliefs and the organization of society: historical and theoretical reflection on collectivist and individual societies. J Polit Econ 102:912–950

Greif A (2000) The fundamental problem of exchange: a research Agenda in historical institutional analysis. Eur Rev Econ Hist 4:251–284

Grosjean G (1977) Der Katalanische Weltatlas von Jahre 1375. Zürich

Haensch R, Mackensen M (2011) Das tripolitanische Kastell Gheriat el-Garbia im Licht einer neuen spätantiken Inschrift: Am Tag, als der Regen kam. Chiron 41:263–286

Halm H (1992) Nachrichten zu den Bauten der Aglabiden und Fatimidenin Libyen und Tunesien. Die Welt des Orients (WdO) 23:129–157

Hamid S The ancient history of the Aujala oasis up to the Ottoman era. Tripoli

Hamid S (1978) Al-ma amil al-islamiyya bil-mathaf al-islami bi-Tarabuls. Tripoli

Hamid S (1978) The Islamic landmarks in the Islamic Museum in Tripoli. Tripoli

Hamid S (1991) The Islamic numismatics in Libya. Arab Antiquities Mag 2

Hamid S (2005) From the history of Nalut. People's Heritage Mag 3 (51)

Hamid S (ed) (2006) The city of Sharba. In: The second international conference about the caravan trade and its political, economic, and social dimensions. Tripoli

Hamid S (2008) Arch of Marcus Aurelius in Tripoli. Tripoli

Hamid S (2008) History and landmarks of civilization and urbanism in Libya. In: A study within the series on the landmarks of the Islamic civilization in Libya. Tripoli

Harding-King WJ (1913) The Libyan Desert from native information. Geogr J 42:277–283

Harmann U (1998) The Dead Ostrich life and trade in Ghadames (Libya) in the nineteenth century. Die Welt des Islams, New Ser. 38 (1):9–94

Harrison RS (1967) Migrants in the City of Tripoli, Libya. Geogr Rev 57(3):397–423

Haynes D (1956) An archaeological and historical guide to the pre-Islamic antiquities of Tripolitania. Tripoli

Haynes EL (1981) The Antiquities of Tripolitania. Tripoli

Hecht F, Fürst M, Klitzsch E (1964) Zur Geologie Libyens. Int J Earth Sci 53:413–470

Heffernan M (1989) The limits of Utopia: Henri Duveyrier and the exploration of the Sahara in the nineteenth century. Geogr J 155 (3):342–352

Helfensteller A (1998) Afrika-Reise: Leben und Werk des Afrikaforschers Gerhard Rohlfs (1831-1896). Gerhard-Rohlfs-Symposium am 1. Juni 1996 in Bremen Vegesack

Herkommer J (1941) Libyen von Italien kolonisiert. Freiburg i, Breisgau

Herzog R (1961) Ein Beitrag zur Geschichte des nordafrikanischen Karawanenhandels. Die Welt des Islam 6(3/4):255–262

Herzog W (1990) Der Maghreb: Marokko, Algerien. Tunesien, München

Hester JJ, Hobler PhM, Russell J (1970) New evidence of early roads in Nubia. Am J Archaeol 74(4):385–389

Hickson CR, Turner JD (2005) Partnership. In: McCusker JJ, Engerman S, Fisher LR, Hancock DJ, Pomeranz KL (eds) Encyclopedia of world trade since 1450. New York

Hildebrand G (1904) Cyrenaica als Gebiet künftiger Besiedlung. Eine Landeskunde mit besonderer Berücksichtigung der wirtschaftlichen Verhältnisse. Bonn

Hinz A (2007) Libyen. In: Weiss W (ed) Die arabischen Staaten: Geschichte, Politik, Religion, Gesellschaft, Wirtschaft, pp 176–193. Heidelberg

Hirschberg HK (1974) A history of the jews in North Africa. Leiden

Hopkins AG (1973) An economic history of West Africa. New York

Hopkins AG (1998) The economic history of Arab Africa (Translated by Ahmad Fuad Balba). Birmingham

Hövermann J, Jensch G, Valentin H, Wöhlke W, Hagedorn H (eds) (1969) Berliner Geographische Abhandlungen. Arbeitsberichte aus der Forschungsstation Bardai/Tibesti, vol 8. Berlin

Hunwick J (1992) Black Slaves in the mediterranean World: introduction to a neglected aspect of the African diaspora. S.5-38. In: Savage E (ed)The human commodity. Perspectives on the Trans-Saharan Slave Trade. London

Hunwick J, Powell ET (2002) The African Diaspora in the Mediterranean Lands of Islam. Princeton

Israel JI (2002) Diasporas within a diaspora. Jews, Crypto-Jews and the World Maritime Empires (1540–1740). Leiden

Issawi C (1966) The economic history of the Middle East, 1800–1914. Chicago

Italiaander R (ed) (1970) Heinrich Barth. Er schloss uns einen Weltteil auf. Unveröffentlichte Briefe und Zeichnungen des großen Afrika-Forschers. Hamburg

Jalloh A, Falola T (eds) (2002) Black Business and Economic Power. Rochester

Jayyusi SK, Holod R, Petruccioli A, Raymond A (eds) (2008) The City in the Islamic World, vol 2. Leiden

Jennings R Public Sitting 21. June 1993 in the Case Territorial Dispute (Libyan Arab Jamahiriya/ Chad). In: The Hague: International Court of Justice

Johnson M (1976) Calico Caravans: the Tripoli-Kano trade after 1880. J Afr Hist 17(1):95–117

Johnson M, Hogendorn J (1986) The shell money of the slave trade. Cambridge

Kassas M (1967) Die Pflanzen der Sahara. In: Krüger C (ed) Sahara. München, pp 162–181

Keenan J (2002) The Tuareg. People of the Ahaggar. London

Kenrick P (2009a) Libya archaeological guides: Tripolitania. London

Kenrick P (2009b) Tripolitania. London

Khaldun I (1867) Geschichte. Band 6 und 7. Kairo

Khalilieh HS (1999) The Ribat system and its role in coastal navigation. J Econ Soc Hist Orient 2(42):212–225

Kieckhefer R (1995) Magie im Mittelalter. München

Kielstra N (1987) The decline of tribal organization in the Souf (S.E. Algeria). Revue de l'Occident musulman et de la 45:11–24

Kirsten E (1961) Nordafrikanische Stadtbilder. Antike und Mittelalter in Libyen und Tunesien, Heidelberg

Klemp E (1968) Africa on maps dating from the twelfth to the eighteenth century. Leipzig

Klingenstein G, Lutz H, Stourzh G (eds) (1980) Europäisierung der Erde?. Studien zur Einwirkung Europas auf die außereuropäische Welt, Wien

Kohl I (2003) Wüstentourismus in Libyen. Auswirkungen, Folgen und lokale Wahrnehmungen. Eine anthropologische Fallstudie aus der Oase Ghat. Berlin

Krause ChL, Mishler C (2001) Standard catalog of world coins, 1801–1900, 3rd edn. Iola, Wis

Krüger C (ed) (1967) Sahara. München

Kulturreferat Bremen-Nord im Ortsamt Vegesack (ed) (1998) Afrika-Reise. Leben und Werk des Afrikaforschers Gerhard Rohlfs. Bonn

Kuper R (2007) Desert parks in the eastern Sahara. In: Bubenzer O, Bolten A, Darius F (eds) Atlas of cultural and environmental change in arid Africa. Köln, pp 118–121

Kuran T (1983) Behavioral norms in the Islamic doctrine of economics: A critique. J Econ Behav Organ 4:353–379

Laabdallaoui M, Rüschoff SI (2005) Ratgeber für Muslime bei psychischen und psychosozialen Krisen. Bonn

Lahr MM, Foley R, Crivellaro F, Fernandez JM, Wilshaw A, Copsey B, Rivera F, Mattingly D (2011) DMP XIV: Prehistoric sites in the Wadi Barjuj, Fazzan, Libyan Sahara. Libyan Stud 42:117–138

Lapworth C (1905) Tripoly and Young Italy. New York

Laronde A (1997) La Libya – à travers les cartes postales 1900–1940. Tunis

Law RCC (1967) The garamantes and trans-Saharan enterprise in classical times. J Afr Hist 8(2):181–200

Le Gall M (1993) The end of the trans-Saharan slave trade in Tripoli. Princeton Pap Near East Stud 2:35–55

Le Houérou HN (2009) Bioclimatology and biogeography of Africa. Berlin, Heidelberg

Leroy S, Costa P (eds) (2004) Environmental catastrophes in Mauritania, the desert and the coast. Mauretanien

Lethielleux J (1946) Le Fezzan, ses jardins, ses palmiers. Tunis

Levtzion N, Hopkins JFP (2000) Corpus of early Arabic sources for West African history. Princeton

Lewicki T (1955) Études Ibadites Nord-Africaines. Partie I, Tasmiya suyuh Gabal Nafusa wa-Qurahum, etc. Warsaw

Lewicki T (1976) Études Maghrébines et Soudanaises. Warsaw

Lewicki T (1983) Le Sahara oriental et septentrional dans le haut moyen age, VIIe-XIIe siècle. Ètudes Maghrébines et Soudanaises 2:42–96

Liverani M (2000) The Garamantes: A fresh approach. Libyen Stud 31:17–28

Liverani M (2000) The Libyan caravan road in Herodotus IV, pp 181–185. Jesho 43(4):496–520

Liverani M, Cremaschi M, Di Lerina S (2000) The "Archaeological Park" of the Tadrart Acacus and Messak Settafet. Sahara 12:121–140

Lovejoy PE (1983) Transformations in Slavery. In: A history of slavery in Africa. Cambridge

Lovejoy PE (1984) Commercial sectors in the economy of the nineteenth-century central Sudan: The trans-Saharan Trade and the Desert-Side salt trade. Afr Econ Hist 13:85–116

Lovejoy PE (2011) Transformations in slavery. In: A history of slavery in Africa, 3rd edn. Cambridge

Lydon G (2008) Contracting caravans: Partnership and profit in nineteenth- and early twentieth-century trans-Saharan trade. JGH 3 (01):89–113

Lydon G (2009) On trans-Saharan Trails. In: Islamic law, trade networks, and cross-cultural exchange in nineteenth-century Western Africa. Cambridge

Lydon G (2009) The organization of caravan Trade in nineteenth and twentieth century Western Africa. In: Ahmida AA (ed) Bridges across the Sahara. Social, economic and cultural impact of the trans-Sahara trade during the 19th and 20th centuries. Newcastle, pp 25–60

Mahadi A (1992) The aftermath of the Jihad in the Central Sudan as a major factor in the volume of the transsaharan slave trade in the nineteenth century. In: Savage E (ed) The human commodity: perspectives on the trans-Saharan slave trade. London, pp 111–128

Mahdi M (1957) Ibn Khaldun's philosophy of history. In: A study in the philosophic foundation of the science of culture. London

Marfaing L, Wippel S (eds) (2004) Les Relations Transsahariennes à l'Epoque Contemporaine. Un Espace en Constante Mutation, Paris, Berlin (Hommes et sociétés)

Martel A (1991) La Libye, 1835–1990. Essai de géopolitique historique, Paris

Martin BG (1962) Five letters from the tripoli archieves. JHSN 2 (3):350–372

Martin BG (1969) Kanem, Bornu and the Fazzan: Notes on the political history of a trade route. J Afr Hist 10(1):15–27

Martin BG (1983) Ahmed Rassim Pasha and the Suppression of the Fazzan Slave Trade. Afr J Int Afr Inst 38(4):545–579

Martin BG (1985) Ahmed Rassim Pasha and the suppression of the Fazzan Slave Trade. In: Willis JR (ed) Slaves and slavery in Muslim Africa: the servile estate. London

Mattes H-P (1982) Die Volksrevolution in der sozialistischen libyschen arabischen Volksjamahiriya. Heidelberg

Mattes H-P (1987) Die innenpolitische Entwicklung 1969–1986. In: Operschall C (ed) Libyen (die verkannte Revolution?). Wien, pp 31–47

Mattes H-P (1989) Die Entkolonisierung der libyschen Geschichtsschreibung: Zur Arbeit des 1978 in Tripolis eröffneten historischen Forschungszentrums. In: Faath S, Mattes H-P (ed) wuqûf 3. Hamburg, pp 259–277

Mattes H-P (1989) Geschichte und Identität: Das Beispiel der politischen Einflussnahme auf den kulturellen Sektor in Libyen seit 1969. In: Faath S, Mattes H-P (eds): wuqûf 3. Hamburg, pp 281–299

Mattes H-P (1995) Außen-, innenpolitisches und soziales Konfliktpotential eines nordafrikanisches Staates: Das Beispiel Libyen. In: Deutsches Orientinstitut (ed): Libyen im 20. Jahrhundert: Zwischen Fremdherrschaft und Selbstbestimmung. Hamburg

Mattes H-P (1999) Wahlen und Wahlrechtsentwicklung in Libyen. In: Bergs R, Schliephake K (eds) Libyen und nordwestliches Ägypten. Aktuelle Elemente der räumlichen Entwicklung, vol 51. Würzburg, pp 213–218

Mattingly D Making the desert bloom. The Garamantian capital and its underground water system. In: Archaeology Odyssey, vol 3, Nr. 2 2000

Mattingly D The Tripoli region in the Roman Era. Edited by Center for the Libyen Archives and Historical Studies. Tripoli

Mattingly D (1995) Tripolitania. London

Mattingly D (2003) Historical summary. In: Mattingly D (ed) The archaeology of Fezzan, vol 1. London, pp 75–106

Mattingly D (ed) (2003) The archaeology of Fezzan. London

Mattingly D (2011) The Garamantes of Fazzan. In: An early libyan state with trans-Saharan connections. In: Dowler A, Galvin ER (eds) Money, trade and trade routes in Pre-Islamic North Africa, pp 40–61

Mattingly D, Sterry M, Edwards DN (2015) The origins and development of Zuwīla, Libyan Sahara: an archaeological and historical overview of an ancient oasis town and caravan centre. Azania: archaeological research in Africa 50(1):27–75

Mattingly D, Hitchner DB (1995) Roman Africa: An archaeological review. J Roman Stud 85:165–213

Mattingly D, Brooks N, Cole F, Dore J, Drake N, Leone A, Hay S (2001) The Fezzan Project 2001: Preliminary report on the fifth season of work. Libyan Stud 32:133–153

Mattingly D, Dore J, Lahr M (2008) DMP II: 2008 fieldwork on burials and identity in the Wadi al-Ajal. Libyan Stud 39:223–262

Mattingly D, Reynolds T, Dore J (2003) Synthesis of human activities in Fezzan. In: Mattingly D (ed) The archaeology of Fezzan. London, pp 327–374

Mauny R (1961) Tableau Géographique De L´ouest Africain Au Moyen Age. D´Aprés Les Sources Écrites, La Tradition Et L´ Archéologie. Ifan-Dakar

McCall DF, Bennett NR (eds) (1971) Aspects of West African Islam. Boston

McCusker JJ, Engerman S, Fisher LR, Hancock DJ, Pomeranz KL (eds) (2005) Encyclopedia of world trade since 1450. New York

McDougall EA (1985) Camel Caravans of the Saharan salt trade. Traders and transporters in the nineteenth century. In: Coquery-Vidrovitch C, Lovejoy PE (eds) The Workers of African Trade. Beverly Hills, pp 99–121

McDougall EA (1992) Salt, Saharans and the Trans-Saharan slave trade: nineteenth century developments. In: Savage E (ed) The human commodity: perspectives on the trans-Saharan slave trade. London, pp 61–88

McLachlan KS (1978) Tripoli and Tripolitania: conflict and cohesion during the period of the Barbary Corsairs (1551–1850). Trans Inst British Geogr New Ser 3(3):285–294

McLachlan KS (2010) Nafusa. In: Bearman P, Bianquis T, Bosworth CE, van Donzel E, Heinrichs WP (eds) Encyclopaedia of Islam, 2nd edn. Leiden: BrillOnline, checked on 6/8/2018 http://dx.doi.org/10.1163/1573-3912_islam_SIM_5744

McLachlan KS (2012) Murzuk. In: Bearman P, Bianquis Th, Bosworth CE, van Donzel E, Heinrichs WP (eds) Encyclopaedia of Islam, 2nd edn. Brill Online, Murzuk

McLaren BL (2006) Architecture and tourism in Italian Colonial Libya. An Ambivalent Modernism, Seattle

Meerpohl M (2007) Mobile Oases in the Libyan Desert. In: Bubenzer O, Bolten A, Darius F (eds) Atlas of cultural and environmental change in Arid Africa. Köln, pp 138–141

Meerpohl M (2009) Camels as trading goods: the transition from a beast of burden to a commodity in the trans-Saharan trade between Chad and Libya. In: Ahmida AA (ed) Bridges across the Sahara.

social, economic and cultural impact of the trans-Sahara trade during the 19th and 20th centuries. Newcastle, pp 157–186
Meerpohl M (2009) Kamele und Zucker. Transsahara-Handel zwischen Tschad und Libyen. In: Meerpohl M (ed) Philosophischen Fakultät der Universität zu Köln. Köln
Meillassoux C (ed) (1971) The Development of indigenous trade and markets in West Africa. London
Metz H (1989) Libya: a country study. Washington
Micara L (2008) The Ottoman Tripoli: A mediteranean medina. In: Jayyusi SK, Holod R, Petruccioli A, Raymond A (eds) The City in the Islamic World, vol 2. Leiden, pp 383–406
Micara L (2013) Tripoli Medina Mediterranea. Tripoli, a Mediterranean Medina. Roma
Miège J-L (1975) La Libye et le commerce trans-saharien au XIXe siègle. ROMM 19:135–168
Miège J-L (1981) Le commerce transsaharien au XIXe siècle. Revue de l'Occident musulman et de la Méditerranée 32:93–119
Miège J-L (1981) Le commerce transsaharien au XIXe siècle: Essai de quantification. Revue de l'Occident musulman et de la Méditerranée 32:90–112
Millington AC (ed) (1994) The environmental change in Drylands. Wiley, Chichester
Möhlig JG, Bubenzer O, Menz G (eds) (2010) Towards interdisciplinarity. Experiences of the Long-term ACACIA project. Köln
Müller M (1952) Afrika verschollen. Leipzig
Müller CC, de Castro I (eds) (2006) Die Wüste. Begleitbuch zur Landesausstellung im Ausstellungszentrum Lokschuppen Rosenheim 2006. Germering/München
Myres JL (1896) An attempt to reconstruct the maps used by Herodotus. Geogr J 8(6):605–629
Nageh A (1989) Zentralität und Stadtwachstum in Libyen. Kassel
Naihoum S (ed) The Tarikhuna Library from pre-history to the sixth century AD, vol 5. Genève
Najem F (2004) Tribe, Islam and State in Libya. In: Analytical study of the roots of the Libyan tribal society and interaction up to the Qaramanli rule (1711–1835). London
Najm M (1973) History of Tarabulus al Gharb. Tripoli
Najm M, Abbas I (1968) Libya in the geography and voyages books. Benghazi
Newbury CW (1966) North African and Western Sudan trade in the nineteenth century: A re-evaluation. J Afr Hist 7(2):233–246
Norris HT (1975) The Tuareg: Their Islamic legacy and its diffusion in the Sahel. Warminster
Obst J (1969) Statik und Dynamik in der Agrarlandschaft des Jabal al Akhdar/Libyen. Ein Beitrag zur Agrargeographie semi-arider Gebiete, Frankfurt a M
Oman G, Christides V, Bosworth CE (2010) Tarabulus al-Gharb. In: Bearman P, Bianquis T, Bosworth CE, van Donzel E, Heinrichs WP (eds) Encyclopaedia of Islam, 2nd edn. Leiden: BrillOnline, checked on 6/8/2018 http://dx.doi.org/10.1163/1573-3912_islam_COM_1176
Operschall C (ed) (1987) Libyen (die verkannte Revolution?). Wien
Orhonlu C (2010) Karwan. In: Bearman P, Bianquis T, Bosworth CE, van Donzel E, Heinrichs WP (eds) Encyclopaedia of Islam, 2nd edn. Leiden: BrillOnline, checked on 6/8/2018 http://dx.doi.org/10.1163/1573-3912_islam_COM_0453
Pasha AH (2006) In the Libyan Desert. In: Al Hawari I (ed). Cairo
Passon J (2009) Zwischen Kulturstätten und Wüstensand. Libyen öffnet sich für den Tourismus, Berlin
Passon J (2011) "Quid novi ex Africa?" Auf Trampelpfaden durch die Sahara - Afrikareisende des 19. Jahrhunderts und ihre Begegnung mit der klassischen Antike im heutigen Libyen. In: Wiegels R (ed) Verschlungene Pfade. Neuzeitliche Wege zur Antike, vol 16. Leidorf, pp 169–202
Pennell CR (1989) Piracy and diplomacy in seventeenth century North Africa. The Journal of Thomas Baker, English Consul in Tripoli, 1677–1685. London
Petermann A (1857) Mittheilungen aus Justus Perthes' Geographischer Anstalt über wichtige neue Erforschungen auf dem Gesammtgebiete der Geographie. Justus Perthes, Gotha
Petersen J (ed) (1998) Faschismus und Gesellschaft in Italien. Köln
Piazza C (1996) La vilayet die Tripoli d'Occidente nel 1300 dell'egira secondo un Almanacco. Studi senesi 108:139–173
Playfair L (1889) The bibliography of the Barbary States. In: Part 1: Tripoli and the Cyrenaica. London
Pliez O (2002) Sebha, une grande ville du Sahara libyen. Méditerranée 99(3/4):91–94
Powers DS (2002) Law, society and culture in the Maghrib, 1300–1500. Cambridge
Prestopino F (1999) Una città e il suo fotografo: La Bengasi coloniale, 1912–1941. Milano
Rahma A (ed) (1999) Sufi Order and Resistance Movement: The Sanusiyya of Libya, 1911–1932. Montreal
Renault F (1982) La traite des esclaves noirs en Libye au XVIIIe siècle. J Afr Hist 23(2):163–181
Rghei AS, Nelson JG (1994) The conservation and use of the walled city of Tripoli. Geogr J 160(2):143–158
Richardson J (1848) Travels in the great desert of Sahara in the years 1845–46, vols I and II. London
Róbert S (2002) Ibn Khaldun. History as Science and the Patrimonial Empire. Budapest
Rosenthal F (1967) Übersetzung, Kommentar von Ibn Chaldun: The Muqaddimah: An introduction to history. Bände, New York
Rossi E Tripoli. In: Houtsma MT, Arnold TW, Basset R, Hartmann R (eds) Encyclopaedia of Islam, 1st edn (1913–1936), checked on 6/8/2018 http://dx.doi.org/10.1163/2214-871X_ei1_SIM_5810
Rossi E (ed) (1968) Storia di Tripoli e della Tripolitania dalla Conquista Araba al 1911. Rom
Rossi E (1974) Libya mundu al-fat al-arabi hatta sanat 1911 (Translated by Khalifa at-Tillisi). Beirut
Rossi E, Nallino M (1968) Storia di Tripoli e della Tripolitania: Dalla conquista araba al 1911. Roma
Rotter G (ed) (1978) Bis zu den Grenzen der Erde. Auszüge aus dem "Buch der Goldwäschen". Tübingen & Basel
Roumani M (2010) Gharian. In: Encyclopedia of Jews in the Islamic World. Leiden: BrillOnline
Roumani, M. (2010): Yefren. In : Encyclopedia of Jews in the Islamic World. Leiden: BrillOnline.
Salama MA (1995) The way to the White City: A guidebook. Tripoli
Savage E (1992) Berbers and Blacks: Ibadi slave traffic in eighth-century North Africa. J Afr Hist 33(3):351–368
Savage E (ed) (1992) The human commodity: Perspectives on the trans-Saharan slave Trade. London
Savage E (1997) A Gateway to hell, a gateway to paradise: The North African response to the Arab Conquest. Princeton
Schacht C (1964) An introduction to Islamic Law. Oxford
Schelhaas B, Wardenga U (2011) Inzwischen spricht die Karte für sich selbst: transformation von Wissen im Prozess der Kartenproduktion. In: Die Werkstatt des Kartographen: Materialien und Praktiken visueller Welterzeugung. München, pp 89–107
Schiffers H (1965) Die grosse Reise. Dr. Heinrich Barths Forschungen und Abenteuer 1850–1855. Minden
Schiffers H (1967) Heinrich Barth. Ein Forscher in Afrika. Leben-Werk-Leistung. Wiesbaden
Schiffers H (1971) Die Sahara und ihre Randgebiete. Darstellung eines Naturgroßraumes. I. Band. Physiogeographie. München
Schiffers H (1975) Libyen: Brennende Wüste, blühender Sand. Berlin

Schliephake K (1979) Tendenzen und Standorte der industriellen Entwicklung in Algerien und Libyen. Würzburger geographische Arbeiten 49:265–286
Schliephake K (1980) Libyan agriculture—natural constraints and aspects of development. Maghreb Rev 5(2–4):51–56
5.Schliephake (1982) Die Oasen der Sahara - ökologische und ökonomische Probleme. Geographische Rundschau 34/6:282–291
Schliephake K (ed) (1988) Konzepte der räumlichen Entwicklung in Tunesien und Libyen. 46. Deutscher Geographentag München, vol 46. München
Schliephake K (ed) (1999) Libyen und nordwestliches Ägypten. Aktuelle Elemente der räumlichen Entwicklung. Würzburg
Schliephake K (2000) Nördliches und nordöstliches Afrika. Informationen zur politischen Bildung (272), 3–4, 13–14, 16–20, 35–40, 41–42, 52, 54–55
Schliephake K, Schulz E (ed) (1991) Tunesien und Tripolitanien. Natur- und kulturgeographische Elemente. Würzburg
Schmidt-Colinet A (1979) Archäologische Forschungen in Libyen. Antike Welt 10(2):25–34
Schmieder O, Wilhelmy H (1939) Die faschistische Kolonisation in Nordafrika. Leipzig
Schroeder F-G (1998) Lehrbuch der Vegetationsgeographie. Wiesbaden
Schliephake (1982) Die Oasen der Sahara - ökologische und ökonomische Probleme. Geographische Rundschau 34(6): 282–291
Schubert R, Wagner G (1988) Pflanzennamen und botanische Fachwörter. Radebeul
Schubert R, Wagner G (2000) Botanisches Wörterbuch. Pflanzennamen und botanische Fachwörter; mit einer Einführung in die Terminologie und Nomenklatur. 12. Aufl. Stuttgart
Sebald P, Musa M (1972) Gottlob Adolf Krause 1850–1938. Leben und Lebenswerk eines antikolonial gesinnten Afrika-Wissenschaftlers unter den Bedingungen des Kolonialismus. Berlin (Ost)
Shboul A (1979) Al-Mas'udi and his world: A muslim humanist and his interest in non-Muslims. London
Sichtermann H (1979) Archäologische Funde und Forschungen in Libyen, Cyrenaika 1959–1961, Tripolitanien 1942–1961. In: Archäologischer Anzeiger, S.439ff
Sieberg H, Schnurer J (1999), Hornemann FK (1772–1801). Begegnungen mit West- und Zentralafrika im Wandel der Zeit. Hildesheim
Simon R (2010) Jebel Nefusa. In: Encyclopedia of Jews in the Islamic World. Leiden: BrillOnline
Sommer M (2011) Trans-Saharan long-distance trade and the Helleno-Punic Mediterranean. In: Dowler A, Galvin ER (eds) Money, trade and trade routes in Pre-Islamic North Africa, pp 61–63
Squarciapino FM (1966) Leptis Magna. Ruinenstädte Nordafrikas, Basel
Stehli P (1978) Sahara—10.000 Jahre zwischen Weide und Wüste. In Rautenstrauch-Joest-Museum für Völkerkunde (ed): Handbuch zu einer Ausstellung des Rautenstrauch-Joest-Museums für Völkerkunde. Köln
Sterry M, Mattingly D, Ahmed M, Savage T, White K, Wilson A (2011) DMP XIII: Reconnaissance survey of archaeological sites in the Murzuq area. Libyan Stud 42:103–116
Sterry M, Mattingly D, Higham T (2012) Desert migrations project XVI: Radiocarbon dates from the Murzuq Region, Southern Libya. Libyan Stud 43:137–147
Strasburger E (2002) Lehrbuch der Botanik für Hochschulen. In: Sitte P, Weiler EW (eds) 35th edn. Berlin
Strunz H (2000) Libyen – zurück auf der Weltbühne. Frankfurt a. M
Subrahmanyam S (1996) Merchant networks in the early modern world. London
Suter K (1967) Zur Geographie der Sahara. In: Gardi R (ed) Sahara. Monographie einer großen Wüste. Bern: Geographischer Verlag, pp 65–123
Swanson J (1975) The myth of trans-Saharan trade during the Roman Era. IJAHS 4:582–600
Talamona M-I (1985) Italienische Agrarsiedlungen in Libyen. In: Frank H (ed) Faschistische Architekturen. Hamburg, pp 139–157
Talbi M (1967) Ibn Khaldun et le sens de l'histoire. In: Studia Islamica (SI), pp 73–148
Tambo DC (1976) The Sokoto Caliphate slave trade in the nineteenth century. Int J Afr Hist Stud 9(2):187–217
Terry J (2004) The history of the Libyan Desert in the middle ages (Translated by Jadallah Azzuz at Talhi). Misrata
Terry J (2008) The history of the Libyan desert in the middle ages (Translated by Jadallah Azzuz at Talhi), 2nd edn. Misrata
Teutsch L (1962) Das Städtetum in Nordafrika: In der Zeit von C. Gracchus bis zum Tod des Kaisers Augustus, Berlin
Tilho J (1920) The exploration of Tibesti, Erdi, Borkou, and Ennedi in 1912–1917. Geogr J 56(2 and 4):81–99, 241–267
The Basic People's Congress (ed) (1985) A Bulletin on Daraj between the Past and the Present. Daraj
Thiry J (1995) Le Sahara libyen dans l`Afrique du Nord médièvale. Leuven
Thiry J (2004) Tarih as-sahra al-lIbiyya fi al-usur al-wusta (Translated by Jad Allah Azuz at-Talhi). Tripoli
Thomas BE (1952) Modern trans-Saharan routes. Geogr Rev 42 (2):267–282
Tillisi HM (1985) Hikayat madina: Libya. 2nd edn. Tunis
Tittoni R (1914) The Italo-Turkish War (1911–12). Kansas
Tracy J (ed) (1990) The rise of merchant empires: Long-distance trade in the early modern world, 1350–1750. Cambridge
Traversi C (1964) L' Italia in Africa. Serie sientifico-culturale, Rom
Trayyeh Sharaf AA (1962) The Geography of Libya. Alexandria
Trivellato F (2007) Merchants' letters across geographical and social boundaries. In: Bethencourt F, Egmond F (eds) Correspondence and cultural exchange in Europe 1400–1700. Cambridge, pp 80–103
Túccari L (1994) I Governi Militari della Libia (1911–1919). Rome
Udovitch AL (1967) Credit as a Means of Investment in Medieval Islamic Trade. J Am Oriental Soc 87:260–264
Vallardi F (1912) Pionieri italiani in Libia. Relazioni dei delegati della Società Italiana di Esplorazioni Geografiche e Commerciali di Milano, 1880–1896. Milano
Vandewalle D (ed) (1995) Qadhafi's Libya, 1969–1994. New York
Vandewalle D (2006) A history of modern Libya. Cambridge
Vasina J (1985) Oral traditions as history. Madison
Verlinden C (1980) Die transatlantische Zwangsmigration afrikanischer Neger und ihre Folgen. In: Klingenstein G, Lutz H, Stourzh G (eds) Europäisierung der Erde? Studien zur Einwirkung Europas auf die außereuropäische Welt. Wien, pp 73–94
Vicari E (1942) L'Ente Turistico ed Alberghiero della Libia. In: Gli Annali dell'Africa Italiana, vol 5, pp 955–975
Vikør K (1987) Al-Sanusi and Qadhafi - continuity of thought? Maghreb Rev 12(1–2):25–28
Vikør K (1987) The early history of the Kawar Oasis. In: A Southern border of the Maghreb or a Northern border of the Sudan? Maghreb Rev 12(3–4), pp 78–83
Vikør K (1996) Sources for Sanusi studies. Bergen
Vikør K (1999) The Oasis of salt. In: The History of Kawar, a Saharan Centre of Salt Production. Bergen
Vikør K, O'Fahey RS (1987) Ibn Idris and al-Sanusi: The teacher and his students. Islam et Sociétés au Sud du Sahara 1:70–83
Vikør KS (2010) Zawila. In: Bearman P, Bianquis T, Bosworth CE, van Donzel E, Heinrichs WP (eds) Encyclopaedia of Islam, 2nd edn. Leiden: BrillOnline, checked on 6/8/2018 http://dx.doi.org/10.1163/1573-3912_islam_SIM_8133

Vireau C (1973) Al-Hawliyyat al-libiyya (Translated by Muhammad Abd al Karim al Wafi), 2nd ed. Tripoli
Voigt I (2012) Die "Schneckenkarte": Mission, Kartographie und transkulturelle Wissensaushandlung in Ostafrika um 1850. Cartographica Helvetic 45–46(45):27–38
von Henneberg K (1996) Imperial uncertainties: Architectural syncretism and improvisation in fascist colonial Libya. J Contemp Hist 31(2):373–395
Ward P (1969) Tripoli. Portrait of a City, Stoughton
Weis H (1996) Murzuch - Blüte und Verfall einer Saharametropole. Graz
Weiss W (ed) (2007) Die arabischen Staaten: Geschichte, Politik, Religion, Gesellschaft. Wirtschaft, Heidelberg
Western R (1989) The flora of the United Arab Emirates—an introduction. Dubai
Westphal W (2002) Tochter des Sultans. Stuttgart
Wiegels R (ed) (2011) Verschlungene Pfade. Neuzeitliche Wege zur Antike, vol 16. Leidorf
Willis JR (ed) (1985) Slaves and slavery in Muslim Africa: The servile estate. London
Winkler G (ed) (1993) Plinius: Naturkunde. Geographie: Afrika und Asien. Düsseldorf 5
Wirth E (2000) Die orientalische Stadt im islamischen Vorderasien und Nordafrika, vol 2. Mainz
Wright J (1989) Libya, Chad and the Central Sahara. London
Wright J (1992) The Wadai-Benghazi slave route. In: Savage E (ed) The human commodity. Perspectives on the Trans-Saharan slave trade. London. pp 174–184
Wright J (1996) The mediterranean middle passage: The nineteenth century slave trade between Tripoli and the Levant. J North Afr Stud 1(1):42–58
Wright J (1997) Diplomacy and tribalism: Consul Warrington and the Awlad Slaiman. Maghreb Rev 22(1–2):96–108
Wright J (ed) (2005) Travellers in Libya. Kent
Wright J (2007) The trans-Saharan slave Trade. London
Wright J (ed) (2008) The emergence of Libya: Selected historical essays. Kent
Zartman, W. (1963): Sahara: Bridge or Barrier? New York.
Zech W (2002) Böden der Welt. Ein Bildatlas, Heidelberg
Zech W, Hintermaier-Erhard G (2002) Böden der Welt - Ein Bildatlas. Heidelberg
Ziadeh NA (1958) Sanusiyah. In: A study of a revivalist movement in Islam. Leiden
Ziegert (1969) Überblick zur jüngeren Besiedlungsgeschichte des Fezzan. In: Hövermann J et al (eds) Berliner Geographische Abhandlungen. Arbeitsberichte aus der Forschungsstation Bardai/Tibesti, vol 8, pp 49–58

Web Publications

Attir M (2005) Modernization & the distraction of the traditional value system: the case of the Libyan Oasis. Available online at www.moatir.com/en/publications.html
Encyclopedia of Jews in the Islamic World (2010). Leiden: BrillOnline
Bearman P, Bianquis T, Bosworth CE, van Donzel E, Heinrichs WP (eds) (2010) Encyclopaedia of Islam, 2nd edn. BrillOnline, Leiden
University of Leicester (2011) Castles in the desert—satellites reveal lost cities of Libya. Available online at http://www2.le.ac.uk/offices/press/press-releases/2011/november/castles-in-the-desert-satellites-reveal-lost-cities-of-libya
Watts-Plumpkin E Garama: An Ancient Civilisation in the Central Sahara. Edited by Taylor & Francis (World Archaeology, Issue 9). Available online at www.world-archaeology.com/features/garamantes-libya/

The manufacturer's authorised representative in the EU is Springer Nature Customer Service Centre GmbH, Europaplatz 3, 69115 Heidelberg, Germany. If you have any concerns regarding our products, please contact ProductSafety@springernature.com

Printed and bound by CPI Group (UK) Ltd, Croydon, CR0 4YY
30/09/2026
02225268-0002